WATER DEFICITS
AND PLANT GROWTH

VOLUME II
Plant Water Consumption and Response

CONTRIBUTORS TO THIS VOLUME

A. S. CRAFTS   A. J. RUTTER
C. T. GATES   P. W. TALBOYS
M. E. JENSEN   R. ZAHNER

WATER DEFICITS
AND PLANT GROWTH

EDITED BY

T. T. KOZLOWSKI

DEPARTMENT OF FORESTRY
UNIVERSITY OF WISCONSIN
MADISON, WISCONSIN

VOLUME II

Plant Water Consumption and Response

1968

ACADEMIC PRESS   New York   San Francisco   London

A Subsidiary of Harcourt Brace Jovanovich, Publishers

ACADEMIC PRESS, INC.
111 Fifth Avenue, New York, New York 10003

United Kingdom Edition published by
ACADEMIC PRESS, INC. (LONDON) LTD.
24/28 Oval Road, London NW1

LIBRARY OF CONGRESS CATALOG CARD NUMBER: 68-14658

PRINTED IN THE UNITED STATES OF AMERICA

LIST OF CONTRIBUTORS

Numbers in parentheses indicate the pages on which the authors' contributions begin.

A. S. CRAFTS (85), Department of Botany, University of California, Davis, California

C. T. GATES (135), Commonwealth Scientific and Industrial Research Organization, Division of Tropical Pastures, Brisbane, Australia

M. E. JENSEN (1), Agricultural Research Service, USDA, Kimberly, Idaho

A. J. RUTTER (23), Botany Department, Imperial College, London, England

P. W. TALBOYS (255), East Malling Research Station, Kent, England

R. ZAHNER (191), Department of Forestry, University of Michigan, Ann Arbor, Michigan

PREFACE

The rapidly proliferating interest in water resources and accelerated research activity on effects of drought on plants indicated the need for this two-volume treatise. Its purpose is to present a comprehensive work on the physical basis of development and control of internal water deficits in plants, the valid measurement of such deficits, and the characterization of various physiological and growth responses of both herbaceous and woody plants to water deficits. The primary concern throughout is with water-stressed plants rather than general water relations of plants which are not undergoing water stress. The treatise is purposely interdisciplinary and should therefore be particularly useful as a text or reference for investigators and students in such fields as plant physiology, soil science, physics, meteorology, plant pathology, horticulture, forestry, agronomy, and hydrology.

An effort was made to standardize terminology throughout the treatise. Because of the large number of different terms used, however, in a few chapters the same symbol is used for different quantities. To avoid difficulties with terminology each author has defined the terms used in his chapter. Special emphasis was given to selection of reference material in an effort to make the volume authoritative, well documented, and up-to-date.

In planning this treatise invitations to prepare chapters were extended to a distinguished group of scientists in universities and government laboratories in the United States and abroad. I wish to express my sincere gratitude to each of these eminent contributors for his scholarly contribution and patient cooperation in revising chapters.

The counsel of Dr. J. E. Kuntz and Dr. R. D. Durbin of the University of Wisconsin in reviewing one of the contributions is acknowledged with appreciation.

T. T. Kozlowski

Madison, Wisconsin
April, 1968

CONTENTS

1. WATER CONSUMPTION BY AGRICULTURAL PLANTS

M. E. JENSEN

2. WATER CONSUMPTION BY FORESTS

A. J. RUTTER

3. WATER DEFICITS AND PHYSIOLOGICAL PROCESSES

A. S. CRAFTS

4. WATER DEFICITS AND GROWTH OF HERBACEOUS PLANTS

C. T. GATES

5. WATER DEFICITS AND GROWTH OF TREES

R. ZAHNER

6. WATER DEFICITS IN VASCULAR DISEASE

P. W. TALBOYS

CONTENTS OF VOLUME I

DEVELOPMENT, CONTROL, AND MEASUREMENT

WATER DEFICITS AND PLANT GROWTH

VOLUME II
Plant Water Consumption and Response

CHAPTER 1

WATER CONSUMPTION BY AGRICULTURAL PLANTS

M. E. Jensen

AGRICULTURAL RESEARCH SERVICE, USDA, KIMBERLY, IDAHO

I. INTRODUCTION

Water consumption by agricultural plants normally refers to all water evaporated from plant and soil surfaces plus that retained within plant tissues. However, the amount of water retained within the tissue of agricultural plants generally is less than 1 % of the total evaporated during a normal growing season. Therefore, water consumption as used in this chapter essentially involves water evaporated from plant and soil surfaces.

Several definitions are presented below to clarify terminology used in this chapter, although it does not differ materially from other terminology in this book.

Transpiration is the loss of water in the form of vapor from plants. All aerial parts of plants may lose some water by transpiration, but most water is lost through the leaves in two stages: (1) evaporation of water from cell walls into intercellular spaces, and (2) diffusion through stomates into the atmosphere. Some water vapor also diffuses out through the epidermal cells of leaves and the cuticle. Small amounts of cuticular transpiration may take place in herbaceous stems, flower parts, and fruits.

Evapotranspiration is the sum of water lost by transpiration and evaporation from the soil or from exterior portions of the plant where water may have accumulated from rainfall, dew, or exudation from the interior of the plant.

Consumptive use is, for all practical purposes, identical with evapotranspiration. It differs by the inclusion of water retained in plant tissues. For most agricultural plants, the amount of water retained by plants is insignificant when compared to evapotranspiration.

II. HISTORICAL ASPECTS

A major stimulus for water requirement studies has been the development of irrigation in the western United States. Irrigation has been practiced for many centuries in other countries, and there is no doubt that some investigations of water requirements can be traced back hundreds of years. Irrigation was also practiced for centuries in the southwestern United States before the Spaniards arrived (Golzé, 1961). The Spaniards began irrigating crops in New Mexico in 1598. Irrigation of small tracts of land began along many rivers of the western United States in the middle of the nineteenth century and expanded throughout this area during the latter part of that century. Numerous studies on water requirements of crops were initiated during this period, reflecting the importance of this information to irrigated agriculture.

The early mechanisms permitting research on water requirements of agricultural crops in the United States were established more than 100 years ago. On May 15, 1862, Lincoln signed "an act establishing the United States Department of Agriculture," and on July 2, 1862, he signed "an act donating public lands to the several States and Territories which may provide colleges for the benefit of American agriculture and the mechanic arts" (Knoblauch et al., 1962). In 1887 the Hatch Act was passed, establishing agricultural experiment stations. This act gave immediate impetus to irrigation research at a number of agricultural experiment stations in the West. These studies frequently involved the assessment of water requirements for agricultural crops.

During the period from 1890 to 1920 the term "duty of water" was used extensively to describe the amount of water being used for irrigation. This term was in general use in the western United States in the 1890's and appears to have originated in Europe, since a number of books on irrigation written in England, France, and Italy during the nineteenth century were cited by Carpenter (1890). Mead (1887) summarized the "duty of water" as determined at Fort Collins, Colorado, where irrigation water applied to wheat, barley, oats, corn, and garden crops was measured during the summer of 1887. Extensive data on water applied to crops such as alfalfa, corn, flax, oats, peas, potatoes, rye, sugar beets, timothy, and wheat, obtained from 1893 to 1898 in Wyoming, were summarized by Buffum (1900). Similar studies were started in Utah in 1890 (Mills, 1895). Such studies expanded throughout the West when funds were provided for irrigation investigations in the Appropriation Act of 1898 (Teele, 1905, 1908). Most of these studies primarily involved the measurement of water delivered to irrigated farms.

Plot and field studies were established during this era to determine the relation between the quantity of water used and crop returns and losses of water by evaporation and percolation through soils (Widstoe *et al.*, 1902; Fortier, 1907; Teele, 1908; Fortier and Beckett, 1912). The primary objective of the plot and field studies was to determine seasonal consumptive use by soil-sampling techniques. Widstoe (1912), e.g., made detailed studies in Utah from 1902 to 1911 on 14 crops. Harris (1920) summarized 17 years of study in the Cache Valley, Utah. Lewis (1919) conducted similar studies near Twin Falls, Idaho, from 1914 to 1916. Hemphill (1922) summarized the studies conducted in the Cache LaPoudre River Valley of northern Colorado. Israelsen and Winsor (1922) made detailed "duty of water" determinations in the Sevier River Valley of Utah from 1914 to 1920. A discussion of the determination of consumptive use by various experimental techniques was presented by Hammatt (1920). An excellent summary of seasonal consumptive use of water can be found in the progress report of the Duty of Water Committee of the Irrigation Division, ASCE "Consumptive Use of Water in Irrigation," presented in 1927 and later published (anonymous, 1930).

Probably the most widely recognized classic investigation of water use by agricultural plants was the transpiration study of Briggs and Shantz (1913, 1914). They initiated extensive experiments at Akron, Colorado, to determine the relative water requirements of crops. These studies were made using small containers in which 44 species and varieties were grown in 1912 and 55 species in 1913. The exposure of these crops was varied. In some experiments a screened area was used to protect the plants from hail and birds. Other experiments were conducted in the open and some with the containers placed in trenches. Because of the observed differences in transpiration, depending on exposure, the data from these studies were not considered as unique values for these plants. Briggs and Shantz (1914) stated that "the water-requirement measurements must therefore be considered relative rather than absolute." This basic relationship between the loss of water through transpiration and the dry matter produced was recognized one-half century ago, but it still is often misinterpreted—even in more recent studies or applications in the 1960's.

Briggs and Shantz also obtained meteorological data, including minimum and maximum air temperatures, wind speed, rainfall, evaporation, sunshine, sun and sky radiation, and wet-bulb depression. They recognized that solar radiation was the primary cause of the cyclic change of environmental factors (Briggs and Shantz, 1916a). Radiation incident on plants exposed to direct sunlight was corrected to an equivalent horizontal area. Advected energy or heat energy extracted from warm air was determined and recognized as a contribution to the energy utilized in transpiration. They stated that "even on bright days, therefore, other sources of energy such as the indirect radiation from the sky and from surrounding objects and the heat energy received directly from the air, contribute materially to the energy dissipated through transpiration" (Briggs and Shantz, 1916b). Other investigators also began studying the influence of various meteorological factors on evaporation and transpiration (Harris and Robinson, 1916; Widstoe, 1902, 1909, 1912).

This summary illustrates the change in the type of studies underway in the western United States from merely the measurement of water delivered to farms in the late 1800's and early 1900's to studies of factors causing and affecting water loss from 1900 to 1920. During the next two decades emphasis was placed on the development of procedures for estimating seasonal consumptive use of water, using available climatological data.

Some of the problems associated with crop yields and consumptive use relationships were recognized in the 1920's. For example, the ASCE Duty of Water Committee recognized that yields may be reduced by plant diseases and insects without significantly affecting seasonal consumptive use. The difficulty in obtaining the same environmental conditions around pots or containers as exist in ordinary cropped land was recognized. Also, the influence of abnormal environmental conditions on consumptive use was recognized as being great enough to render questionable the pot method of determining consumptive use. Drainage from the soil profile following infrequent heavy irrigations or frequent light irrigations was recognized as a probable source of error when soil sampling methods were used. Many consumptive use data in the literature determined by soil sampling or using the neutron moisture meter obviously include a significant drainage component.

The emphasis on factors controlling transpiration expanded extensively during the middle of the twentieth century. The energy balance concept— applied in estimating evaporation from water surfaces in the 1920's and the 1930's (Bowen, 1926; Cummings and Richardson, 1927; McEwen, 1930;

Richardson, 1931; Cummings, 1936, 1940; Kennedy and Kennedy, 1936)—
was applied to crop surfaces in the 1940's by Penman (1948) and Budyko
(1948). Penman combined the energy balance equation and an aerodynamic
equation into what now is commonly referred to as the combination method
(see Chapter 4, Volume I). The Penman equation, or equations of the Penman type, have been evaluated throughout the world. The meteorological
measurements required for the combination method are mean air temperature, dew-point temperature, mean wind speed, and net radiation. On a field
basis, water losses from soil and plant surfaces can be conservatively approximated more readily by using an energy balance or a combination approach
than by any of the other available methods relating the evaporation and
transfer of water to the atmosphere, such as the Dalton equation or aerodynamic equations. A detailed discussion of the characteristics of these
methods is presented in Chapter 4, Volume I.

This brief summary of progress in assessing water requirements by agricultural crops, beginning with crude measurements of water applied to fields, is
not all inclusive, but it illustrates the trend of early studies and progress made
throughout the world. The general relationships of consumptive use to climate still need refinement for developing efficient irrigated agriculture and
for maximizing the development of water resources. Detailed studies involving the biochemistry and internal processes within the plants as influenced by
the state of water within the plant are currently underway.

III. DETERMINING EVAPOTRANSPIRATION

A detailed discussion of evaporation from plant and soil surfaces as
related to micrometeorological parameters and a summary of various methods
of calculating evaporation using energy balance or mass transfer concepts are
presented in Chapter 4, Volume I. Other methods are also used. The most
common method of determining water requirements of agricultural plants
under natural environmental conditions for 5- to 20-day periods is by soil
moisture depletion. This method has been used extensively in irrigated areas of
the world and in the western United States for more than 70 years. The major
problems encountered in soil sampling are summarized in Chapter 5, Volume
I. The precautionary measures needed to minimize errors in evapotranspiration determinations using soil moisture depletion techniques follow: (1) the
sampling sites must be representative of the general field conditions; (2) depth
to a saturated zone should be much greater than the root zone depth; (3) only
those sampling periods where rainfall is light should be used—all others are
questionable because drainage may be excessive; and (4) drainage should be
minimized by (a) applying the preplant irrigation at least 10 to 30 days before
planting, (b) controlling irrigation so as to apply less water than can be

retained within the effective root zone, (c) waiting at least 2 days after normal light irrigations before taking the first sample (longer periods are required if excessive irrigations or high soil moisture levels are involved and when evapotranspiration rate is small), and (d) using only the effective root zone depth or the depth to the plane of zero hydraulic gradient (Jensen, 1967a).

A summary of lysimeters for measuring evapotranspiration is also presented in Chapter 4, Volume I. The major sources of unreliable data obtained with lysimeters are as follows: (1) the vegetative and soil moisture conditions in the lysimeter may not be comparable to those of the surrounding crop; (2) the effective leaf area for the interception of radiation and transpiration may be greater than the surface area of the lysimeter, i.e., the foliage may extend beyond the perimeter of the lysimeter or extend above the surrounding crop; and (3) the edge of the lysimeter may represent an excessively large proportion of the surface area of the lysimeter, resulting in unrealistic border effects caused by the lysimeter itself, which can influence the microclimate in the plant–air zone. When properly installed, operated, and instrumented, lysimeters provide the most accurate measurement of evapotranspiration. This is especially true under high rainfall conditions, because the probable error resulting from drainage increases with other methods such as soil moisture depletion techniques.

Meteorological methods for determining evapotranspiration are being used increasingly as a result of the tremendous development of electronic instrumentation during the 1950's and 1960's. A thorough discussion of these techniques is presented by Webb (1965) and in Chapter 4, Volume I. In general, the instrumentation and technical skill requirements limit these methods to detailed research studies or comprehensive operational studies at a few locations.

IV. CLIMATIC REGIMES

A. POTENTIAL GROWING SEASONS

The climatic regime and the potential growing season control the type of agricultural crops that are grown and, consequently, greatly influence the annual or seasonal water use by agricultural crops. In general, seasonal water use is greater with long growing seasons than with short ones (Milthorpe, 1960; Penman, 1963).

A detailed classification of climates of the world and their agricultural potential is presented by Papadakis (1966). Three climatic regimes are presented in Fig. 1 to illustrate the range of climatic conditions encountered in agricultural areas. Obviously, the potential growing season at Maiquetia, Venezuela, is all year long primarily because of proximity to the Equator and

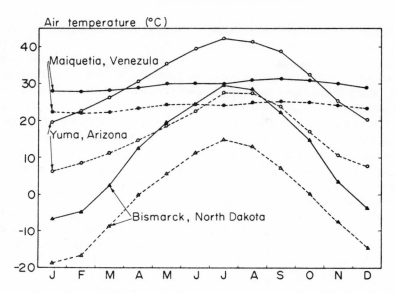

Fig. 1. Monthly mean maximum and mean minimum air temperatures.

the Caribbean Sea. Mean monthly solar radiation at this location varies from a low of 400 cal cm^{-2} day^{-1} in December to 530 in July.

Papadakis classified the climate at Maiquetia, Venezuela, as dry, semi-hot tropical and at Yuma, Arizona, as hot subtropical desert. The latitude at Yuma is 33.7°N as compared to 10.5°N at Maiquetia. Clear skies are common at Yuma; consequently, there is a much greater variation in monthly mean daily solar radiation—varying from a low of 270 cal cm^{-2} day^{-1} in December to 700 in June. The mean minimum temperature at Yuma would indicate that crops may be grown all year long. However, winter temperatures have reached -5°C and, thus, some crops may be subject to frost damage. Monthly mean maximum and mean minimum temperatures at Bismarck, North Dakota (Fig. 1), obviously restrict the potential growing season to the period from mid-April to mid-October. The probability of a late frost in the spring or an early frost in fall further limits the growing season for many farm crops in that area. The usual planting dates for barley and oats in North Dakota range from April 15 to June 5, whereas harvest dates for sugar beets generally range from September 10 to October 20 (Burkhead *et al.*, 1965).

B. Potential Evapotranspiration

Potential evapotranspiration, as used in this chapter, represents the upper limit of evapotranspiration that occurs with a well-watered agricultural crop that has an aerodynamically rough surface such as alfalfa with 30–50 cm of top growth. Potential evapotranspiration so defined occurs in either humid or arid areas in fields that are surrounded by sufficient buffer area so that the edge or "clothes line" effect is small or negligible. The width of the buffer strip required to minimize the edge effect may be only 30 m or less for most short, closely spaced field crops. A detailed theoretical discussion, supported by experimental data, of the horizontal transport of heat and moisture in the 16-m zone is presented by Rider *et al.* (1963). The effect of regional advection of heat or the "oasis" effect would be included in the term as defined because most irrigation projects and most farm fields are subjected to these conditions during parts of the growing season. For comparative purposes, evapotranspiration from well-watered short grass generally would be less than potential evapotranspiration as used here.

Estimates of potential evapotranspiration E_o for the three locations previously mentioned using solar radiation and mean air temperature are presented in Fig. 2. The curves approximate the mean daily upper limit of

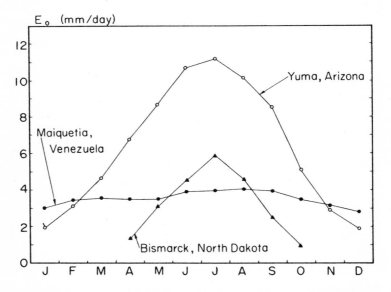

Fig. 2. Monthly mean potential evapotranspiration (E_o). From Jensen (1967b).

evapotranspiration that can occur from a well-watered, aerodynamically rough crop that is actively growing throughout the season. Potential evapotranspiration for a 10-day period in any one year may exceed these mean values for such locations as Bismarck, North Dakota, where climatic conditions may vary widely from one week to the next. In contrast, during the summer months at Yuma, Arizona, there is very little variation in climatic conditions from week to week. In general, larger variability in day-to-day solar radiation occurs more often in semihumid and humid areas (oftentimes with only small changes in air temperature) than in arid areas. An analysis of the frequency of daily solar radiation in July for three locations by Jensen (1967b) indicated that daily solar radiation deviated only $\pm 10\%$ from the long-time mean 20 days of the month at Phoenix, Arizona. In contrast, the deviation was $\pm 24\%$ at a Florida location and $\pm 32\%$ in Wisconsin.

Seasonal evapotranspiration for most common farm crops will be less than the potential because the soil may be completely bare for some time prior to planting, leaf area is limited as the seedlings emerge and develop, and the effective resistance to transpiration increases as the crop begins to mature.

Meteorological parameters controlling potential evapotranspiration, as illustrated in Fig. 2, indicate that crops such as small grains would not necessarily require the same amount of water when grown in different regions under widely different climatic conditions or when grown at different times during the year at a given location. Thus water requirements of a crop cannot be discussed without considering the crop season and potential evapotranspiration at various stages of plant growth.

C. AGRICULTURAL CROPS AND POTENTIAL EVAPOTRANSPIRATION

A simple analytical model of evaporation from soil and plant surfaces would be very useful in a discussion of water consumption by agricultural crops. However, even the most elementary model of this system, involving parallel and series hydraulic and diffusive resistances to water flux, heat sinks, and heat and vapor sources, and operating under conditions involving both spatial and time changes as well as "feedback" effects, can be unwieldy, as evidenced by the simplified ohm's law model presented by Cowan (1965). Models of this system, such as those discussed by Monteith (1965) and Tanner in Chapter 4, Volume I, are extremely useful in describing the general influence of diffusive resistance to water vapor transfer from the leaves to the atmosphere on evapotranspiration. As long as the limitations and pitfalls of using simplified one-dimensional models for quantitative inferences and estimates discussed by Philip (1966) are recognized, models can be effectively used for many purposes.

Individuals forced to manage this complex system often must rely on extremely simple but rational relationships for estimating evapotranspiration. One such relationship, which is based on the concept of potential evapotranspiration and a crop coefficient, is presented here. The basic meteorological parameter required is potential evapotranspiration as previously defined. The reference crop must have a root system that is fully developed, sufficient leaf area so as not to materially limit transpiration, and adequate soil moisture so that evapotranspiration is essentially limited by meteorological conditions. Potential evapotranspiration could be calculated by one of several methods discussed in Chapter 4, Volume I, or it could be measured with one or two good lysimeter installations in the general region under consideration. Because of the conservativeness of potential evapotranspiration and a real uniformity (with the exception of areas adjacent to large water bodies or major orographic changes), accurate determinations at a few locations would be preferred over numerous crude determinations throughout the area of interest. Evapotranspiration for a given agricultural crop can be related to potential evapotranspiration E_o as follows:

$$E_t = K_c E_o \tag{1}$$

in which K_c is a dimensionless coefficient, similar to that proposed by van Wijk and de Vries (1954), representing the combined relative effects of resistance to water movement from the soil to the evaporating surfaces, resistance to diffusion of water vapor from the evaporating surfaces through the laminar boundary layer, resistance to turbulent transfer to the free atmosphere, and the relative amount of radiant energy available as compared to the reference crop.

From an energy balance viewpoint, the crop coefficient represents the relative heat energy converted to latent heat. Thus, K_c is related to the major energy terms of the soil–plant–air continuum as follows:

$$K_c = (R_n + A + G)/(R_{no} + A_o + G_o) \tag{2}$$

in which R_n is net radiation, A is sensible heat flux to or from the air, and G is sensible heat flux to or from the soil. The subscript o designates concurrent values for the reference crop in the immediate vicinity (in this case alfalfa). The energy terms are positive for input to the crop air zone and negative for outflow. Of the energy terms, only sensible heat flux is difficult to determine or predict. However, it is related to the overall effective hydraulic and diffusive resistance of the soil–plant–air system. The energy terms of the energy balance equation can be rewritten using the Bowen ratio approach from which

$$K_c = \frac{1 + \beta_o}{1 + \beta} \frac{(R_n + G)}{(R_{no} + G_o)} \tag{3}$$

where β represents the partitioning of latent and sensible heat flux (the ratio of sensible heat flux to latent heat flux) or the Bowen ratio, A/LE (see Chapter 4, Volume I). The magnitude of $(1 + \beta)$ is largely controlled by the overall resistance to the transfer of soil water to water vapor in the free atmosphere.

The overall resistance to water flow from the soil to plant roots and through the plant, as well as to the soil surface, the resistance to diffusion of water vapor through the leaves and the dry surface layer of soil, the laminar boundary layer, and the resistance to turbulent transfer of water vapor to the free atmosphere all indirectly affect the magnitude of G and to some extent R_n. When evaporation from the soil is negligible compared to transpiration, the hydraulic resistance to water flow to the roots would be inversely proportional to hydraulic conductivity of the soil and the length of roots per unit volume of soil (Gardner, 1966). Visser (1965) and Gardner and Ehlig (1963) presented similar analogies of soil resistance. Monteith (1965) presented a detailed analysis of the influence of the diffusive resistance of a crop on transpiration. Until quantitative values are available for calculating the influence of various soil and crop resistances, crop coefficients K_c can be determined from experimental data for estimating purposes.

A typical example of the influence of growth stage on the crop coefficient where soil water is not limiting is shown in Fig. 3 (Jensen, 1967b). The data in

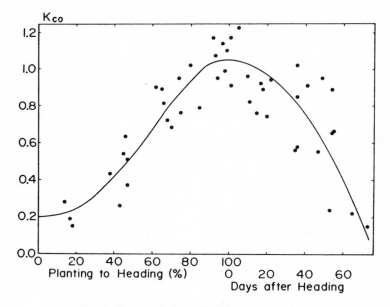

Fig. 3. Crop coefficients (K_{co}) for grain sorghum.

Fig. 3 illustrate primarily the relatively large diffusive resistance of bare soil immediately after planting. The resistance decreases during the period of rapid leaf-area development, approaching an effective overall resistance (and net radiation) similar to that of alfalfa (estimated potential evapotranspiration in this case) near heading and then increases as the crop matures after heading. A similar curve for corn was presented by Denmead and Shaw

(1959) and a curve for soybeans by Laing (1965, from Shaw and Laing, 1966) using evaporation from a pan as an estimate of evaporative demand.

D. Crop Evapotranspiration (Water Nonlimiting)

Since most farm crops do not require as much water during the season as would be needed to meet potential evapotranspiration, even though adequate soil moisture is provided, an additional term is desired to differentiate water requirements of agricultural crops when water is not limiting from water use when soil moisture may be limiting during a portion of the season. This term can be referred to as "crop potential evapotranspiration," E_{oc}. The magnitude of this term generally will be less than potential evapotranspiration during much of the season, as previously indicated, primarily because of limited plant canopy during a portion of the season and the overall increase in resistance to evaporation as the crop matures. Crop potential evapotranspiration, as defined, can be represented by the following equation:

$$E_{oc} = K_{co} E_o \tag{4}$$

where K_{co} is the crop coefficient when soil water is *not* limiting. Thus, crop potential evapotranspiration, E_{oc}, represents the rate of evapotranspiration for a given crop at a given stage of growth when water is not limiting and other factors such as insects, diseases, and nutrients have not materially restricted plant development.

Other terms that become important in planning and discussing water requirements of agricultural crops are seasonal totals of potential evapotranspiration, crop potential evapotranspiration, and actual evapotranspiration with limited soil moisture at some stages of growth or other factors that significantly influence the characteristics of the crop itself, such as diseases, defoliation, etc. Seasonal totals for these three values are represented by Eqs. (5)–(7):

$$W_o = \int_{t_1}^{t_2} E_o \, dt \tag{5}$$

$$W_{oc} = \int_{t_1}^{t_2} E_{oc} \, dt = \int_{t_1}^{t_2} K_{co} E_o \, dt \tag{6}$$

$$W_{ct} = \int_{t_1}^{t_2} E_t \, dt = \int_{t_1}^{t_2} K_c E_o \, dt \tag{7}$$

When considering seasonal totals, t_1 and t_2 in Eqs. (5)–(7) represent the dates of planting and harvest. The totals for a portion of a growing season, or for periods including evaporation from the soil before and after the crop growing season, can be obtained by integrating over the entire time interval. The seasonal total potential evapotranspiration will vary with the meteorological conditions during a given growing season. The total crop potential evapotranspiration will also vary with seasonal meteorological conditions, but in addition it will be influenced by the date of planting within the growing season and date of harvest or freeze near the end of the season. Water use by barley, e.g., is about 40 cm in North Dakota and about 55 cm in Arizona (Haise, 1958; Woodward, 1959).

V. CROP CHARACTERISTICS AND EVAPOTRANSPIRATION

A. PLANT EFFECTS

Characteristics of agricultural plants, such as number and distribution of stomata, leaf coatings, etc. discussed in Chapters 6 and 7, Volume I, can affect evapotranspiration for a given crop primarily by influencing diffusive resistance. For example, in a recent analysis and estimate of water use by an orange orchard, van Bavel *et al.* (1967) concluded that the canopy resistance to evaporation is considerable in citrus orchards, resulting in evapotranspiration much below the potential rate. Similarly, other plant characteristics such as wilting and maturation can influence the effective resistance to evaporation.

B. ROOT AND SOIL EFFECTS

The volume of soil occupied by plant roots and the number of roots within this volume can significantly influence effective soil resistance to water movement. This subject is discussed in detail in Chapter 5, Volume I. Obviously, evapotranspiration with a deep-rooted crop such as alfalfa would not be influenced as much by the removal of a given amount of soil water as with a shallow-rooted crop such as grass. Characteristics of the soil profile that may severely restrict the volume of soil occupied by a dense root system would also result in greater changes in the crop coefficient as soil moisture decreased. Shaw and Laing (1966) presented data illustrating the relative transpiration rate as a function of soil suction in Colo silty clay loam. These data indicate that the relative transpiration rate, or K_c, for corn was also influenced by the evaporative demand. This effect is also represented in Eq. (3). For example, the turgor loss point at low potential evapotranspiration rates occurred at higher soil water matric potentials than at high potential rates. Loss of turgor would increase the effective diffusive resistance and the Bowen ratio. Also, if the evaporative demand were large, the effective soil resistance would increase more rapidly with a limited root system as compared to a crop with a deep, dense root system such as alfalfa.

C. Precipitation Effects

For a row crop with a partial canopy, the effective diffusive resistance would be large when the soil surface is dry. However, immediately following a light rain, the plant surfaces and the surface of the soil between plant rows would be wet and effective diffusive resistance would be greatly decreased. In addition, the albedo of the wet soil surface would decrease, thereby increasing relative net radiation. Obviously, the crop coefficient as given by Eq. (3) would not remain constant at a given stage of growth under these conditions. Under arid conditions, experimentally determined mean crop coefficients at given stages of growth can be effectively used to estimate water use by various crops because of infrequent light rains. Even when rains occur, the soil surface dries within 3 days or less. Crop curves representing K_{co} at various stages of growth would be similar for semihumid areas except for the mean value under partial cover, and as the crop matures K_{co} might be larger than those under arid conditions because of more frequent rains and smaller variations in the Bowen ratio.

D. Perennial Crops

The effective diffusive resistance for perennial crops decreases in the spring when new growth begins, or when new growth begins after a period of dormancy. The effective diffusive resistance during dormancy, or when climatic conditions are such that growth does not take place, will generally be much higher. Consequently, the potential crop coefficient for a perennial crop would be small during a period of dormancy.

Other factors also influence water use by perennial crops. A common cultural practice, e.g., that significantly affects water use is cutting. Cutting alfalfa drastically changes the effective diffusive resistance, although for a short time period after cutting evaporation from the soil surface may compensate for the decrease in transpiration. Bahrani and Taylor (1961) found that evapotranspiration decreased following the cutting of alfalfa. The decrease in this case was attributed to less net radiation as surface soil temperature increased. Irrigation of alfalfa that had previously been cut increased net radiation.

Another situation that is encountered is a two-stage agricultural crop such as a deciduous orchard with an alfalfa cover crop. The potential crop coefficient K_{co} would be small because of high effective diffusive resistance before the trees leaf out and before alfalfa begins to grow in the spring. However, as the season progresses the effective diffusive resistance would decrease markedly with the development of leaves and growth of the cover crop. The two-crop combination would probably increase the effective aerodynamic roughness of the surface, thus decreasing the effective resistance to turbulent transfer of water vapor. Analysis of evapotranspiration data from an apple orchard with an alfalfa cover crop indicates that the crop coefficient may be as high as 1.2. The increase in the crop coefficient above 1 can be attributed largely to a smaller resistance to turbulent transfer, resulting from greater aerodynamic roughness of the two-level crop, and a smaller effective diffusive resistance because of the larger leaf-area and distribution of evaporating surfaces above the soil surface as compared to alfalfa alone.

E. ANNUAL CROPS

The discussion in Section V, A–D covered various aspects of perennial crops and their influence on resistance to evapotranspiration. These effects can be grouped into three broad categories: (1) The influence of degree of crop cover or canopy that influences diffusive resistance. (2) The maturation of the crop, including the development of seed heads above a crop that can influence evapotranspiration by decreasing the proportion of net radiation converted to latent heat of vaporization. For example, Fritschen and van Bavel (1964) reported that absorption of net radiation by the seed head protruding above a crop of Sudan grass resulted in a greater portion of net radiation being converted to sensible heat. (3) Cultural practices such as the frequency of irrigation can influence the effective diffusive resistance. Frequent light irrigations of a row crop that keep the soil surface moist decrease the effective diffusive resistance. One would expect the crop coefficient to be larger on widely spaced row crops under these conditions as compared with infrequent heavy irrigations where the soil surface may remain dry for extensive periods of time.

The spacing of row crops can also affect the crop coefficient during early stages of leaf development. Porter *et al.* (1960) reported that more water was used early in a season with narrow row spacings than with wide row spacings. Differences in total water use among either the spacing or planting rate means for the entire season were insignificant.

VI. EVAPOTRANSPIRATION WITH SOIL WATER NONLIMITING

A. FULL-SEASON CROPS

A brief summary of water use by alfalfa in two widely different climatic regimes is presented in Table I to illustrate the small effects of limited leaf area early in the spring and cuttings during the season on seasonal evapotranspiration. The sample data indicate that seasonal water use by a perennial crop such as alfalfa may be about 90% of potential evapotranspiration.

The decrease primarily results from higher diffusive resistance when growth first begins in the spring and immediately following a cutting as compared to a reference crop without cutting.

TABLE I

MEAN WATER USE BY ALFALFA AND POTENTIAL
EVAPOTRANSPIRATION

Location	W_{et} (cm)	W_o (cm)	$\dfrac{W_{et}}{W_o}$
Bismarck, North Dakota[a]	58	66	0.88
Phoenix, Arizona[b]	188	214	0.88

[a] Haise (1958).
[b] Erie et al. (1965).

B. PART-SEASON CROPS

Grain sorghum is an example of a crop that is planted after the potential growing season begins and is harvested before the potential growing season ends. Jensen and Sletten (1965) reported mean seasonal evapotranspiration (from planting to harvest) for grain sorghum at Bushland, Texas, to be 55 cm, whereas potential evapotranspiration for the same period averaged 84 cm. The data for a row crop such as grain sorghum, which undergoes large changes in leaf area along with changes in crop characteristics as the seeds develop, fill, and mature, indicate that seasonal evapotranspiration may be only 65% of potential evapotranspiration. Similarly, data presented by Ripley (1966) indicate that seasonal water use by many farm crops may range from 55 to 75% of potential evapotranspiration if alfalfa water use is 90% of the potential.

C. CITRUS CROPS

Water requirements of citrus orchards are generally much lower than for crops such as alfalfa, providing the soil surface is kept bare (Jensen and Haise, 1963). Van Bavel et al. (1967) evaluated canopy resistance of an orange orchard near Tempe, Arizona, by direct and indirect techniques. They found that the resistance may vary from 3.4 to 7.6 sec cm^{-1} as compared with 0.3 to 0.5 sec cm^{-1} for field crops reported by Monteith (1965). Mean annual evapotranspiration for clean-tilled grapefruit orchards in Arizona was estimated as 115 cm (Jensen and Haise, 1963). Thus, annual evapotranspiration for a high-resistance canopy crop may be only 55 % of mean annual potential evapotranspiration.

D. OFF-SEASON LOSSES

Loss of soil water before planting and after harvest of an annual crop may vary greatly, depending on cultural practices involved. For example, weeds allowed to grow following the harvest of a crop such as winter wheat (approximately June in the southern Great Plains) can significantly increase the amount of water lost prior to preparation of a seed bed and planting of the next crop. The causes for these losses are obvious and are discussed in Section V. The loss of water with weed growth is normally greater than under fallow conditions as a result of (1) smaller diffusive resistance and (2) smaller effective soil hydraulic resistance because of weed roots. Losses of water under fallow conditions also may be high because of high rates of evaporation from the soil surface immediately following light rains. Numerous data in the Great Plains from Texas to Canada indicate that the amount of water remaining in the soil after a fallow period may be only 15–30 % of the total precipitation received during the period.

VII. EVAPOTRANSPIRATION WITH LIMITED SOIL WATER

As the soil water content decreases and is not replenished by rainfall or irrigation the effective hydraulic resistance increases greatly. This increase in hydraulic resistance results in various degrees of plant water stress, depending on the evaporative demand and plant characteristics. At high plant water stress, diffusive resistance also increases, illustrating the interdependence of hydraulic and diffusive resistances. Seasonal water use is related to the soil water available at planting, seasonal precipitation, and root penetration. Dreibelbis and Amerman (1964) presented data illustrating the importance of root penetration on water use under dryland conditions. These aspects are covered in Chapters 5 and 7, Volume I and are discussed by Black (1966) and Viets (1962, 1966).

A simple expression linearly relating effects of plant water stress caused by inadequate soil water on yields would be desirable for two general types of crops: (1) those having a determinate type of flowering such as a grain crop and (2) crops such as grass that can tolerate severe stress for a period of a week during the growing season and completely recover following application and maintenance of adequate soil water during the remainder of the season with only a small decrease in total dry matter production. A detailed discussion of the effects of water stresses on physiological processes within the plant is presented in Chapter 3.

A. Determinate Crops

The effects of limited soil moisture (resulting in reduced water use during a growth stage) on the development of the marketable product of a determinate flowering crop can be linearly related to yields by the following expression, providing other factors such as plant nutrients are not limiting:

$$\frac{Y}{Y_o} \simeq \prod_{i=1}^{n} \left(\frac{W_{et}}{W_{oc}}\right)_i^{\lambda_i} \tag{8}$$

where Y/Y_o represents the relative yield of the marketable product from an agricultural crop (Y_o is the yield when soil moisture is not limiting); $(W_{et}/W_{oc})_i$ represents the relative total evapotranspiration during a given stage of physiological development, e.g., the boot stage or heading stage of a crop such as small grain or sorghum (W_{et} is the actual use of water and W_{oc} is the use if soil moisture was not limiting); and λ_i represents the relative sensitivity of the crop to water stress during the stage of growth i. The right side of Eq. (8) is a product. Therefore, severe water stress, as indicated by reduced water use, during a single growth stage could reduce the yield of the marketable product severely. The magnitude of λ for specific growth stages would depend primarily on the sensitivity of plant growth to water stress during each growth period. The primary implication of Eq. 8 is that the yield of the marketable product of a farm crop may not be linearly related to total water use when plants are stressed. Jensen and Sletten (1965) found that delaying irrigations of

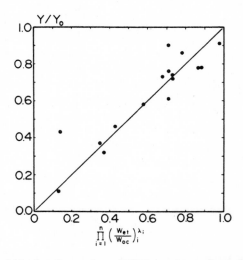

Fig. 4. Relative yield of grain sorghum (Y/Y_o) vs. the product of relative water use (W_{et}/W_{oc}) during various growth periods.

grain sorghum reduced yields an average of 35% but reduced water use only 20%. Similarly, delaying irrigations so that yields were reduced 70% reduced water use only about 40%. Equation (8) was evaluated using the same data with values of W_{et}/W_{oc} approximated from soil moisture data, precipitation, and irrigations (see Fig. 4). Three periods were used: (1) emergence to boot stage with $\lambda = 0.5$, (2) boot to milk stage with $\lambda = 1.5$, and (3) milk to harvest with $\lambda = 0.5$. The results indicate that Eq. (8) may adequately represent the effects of water stress as indicated by reduced water use on yields of a determinate crop. A detailed discussion of water stress during various physiological stages of growth on reproduction and grain development is presented by Shaw and Laing (1966).

The yield of other crops that must meet minimum quality characteristics, e.g., potatoes and lettuce, and for which specific growth stages may have significant effects would be also, probably, linearly related to relative water use by Eq. (8). Detailed discussions of water stress and physiological processes are also presented in Chapter 3.

B. INDETERMINATE CROPS

An expression linearly relating the marketable yield or dry matter produced by an indeterminate crop, such as grass, to water use when soil moisture is limiting, providing other factors such as plant nutrients are not limiting, is as follows:

$$\frac{Y}{Y_o} \simeq \frac{\sum_{i=1}^{n} \lambda_i (W_{et})_i}{\sum_{i=1}^{n} \lambda_i (W_{oc})_i} \tag{9}$$

DeWit (1958) presented a detailed analysis of dry matter production vs. relative transpiration (transpiration/free water evaporation) that generally would substantiate Eq. (9) even though W_{oc} is used in Eq. (9) instead of free water evaporation. The primary difference between Eqs. (8) and (9) is that in Eq. (9) the effects of water stress on yield during a specific growth stage are independent of other growth stages.

REFERENCES

Anonymous (1930). Consumptive use of water in irrigation. *Trans. Am. Soc. Civil Engs. Prog. Rep. Duty Water Comm. Irrigation Div.* **94**, 1349.

Bahrani, B., and Taylor, S. A. (1961). Influence of soil moisture potential and evaporative demand on the actual evapotranspiration from an alfalfa field. *Agron. J.* **53**, 233.

Black, C. A. (1966). Crop yields in relation to water supply and soil fertility. *In* "Plant Environment and Efficient Water Use" (W. H. Pierre, D. Kirkham, J. Pasek, and R. Shaw, eds.) p. 177. Am. Soc. Agron.; Soil Sci. Soc. Am.

Bowen, I. S. (1926). The ratio of heat losses by conduction and by evaporation from any water surface. *Phys. Rev.* **27**, 779.

Briggs, L. J., and Shantz, H. L. (1913). The water requirements of plants. I. Investigations in the Great Plains in 1910 and 1911. *U. S. Dept. Agr. Bur. Plant Ind. Bull.* **284**, 49 pp.

Briggs, L. J., and Shantz, H. L. (1914). Relative water requirements of plants. *J. Agr. Res.* **3**, 1.

Briggs, L. J., and Shantz, H. L. (1916a). Hourly transpiration rate on clear days as determined by cyclic environmental factors. *J. Agr. Res.* **5**, 583.

Briggs, L. J., and Shantz, H. L. (1916b). Daily transpiration during normal growth period and its correlation with the weather. *J. Agr. Res.* **7**, 155.

Budyko, M. I. (1948). Evaporation under natural conditions. (Translated from Russian). *U. S. Dept. Comm. Israel Program Sci. Transl.* **751**, 130 pp.

Buffum, B. C. (1900). The use of water in irrigation in Wyoming and its relation to the ownership and distribution of the natural supply. *U. S. Dept. Agr. Office Expt. Sta. Bull.* **81**, 56 pp.

Burkhead, C.E., Kirkbridge, J. W., and Losleben, L. A. (1965). Usual planting and harvesting dates (field and seed crops). *U.S. Dept. Agr., Agr. Handbook* **283**, 84 pp.

Carpenter, L. G. (1890). Section of meteorology and irrigation engineering. *Colo. Agr. Expt. Sta. Ann. Rept.* **3**, 57.

Cowan, I. R. (1965). Transport of water in the soil-plant-atmosphere system. *J. Appl. Ecol.* **2**, 221.

Cummings, N. W. (1936). Evaporation from water surfaces. *Trans. Am. Geophys. Union 16th Ann. Meeting* Part 2, *Pasadena, California, January, 1936*, p. 507.

Cummings, N. W. (1940). The evaporation-energy equations and their practical application. *Trans. Am. Geophys. Union* **21**, 512

Cummings, N. W., and Richardson, B. (1927). Evaporation from lakes. *Phys. Rev.* **30**, 527.

Denmead, O. T., and Shaw, R. H. (1959). Evapotranspiration in relation to the development of the corn crop. *Agron J.* **51**, 726

DeWit, C. T. (1958). Transpiration and crop yields. *Mededel. Inst. Biol. Scheik. Onderzoek. Landbouwgewassen, Wageningen*, **59**, 88 pp.

Dreibelbis, F. R., and Amerman, C. R. (1964). Land use, soil type and practice effects on the water budget. *J. Geophys. Res.* **69**, 3387.

Erie, L. J., French, O. F., and Harris, K. (1965). Consumptive use of water in Arizona. *Univ. Ariz. Tech. Bull.* **169**, 41 pp.

Fortier, S. (1907). Evaporation losses in irrigation and water requirements of crops. *U. S. Dept. Agr. Office Expt. Sta. Bull.* **177**, 64 pp.

Fortier, S., and Beckett, S. H. (1912). Evaporation from irrigated soils. *U. S. Dept. Agr. Bull.* **248**, 77 pp.

Fritschen, L. J., and van Bavel, C. H. M. (1964). Energy balance as affected by height and maturity of sudangrass. *Agron. J.* **56**, 201

Gardner, W. R. (1966). Soil water movement and root absorption. *In* "Plant Environment and Efficient Water Use" (W. H. Pierre, D. Kirkham, J. Pasek, and R. Shaw, eds). p. 127. Am. Soc. Agron.; Soil Sci. Soc. Am.

Gardner, W. R., and Ehlig, C. F. (1963). The influence of soil water on transpiration by plants. *J. Geophys. Res.* **68**, 5719.

Golzé, A. R. (1961). "Reclamation in the United States," 486 pp. Caxton, Caldwell, Idaho.

Haise, H. R. (1958). Agronomic trends and problems in the Great Plains. *Advan. Agron.* **10**, 47.

Hammatt, W. C. (1920). Determination of the duty of water by analytical experiment. *Trans. Am. Soc. Civil Engrs.* **83**, 200

Harris, F. S. (1920), The duty of water in the Cache Valley, Utah. *Utah Agr. Expt. Sta. Bull.* **173**, 16 pp.

Harris, F. S., and Robinson, J. S. (1916). Factors affecting the evaporation of moisture from the soil. *J. Agr. Res.* **7**, 439.

Hemphill, R. G. (1922). Irrigation in northern Colorado. *U. S. Dept. Agr. Bull.* **1026**, 85 pp.

Israelsen, O. W., and Winsor, L. M. (1922). The net duty of water in the Sevier Valley, Utah. *Utah Agr. Expt. Sta. Bull.* **182**, 36 pp.

Jensen, M. E. (1967a). Evaluating irrigation efficiency. *J. Irrigation Drainage Div. Am. Soc. Civil Engrs.* **93**, 83.

Jensen, M. E. (1967b). Empirical methods of estimating or predicting evapotranspiration using radiation. *Proc. Evapotranspiration Its Role Water Res. Management Conf., Am. Soc. Agr. Engrs., Chicago, 1966*, 49.

Jensen, M. E., and Haise, H. R. (1963). Estimating evapotranspiration from solar radiation. *J. Irrigation Drainage Div. Am. Soc. Civil Engrs.* **89**, 15.

Jensen, M. E., and Sletten, W. H. (1965). Evapotranspiration and soil moisture-fertilizer interrelations with irrigated grain sorghum in the Southern High Plains. *U. S. Dept. Agr. Conserv. Res. Rept.* **5**, 27 pp.

Kennedy, R. E., and Kennedy, R. W. (1936). Evaporation computed by the energy-equation. *Trans. Am. Geophys. Union* **17**, 426.

Knoblauch, H. C., Law, E. M., Meyer, W. P., Beacher, B. F., Nestler, R. B., and White, B. S., Jr. (1962). State experiment stations. *U. S. Dept. Agr. Misc. Publ.* **904**, 262 pp.

Laing, D. R. (1965). The water environment of soybeans. Unpublished Ph.D thesis, Iowa State Univ. Library.

Lewis, M. R. (1919), Experiments on the proper time and amount of irrigation, Twin Falls Expt. Sta. 1914, 1915, and 1916. *U. S. Dept. Agr., cooperating with Twin Falls County Comm., Twin Falls Canal Co., Twin Falls Com. Club.*

McEwen, G. F. (1930). Results of evaporation studies. *Scripps Inst. Oceanog. Tech. Ser.* **2**, 401

Mead, E. (1887). Report of Experiments in irrigation and meteorology. *Colo. Agr. Expt. Sta. Bull.* **1**, 12 pp.

Mills, A. A. (1895). Farm irrigation. *Utah Agr. Expt. Sta. Bull.* **39**, 76 pp.

Milthorpe, F. L. (1960). The income and loss of water in arid and semi-arid zones. *In* "Plant-Water Relations in Arid and Semi-Arid Conditions. Reviews of Research," p. 9. UNESCO, Paris.

Monteith, J. L. (1965). Evaporation and the environment. *Symp. Soc. Exptl. Biol.* **19**, 205.

Papadakis, C. J. (1966). "Climates of the World and Their Agricultural Potentialities." Available through FAO Rome.

Penman, H. L. (1948). Natural evaporation from open water, bare soil and grass. *Proc. Roy. Soc.* **A193**, 120

Penman, H. L. (1963). Vegetation and hydrology. *Commonwealth Bur. Soil Sci. (Gt. Brit.) Tech. Commun.* **53**, 124.

Philip, J. R. (1966). Plant water relations: some physical aspects. *Ann. Rev. Plant Physiol.* **17**, 245.

Porter, K. B., Jensen, M. E., and Sletten, W. H. (1960). The effect of row spacing, fertilizer and planting rate of the yield and water use of irrigated grain sorghum. *Agron. J.* **52**, 431

Richardson, B. (1931). Evaporation as a function of insolation. *Trans. Am. Soc. Civil Engineers.* **95**, 996.

Rider, N. E., Philip, J. R., and Bradley, E. F. (1963). The horizontal transport of heat and moisture—a micrometeorological study. *Quart. J. Roy. Meteorol. Soc.* **89**, 507.

Ripley, P. P. (1966). The use of water by crops. *Proc. Intern. Comm. Irrigation Drainage, New Delhi, India, January, 1966, Spec. Session Rept.* **2**, S9.

Shaw, R. H., and Laing, D. R. (1966). Moisture stress and plant response. *In* "Plant Environment and Efficient Water Use" (W. H. Pierre, D. Kirkham, J. Pasek, and R. Shaw, eds.), p. 73. Am. Soc. Agron.; Soil Sci. Soc. Am.

Teele, R. P. (1905). Annual report of irrigation and drainage investigations, 1904. *U. S. Dept. Agr. Office Expt. Sta. Separate* **1**, 75 pp. (reprint *Office Expt. Sta. Bull.* **158**).

Teele, R. P. (1908). Review of ten years of irrigation investigations. *U. S. Dept. Agr. Office Expt. Sta., Ann. Rep., year ending June 30,* **1908**, p. 35.

van Bavel, C. H. M., Newman, J. F., and Hilgeman, R. H. (1967). Climate and estimated water use by an orange orchard. *Agr. Meteorol.* **4**, 27.

van Wijk, W. R., and de Vries, D. A. (1954). Evapotranspiration. *Neth. J. Agr. Sci.* **2**, 105.

Viets, F. G., Jr. (1962). Fertilizers and efficient use of water. *Advan. Agron.* **14**, 223.

Viets, F. G., Jr. (1966). Increasing water use efficiency by soil management. *In* "Plant Environment and Efficient Water Use" (W. H. Pierre, D. Kirkham, J. Pasek, and R. Shaw, eds.), p. 259. Am. Soc. Agron.; Soil Sci. Soc. Am.

Visser, W. C. (1965). The moisture consumption of plants described as a hydrological phenomenon. *Inst. Land Water Management, Wageningen, Netherlands, Res. Bull.* **40**; *In* "Water Stress in Plants" (B. Slavik, ed.), Proc. Symp. Prague, 1963, p. 257. Czech. Acad. Sci., Prague.

Webb, E. K. (1965). Aerial microclimate, *Meteorol. Monographs* **6**, 27.

Widstoe, J. A. (1909). Irrigation investigations: factors influencing evaporation and transpiration. *Utah Agr. Expt. Sta. Bull.* **105**, 64 pp.

Widstoe, J. A. (1912). The production of dry matter with different quantities of water. *Utah Agr. Expt. Sta. Bull.* **116**, 64 pp.

Widstoe, J. A., Swendsen, G. L., Merrill, L. A., McLaughlin, W. D., Beers, W. O., and Widstoe, O. (1902). Irrigation investigations in 1901. *Utah Agr. Expt. Sta. Bull.* **80**.

Woodward, G. O. (ed.) (1959). "Sprinkler Irrigation" 2nd ed., 377 pp. Sprinkler Irrigation Assoc., Washington, D. C.

CHAPTER 2

WATER CONSUMPTION BY FORESTS

A. J. Rutter

BOTANY DEPARTMENT, IMPERIAL COLLEGE, LONDON, ENGLAND

I. INTRODUCTION

Many investigations of the water consumption of forests have been undertaken in relation to water supplies for human populations. The objective of such studies has been to determine, directly or indirectly, the effects of forests on the flow of streams arising in them. However, the water consumption of a forest is also of importance to the forest itself and to all the plants in it.

Terrestrial plants obtain most of their water from the soil, and the amount of available water that soils can store is limited to 50–250 mm per meter depth of soil within the root range. This store is only intermittently renewed by precipitation, but it is continuously depleted during the growing season by transpiration and evaporation from the soil. As the available water is reduced there is a corresponding fall in the potential of the soil water and hence in the potential of water in the plants dependent on it. Furthermore, at any given time, the gradient of water potential between soil and plant is related to transpiration rate, an increase in transpiration rate lowering the water potential of the plant. The general relation of plant water potential to soil water potential, and the diurnal fluctuation of plant water potential with changes in transpiration rate, are illustrated in Gardner and Nieman's (1964) experiment on pepper. If leaf water deficit is taken as an indirect measure of leaf water potential, similar relations have been shown in a plantation of *Pinus radiata* by Johnston (1964), and it may be assumed that transpiration plays as important a part in the ecology of trees and forest plants as in any other vegetation type.

Not all the water absorbed by trees is transpired; some is combined with carbon dioxide in photosynthesis and some is retained within the plant body without being chemically changed. If the annual dry matter production of forests is about 2×10^4 kg ha^{-1} (Ovington, 1962), this represents the annual chemical combination of about 1.2 mm depth of water. Ovington (1962) also reported that the liquid water content of a number of plantations of different species just under 50 years old was between 1 and 2×10^5 kg ha^{-1}. This represents a mean annual increase of water content of 0.2–0.4 mm. Reynolds (1966) discussed the difference between transpiration and uptake of water by trees and the possibility that some transpiration in summer may be met by a temporary depletion of water in the trunk or foliage. However, he concluded that the seasonal fluctuation of water content in the trunks of *Acer pseudoplatanus* was equivalent to no more than 0.6 mm and in the leaves of *Pinus radiata*, 0.4 mm. If trees transpire 1–5 mm per day during the growing season, these small differences between uptake and transpiration may be neglected for present purposes.

A distinction must be drawn between water absorbed by trees and other forest plants and that evaporated from the forest. The latter includes not only transpiration but also the evaporation of precipitation intercepted by the surfaces of plants and evaporation from the soil. It is clear that both transpiration and evaporation from the soil constitute a drain on stored water. The significance of intercepted water is more controversial. Since it never enters the soil, many authors regard it as a complete loss to the trees, unless some of it is absorbed by leaves before it is evaporated. Others point out that a leaf with a wet surface is unlikely to transpire and that during evaporation of intercepted water there will be a saving in transpired water. The quantity of water involved is too large to ignore, and the significance of interception, and its evaporation, is discussed in Section IV,D.

The term "water consumption" is commonly equated with the total of all water evaporated from an area of vegetation, and all paths for evaporation of water are considered in this chapter. However, from a physiological viewpoint, it is important to distinguish between them. While both transpiration and evaporation from the soil deplete soil water, the temperature (Gates 1964) and water potential of leaves are in especially close relation to their transpiration rates. Moreover, while intercepted water may result in a saving of transpiration in trees that otherwise would be actively transpiring, it may be almost useless to trees that have exhausted the available water in the soil.

II. DISCUSSION OF METHODS

The size, longevity, and deep rooting habit of many trees make the measurement of their water consumption particularly difficult and introduce considerable error into standard methods used with other kinds of vegetation. The results of investigations reported in Section III must be viewed in relation to the limitations of the methods employed, and these will be discussed first.

Evaporation can be estimated from micrometeorological profiles above the vegetation, from the loss in weight of parts of plants (transpiration and the rate of loss of intercepted water), or indirectly from the hydrological equation:

$$E = P \pm \Delta S - R \qquad (1)$$

where E is evaporation, P is precipitation, ΔS is the change in storage of the soil or other defined system (e.g., a complete watershed), and R is run-off and drainage from the same system.

Many methods of estimating E rely on measurements of the terms on the right-hand side of Eq. (1) and especially on finding or making circumstances in which ΔS and R can either be accurately measured or else neglected. If all the terms in Eq. (1) are independently measured, a check on the accuracy of the observations is obtained, but such checks have rarely been made.

A. MICROMETEOROLOGICAL ESTIMATES OF VAPOR FLUX

Estimates of the vapor flux away from the upper surface of vegetation can be calculated, in the mass transport method, from measurements of the vertical gradients of wind speed and humidity. Alternatively, in the energy balance method, observations on vertical gradients of temperature and humidity can be used to calculate the partition of the available energy, which must be measured, between heat flux and vapor flux. Examples of the comparison of these two methods with accurate lysimeters are provided respectively by Penman and Long (1960) and Fritschen (1965). In both there are technical and theoretical difficulties to be overcome and as yet they hardly have been applied to measuring evaporation from forests, although Denmead (1964) used the energy balance method to measure apparent diffusivities within a canopy of *Pinus radiata* and to investigate the distribution of evaporation sources with height in the canopy. Both methods assume a horizontal uniformity of meteorological conditions, which seems unlikely just above the irregular surface of a forest canopy.

Observations on vapor flux or energy balance seem very suitable for measuring evaporation from the soil under trees. So far, only crude measurements of this component have been made. Some results are mentioned at the end of Section IV,B,4.

B. Transpiration

1. The Cut-Leaf or Quick-Weighing Method

Many ecologists have investigated the transpiration of forest trees by observing the loss in weight of leaves, twigs, or branches for a short time after their removal from the tree. It is well known that transpiration rates may change considerably at the moment of cutting and more slowly but steadily after cutting (Ivanov *et al.*, 1950; Andersson *et al.*, 1954). The cut-leaf method may have some comparative value, but the results can have absolute value only if it is first shown that the difference between the transpiration of a detached part and a whole plant is either negligible or reasonably constant.

Investigations on the effect of excision on the transpiration of the leaves and shoots of trees have been made by Ringoet (1952), Rawitscher (1955), Halevy (1956), Parker (1957), Stålfelt (1963), and Rutter (1966). The effect is variable between species and is dependent on environmental conditions. In many cases, however, it is small for a few minutes after excision and may be corrected if its magnitude is first determined.

The cut-leaf method can certainly be used, as it has been, e.g., by Pisek and Tranquillini (1951) and Parker (1957) to compare transpiration at different heights and on different sides of trees, but there are many difficulties in using it to estimate the transpiration of forest stands over long periods of time. Both the variations in space and the diurnal and long-term variations in time are large. Many early investigators neglected the high sensitivity of transpiration rate to climatic conditions and confined observations to a limited period around midday or made them on leaves suspended, between weighings, in the open near the balance. Some used the transpiration data of other investigators in other places and combined them with estimates of the weight or area of foliage per unit area of forest that they themselves made.

The estimation of weight of canopy per unit forest area presents further difficulties that will not be discussed.

The transpiration rate per unit leaf weight or area does not even have comparative significance for the transpiration rate of unit area of forest. Several European workers placed the transpiration rate per unit leaf weight of seven tree species in an order that was very similar in each investigation and showed a five- to sevenfold difference between *Picea* and *Betula* (Polster, 1950). As Polster pointed out, the species with highest transpiration rates are those with the least dense canopies, and the combination of transpiration per unit leaf weight with canopy weight per unit forest area gives estimates of the transpiration of forest stands that do not vary much between species, except that *Pinus sylvestris* is lower than the other species (Table I).

TABLE I

DAILY TRANSPIRATION OF LEAVES AND STANDS[a]

	Weight of foliage (kg ha^{-1})	Daily transpiration of leaves (gm gm^{-1} fresh wt)	Daily transpiration of stands (mm)
Birch	4,940	9.50	4.7
Beech	7,900	4.83	3.8
Larch	13,950	3.24	4.7
Pine	12,550	1.88	2.35
Spruce	31,000	1.39	4.3
Douglas fir	40,000	1.33	5.3

[a] Cited by Polster (1950), from data of Hartig (1865) and Burger (1935–1948), for stands aged 40–50 years.

Rutter (1966) attempted to solve the problems of adequately sampling the variations of transpiration in time and space and combining them with estimates of canopy weight. He checked his estimates of transpiration (plus interception and evaporation from the soil) against the sum of rainfall and soil water depletion and obtained reasonable agreement. It appears that fairly accurate estimates of the transpiration of the forest can be made in favorable circumstances by the cut-leaf method, but results so obtained should be examined very critically. It is desirable to separate transpiration from other paths of evaporation, and the cut-leaf method allows comparisons of transpiration at different times and on different parts of the tree. This method, however, is very liable to error and cannot be operated on any automatic basis or recorded. It is hoped that the method of investigating vapor fluxes at different heights of the canopy already tested by Denmead (1964) will be developed sufficiently to achieve the same analysis in a more convenient way.

2. The Heat Pulse Method

This method, developed by Huber (1932), has been used by a number of other workers (Ladefoged, 1960, 1963; Skau and Swanson, 1963; Doley and Grieve, 1966). Heat is applied for a few seconds at a fixed level on a trunk, and the velocity of the transpiration stream is calculated from the time required for the heat to reach a thermojunction or thermistor a short distance above the point of application. Thus, variation in flow rate up the trunk at various times of the day may be obtained, but the total volume flowing cannot be calculated unless the cross-sectional area of flow is known. Marshall (1958) suggested a method by which the cross-sectional area of flow could be estimated in conifer trunks. Ladefoged (1960, 1963) calibrated the method at the end of a season's observations by sawing through the trunks of experimental trees under water and then observing the relation between the rate of movement of the heat pulse and the rate of water uptake into the trunk. It is not certain that such calibration is valid for an entire season's measurements. Swanson and Lee (1966) drew attention to the considerable diurnal and seasonal fluctuations in the water content of trunks and suggested that the cross-sectional area of sap flow may well vary in the same way. Reynolds (1966) found that diurnal variations in water content caused variability in the relation between the quantity of water ascending *Picea abies* stems and transpiration into a plastic bag enveloping the crown.

3. The Transpiration Tent

Although transpiration of a leaf or shoot enclosed in a transparent chamber can be calculated from the increase in water content of air that is passed through the chamber, conditions inside the chamber may be changed so much, relative to ambient conditions, that transpiration rates must be altered. Decker *et al.* (1962) constructed a "tent," 3 m in diameter, of polyester film for measuring the transpiration of shrubs of *Tamarix*. Both temperature and vapor pressure were higher inside the tent than outside, and the net effect of the altered environment was to depress transpiration of test plants by 15–20%. Investigation of this technique is being continued at the Institute of Atmospheric Physics, Tucson, Arizona (van Bavel, in discussion of Shachori *et al.*, 1965). Shachori *et al.* (1965) described a tent of polyvinylchloride, large enough to enclose 10 m^2 of maqui vegetation, with ancillary equipment able to maintain normal wind speeds over the canopy and hold the temperature no more than 1–1.5°C above that of ambient air. Preliminary observations with small potted trees showed that transpiration in the tent was increased by only 1–3.5%.

C. INTERCEPTION

In this review, "interception" means precipitation retained by the aerial part of vegetation and either absorbed by it or evaporated. Therefore, it is synonymous with "interception loss" of Hamilton and Rowe (1949) and with "gross interception loss" of Burgy and Pomeroy (1958). Burgy and Pomeroy also distinguish "net interception loss" as that part of precipitation retained on the aerial part of vegetation that has no effect on the soil water consumption of the plant. Net interception loss is considered in Section IV,D.

Interception is measured by gauging the precipitation that falls through the canopy and runs down the trunks and then subtracting the sum of these from precipitation measured in clearings or outside the forest. There may be some error in assuming that the catch outside the forest or in a clearing is identical with that falling on the canopy, but this error appears to be less than what arises if gauges are exposed above the canopy. Sampling procedures for throughfall and stem flow have been discussed by Wilm (1943a), Reynolds and Leyton (1963), Reigner (1964), and Helvey and Patric (1965b). The present discussion will be limited to the question of whether all intercepted water is reevaporated from the intercepting surfaces or some of it is absorbed by trees.

Stone (1957, 1958) showed that the survival of *Pinus ponderosa, Libocedrus decurrens,* and *Abies concolor* in soil at the permanent wilting percentage was extended if their foliage was artificially wetted at night. He considered the effect on survival to be the result of a partial resaturation of leaf tissues. Leyton and Juniper (1963) found that the adaxial surface of *Pinus sylvestris* needles beneath the sheath absorbed much more water than the rest of the leaf surface and that tests of absorption of water through the cuticle of detached leaves are inadequate if the basal parts remain in air. A similar result was obtained with leaves of *Pinus radiata* by Johnston (1964), who also found that in a plantation in which the soil water was near the permanent wilting percentage, rain appreciably reduced the leaf water deficit even when a plastic cover on the surface of the soil prevented this rain from reaching tree roots. Stålfelt (1963) showed that the above-ground parts of *Picea abies* absorb water, and Vaadia and Waisel (1963) found that tritiated water entered leaves of *Pinus halepensis.* Little idea of the quantity of water involved can be obtained from these investigations. However, a calculation can be made from Johnston's observation that an overnight fall of rain of about 32 mm reduced the leaf water deficit in *Pinus radiata* by about 6%. Assuming that the leaves were approximately half water and half dry matter and that, as Rutter (1963) found in *Pinus sylvestris,* the maximum amount of water that could be retained on the leaf surfaces was about equal to their dry weight, the change in leaf water deficit was about 6% of the intercepted water. This suggests that the assumption that all intercepted water is evaporated is not seriously in error, although it needs quantitative examination for other species. Absorption of intercepted water however, may be of physiological significance to trees in dry soil when they are wetted by dew or by rain too light to penetrate the soil.

D. SOIL WATER DEPLETION

When the soil water has been reduced below field capacity, evaporation is frequently estimated from the sum of rainfall and soil water depletion,* less any surface run-off that occurs. Apart from random sampling errors, and uncertainty attached to calibration curves of soil-moisture resistance units when these are used to follow depletion, the main errors appear to arise from failure to sample the full depth of rooting and from the assumption that percolation is negligible. Tree roots in soil resting on rock frequently enter rock fissures. If the rock is crumbling or porous, appreciable quantities of water, which cannot readily be measured, may be obtained from it. Shachori *et al.* (1967) drilled the access holes for a neutron-scattering probe to a depth of 9 m in limestone beneath 1 m of soil and found that maqui scrub withdrew water to a depth of about 8m.

While percolation in many soils may be negligible when water content is below field capacity, this is not true of all soils (Russell, 1961). The lower horizons of deep soils still may be losing water by drainage when the upper ones have been reduced below field capacity.

E. LYSIMETERS

Because of instrumental and theoretical difficulties of calculating vapor flux by micrometeorological methods, and the difficulty of separating the downward flux of water from the upward flux of water vapor in observations on soil water depletion, van Bavel (1961) considered that the study of moisture changes in a body of soil confined in a lysimeter was the only practicable method for measuring evaporation with adequate precision. He pointed out, however, that lysimeters gave valid estimates of actual evaporation only if their construction and operation met certain requirements. The main considerations were that lysimeters should contain an undisturbed and representative soil profile, that the bottom should be well below the deepest roots or that there should be some device for maintaining water tension at the bottom equivalent to the tension in the surrounding soil at the same depth, and that the lysimeter should be surrounded by an adequate area with the same soil

---

* This is more correctly defined as the sum of rainfall and net change of soil water storage. Many studies cited in this chapter as being based on observations of soil water depletion actually include both depletion and some repletion in the net change in storage.

and vegetation as in the lysimeter. There are few tree-covered lysimeters in the world, and of the three from which results are quoted (Deij, 1948, 1954; Law, 1958; Patric, 1961) perhaps none is entirely free from criticism in these respects.

F. EXPERIMENTAL WATERSHEDS

The difficulty of establishing forest-covered lysimeters with satisfactory design and surroundings makes the proposition attractive that an entire valley may be regarded as a natural lysimeter. The drawbacks of this method for estimating evaporation lie first in the possibility of large errors of measurement in any one watershed and second, in the difficulty of establishing adequate controls.

The basic measurements made on experimental watersheds are precipitation and streamflow. The calculation of evaporation as the difference between these measurements assumes first that they can be accurately measured and second that no liquid water enters the watershed except as precipitation or is lost except through the stream gauge. Hence, the use of watershed studies to estimate evaporation requires either geological evidence of watertightness, or a check made by estimating the water percolating from the root zone and comparing it with stream flow, over a period long enough to allow the lag between them to be neglected. Rowe and Colman (1951) made such a comparison in Monroe Canyon near San Dimas, California, where they found that estimated annual percolation from the root zone was about 450 mm, whereas annual stream flow was in the range of 80–120 mm. In view of the extensively faulted nature of the underlying rock, this result did not surprise them. An essentially similar but more highly developed check is described by Pereira et al. (1962a). They showed catchments at Kericho (Pereira et al., 1962b) and Kimakia (Pereira et al., 1962c) in Kenya to be essentially watertight.

Since measurement of evaporation is not the primary purpose of most watershed experiments, and since Dr. J. E. Douglass justifiably indicated in correspondence that workers at the Coweeta watersheds are not prepared to equate evaporation with the difference between precipitation and stream flow, the term "apparent evaporation" from watersheds will be used. Examples have been excluded where clear evidence of leakage has been pointed out by authors. For the rest, it is doubtful whether the errors attached to "apparent evaporation" are any greater than those that enter estimates made by other methods.

If different kinds of vegetation are to be compared by a watershed experiment it is clearly desirable that the watersheds should be as similar as possible in all respects other than their vegetation. Among the important properties of watersheds that may affect evaporation are aspect and shape, which influence exposure to radiation and wind, and soil and drainage conditions, which affect the amount of stored water available for evaporation. Bates and Henry (1928) first attempted to overcome the difficulty of obtaining complete comparability by comparing two watersheds with the same vegetation for a number of years and then changing the vegetation on one and using the other as a control. This method has been improved by fitting the regressions of run-off on climatic variables for each watershed and testing the significance of the difference between before- and after-treatment regressions (Wilm, 1943b; Kovner and Evans, 1954). Goodell (1951) developed a still more sensitive technique capable of predicting flow from one watershed in selected months or natural segments of the hydrograph from the behavior of its control. Reinhart (1967) reviewed watershed calibration methods.

G. CONCLUSIONS ON METHODS

Methods of measuring evaporation have been described in a series that began with microclimatic observations above and among trees and progressed through observations on the trees themselves, to measurements of uptake from the soil, percolation from lysimeters, and streamflow from watersheds. The watershed method has the advantage of providing data from a large area and integrating the effects of all irregularities that may occur in forested land. This advantage, however, is offset by the disadvantage that in many cases it is impossible to obtain valid estimates of evaporation for less than a year, or at least a growing season. Consequently, it is impossible to analyze the relation between evaporation and environment or to extrapolate with confidence from the results obtained to what would be expected in other environments.

At the other extreme, micrometeorological measurements can show hour-by-hour changes in evaporation rate and allow a detailed analysis of their relation to environment. Normally they are used to measure the vapor flux over an extremely small area of vegetation, although a wider sampling would appear to be prevented by nothing but expense.

In between these two methods, observations on soil water depletion can give estimates of evaporation for periods of as little as one to a few weeks when soil water is below field capacity, and well-constructed draining lysimeters (weighable lysimeters are excluded from consideration) do the same in more humid conditions. In both these methods some analysis of the relation of evaporation to environmental conditions is possible.

Watershed experiments are the only means of investigating effects of watershed management on the behavior of streams, but understanding of evaporation can be advanced only through methods that allow for analysis of its relation to seasonal, or even diurnal, changes in environmental conditions.

III. REVIEW OF RESULTS

A. COMPILATION OF THE DATA

Before reviewing measurements of the water consumption of forests, the method of constructing Table II, in which much of the available data is summarized, will be described. In order to generalize from the results, the conditions under which they were obtained needed to be known, and a system had to be devised for specifying the conditions into which results obtained and reported in diverse ways could be fitted.

The system adopted assumes that the environmental factors controlling evaporation in a given period of time are the supply of energy and the availability of water. The supply of energy is made up of net radiation, i.e., the difference between incoming solar (shortwave) radiation and outgoing long-wave and reflected shortwave radiation, and of advected energy which may be defined as heat obtained from the air and derived from energy exchanges outside the area considered. Advected energy had to be neglected in Table II but net radiation was obtained from Budyko's (1955) *Atlas of Heat Balance*. This atlas contains, among other things, maps of net radiation over the earth's surface for the year and for each month separately. The theoretical bases of its construction have been described by Budyko (1956). The maps of net radiation show lines of equal radiation that were drawn by interpolation between observations made at 770 stations, 420 of them on land. Clearly, large errors must arise when the values for particular locations, defined by latitude and longitude in Table II, are read by interpolation between the lines on the map. This is particularly the case in coastal or mountain regions and, in fact, many mountain regions are left unmapped in the atlas. Rough interpolations for some mountain regions have been included in Table II, and they are duly marked. Approximate altitudes are entered in the table, but slope and aspect which are obviously important in the radiation balance have not been included. Further inaccuracy must arise from the fact that one component of the net radiation, viz, the reflected solar radiation, varies with the vegetation. In constructing the maps Budyko necessarily assumed a coefficient of reflection appropriate to the main vegetation type over a large area. Reflection coefficients for the particular forest plots of Table II may well have differed from those assumed for their region when the map was made. Nevertheless, the maps probably provide the most satisfactory index available of evaporating conditions over the globe and, as will be seen later, the values obtained from them account for a large part of the variation of the recorded water consumption of forests.

Water supply is initially dependent on precipitation and also on the capacity of the soil to store water for use in periods when evaporation exceeds

TABLE II

Evaporation from Forests Arranged in Classes According to Soil Water Deficit in the Growing Season and Within Each Class by Increasing Net Radiation[a]

No.	Location and vegetation	Latitude	Longitude	Altitude (m)	Precipitation Annual (mm)	Precipitation Growing season[b] (mm)	Rooting depth (m)[c]	Available water (mm)	Maximum soil water deficit (mm)	Evaporation Annual (mm)	Evaporation Growing season (mm day⁻¹)[g]	Net radiation Annual (mm)	Net radiation Growing season (mm day⁻¹)	Author
NEGLIGIBLE SOIL WATER DEFICIT														
1	Northern taiga, USSR Mainly *Pinus sylvestris* and *Picea abies*	62–65 N	40–45 E	< 200	525	(50%)								Molchanov (1960)
	Mean of 7 studies								(25)	286		330		
	Range of 7 studies									209–370				
2	Middle taiga, USSR Mainly *Picea abies* and *Betula*	c.59 N	c 40 E	c 200	600	(50%)								Molchanov (1960)
	Mean of 5 studies								(25)	329		420		
	Range of 5 studies									210–449				
3	Yorkshire, England *Picea sitchensis*	54 N	2 W	200	1350	50%			(50)	800		500		Law (1958, and personal communication)
4	Harz Mountains, West Germany *Picea abies*	52 N	9 E	600	1250	50%			(50)	579		580		Delfs et al. (1958)
5	Carpathian Mountains, Czechoslovakia Mixed conifer and deciduous	49 N	18 E	750	997	(66%)			(0±)	525		630		Valek (1959)
6	Emmental, Switzerland Mixed conifer and deciduous	47 N	8 E	1100	1650	950			(0±)	861		670		Burger (1943, 1954)
7	Northern Japan Kamikawa-Kitatani 40% conifer, 60% deciduous	c 38 N	c 140 E		1437	(60%)			(0)	630		670		Nakano (1967)
	Kamabuchi, No.2 Mixed conifer and deciduous				2617	(66%)			(0)	542		670		
8	Ota, Japan Mixed conifer and deciduous	37 N	141 E	350	1600	(66%)			(0)	658		750		Hirata (1929, cited by Penman (1963)

No.	Location and forest type	Lat.	Long.												Reference
9	Shackham Brook, Central New York, U.S. 57% conifer, 27% deciduous, 16% pasture and crops 1952, 1953, 1955, 1956	43 N	75 W	450	1050	500			(0±)	504			750		Schneider and Ayer (1961)
10	Mt. Koskiusko, New South Wales, Australia *Eucalyptus niphophila*	36 S	148 E	1650	2500	50%	0.6	175+	27		XI–III[g]	3.8	950	3.9	Costin et al. (1964)
11	Coweeta, North Carolina, U.S. Mixed hardwoods, (Watershed 13)	35 N	83 W	900	1730	(50%)			(< 50)	875			1000		Hoover (1944)
12	Coweeta, North Carolina, U.S. Mixed hardwoods	35 N	83 W	c 800	2000	(50%)			(0±)	700			1000		Johnson and Kovner (1956)
	SMALL SOIL WATER DEFICIT														
13	Southern taiga, USSR Mainly *Picea abies* and *Betula* Mean of 4 studies	58–60 N	30–40 E		(600)	(50%)			(100)	412			420		Molchanov (1960)
	Range of 4 studies									330–550					
14	Bregentved, Denmark *Picea abies*	56 N	9 E	(< 150)	c 680	50%			(130)	470			500		Holstener-Jorgensen (1959)
										449					
15	Bregentved, Denmark *Fagus sylvatica*	56 N	9 E	(< 150)	520	60%				c 450			500		Holstener-Jorgensen (1961)
	Fagus sylvatica, Quercus species and other deciduous									c 400					
16	Scania, Sweden *Picea abies* Wet site	55 N	14 E	< 150	793	50%			(125)	883	V–VIII	4.8	500	3.5	Stålfelt (1963)
	Dry site									716		3.4			
17	Castricum, Holland, 1955–1964 *Pinus nigra*	53 N	5 E	0	840	50%		(250)	(100)	655			550		Deij (1954 and personal communication)
	Mixed deciduous							(250)	(50)	500					
18	Berkshire, England *Pinus sylvestris*	51 N	1 W	100	686	378	1.8	300–370	150	655			580		Rutter (1964), Rutter and Fourt (1965), and unpublished calculations
19	Lower Michigan, U.S. *Pinus resinosa* and *Quercus* species in pure stands	43 N	85 W	(300)	550	550		147	100		IV–X	2.8	670	3.1	Urie (1959)
20	Fernow, West Virginia, U.S. Mixed deciduous	38 N	80 W	750	1490	50%			(100–150)	865			880		Reinhart and Eschner (1962)
21	Okayama, Japan Tatsuno Kuchiyana-Minamitani *Pinus densiflora*	35 N	134 E		1153	(66%)			(100)	847			920		Nakano (1967)

TABLE II—Continued

No.	Location and vegetation	Latitude	Longitude	Altitude (m)	Precipitation (mm) Annual	Precipitation Growing season[b]	Rooting depth (m)[c]	Available water (mm)	Maximum soil water deficit (mm)	Evaporation Annual (mm)	Evaporation Growing season (mm day^{-1})[g]	Net radiation[d] Annual (mm)	Net radiation[d] Growing season (mm day^{-1})	Author
	Tats.-Kitatani Pinus densiflora				1113	(66%)			(100)	823	VI–IX	920		
22	Gunnison, Colorado, U.S. Populus tremuloides	39 N	107 W	3000	530	25%	1.8	730	350		4.2	950[e]	3.9[e]	Brown and Thompson (1965)
	Picea engelmannii						2.1	480	240		3.3			
23	Fraser, Colorado, U.S. Pinus contorta	40 N	106 W	2900	620	c 150			55+	360	VI–IX	950[e]		Wilm and Dunford (1948)
24	Wagon Wheel Gap, Colorado, U.S. Pseudotsuga taxifolia and Populus tremuloides	38 N	107 W	3050	530	< 50%			(< 150)	383	1.5	950[e]	3.9[e]	Bates and Henry, (1928)
25	Kericho, Kenya Evergreen rain forest	0	35 E	2200	1950	Well distributed	>3.0	550+	250	1570		1330[e]		Pereira et al. (1962b)
26	Aberdare Mountains, Kenya Bamboo forest	1S	36 E	2400	2160	Well distributed	(> 3.0)	750+	100+	1150		1330[e]		Pereira et al. (1962c)
MODERATE SOIL WATER DEFICIT														
27	Moscow region, USSR Mainly Pinus sylvestris and Picea abies	c 56 N	c 38 E		480–575	50%			(180)			450		Molchanov (1960)
	Mean of 25 studies									437				
	Range of 25 studies									346–537				
28	Forest-steppe, Russia Deciduous forests	51–52 N	35–39 E		510–525	50%			(150)			500		Molchanov (1960)
	Mean of 34 studies									405				
	Range of 34 studies									309–528				
	Pine forests								(150)			500		
	Mean of 10 studies									412				
	Range of 10 studies									287–698				
29	Steppe zone, European USSR Quercus, Fraxinus, Betula, and other deciduous	47–50 N	30–40 E		375–457	50%								Molchanov (1960)

No.	Location and species	Lat.	Long.	Elev. (m)								Growing season / value	Ratio	Reference
	Mean of 21 studies								(225)	424	580			
	Range of 21 studies									212–568				
30	Lebanon State Forest, New Jersey, U.S. *Pinus echinata* with understory of *Quercus* species	40 N	75 W	c 50	(1200)	50%	3.6	140	125		630	IV–X / 2.97	3.05	Lull and Axley (1958)
31	Farmington, Utah, U.S. *Populus tremuloides*	41 N	112 W	(1000)	1340	100–250	1.8+	320	293	567	830[e]	VI–IX / 3.92	3.7[e]	Croft and Monninger (1953)
32	Vinton, Ohio, U.S. *Pinus echinata* / *Quercus* species / Shrubs	39 N	83 W	c 200	1000	30%	{All to rock at 0.9 m}	160 / 160 / 160	135 / 125 / 108		830	IV–IX / 3.5, 3.5, 3.15	3.65	Marston (1962)
33	Union, South Carolina, U.S. *Pinus taeda*	35 N	82 W	200	1000	50%	1.7	180	150		1000	3.7 (40 rainless days)	4.3	Metz and Douglass (1959)
34	Union, South Carolina, U.S. *Pinus taeda* (1952 only)	35 N	82 W	200		530	2.4	(260)	206		1000	IV–X / 4.0	4.1	Hoover *et al.* (1953)
35	Crossett, Arkansas, U.S. *Quercus* species / All aged, to 40 years / Even aged	33 N	92 W	> 150	(1200)	460	> 1.2 / > 1.2 / > 1.2	345 / 345 / 345	320 / 280 / 320		1000	VI–IX / 4.20, 3.89, 4.49	4.45	Moyle and Zahner (1954)
	Pinus echinata + *P. taeda*													
36	Aberdare Mountains Kenya / Bamboo forest / *Pinus radiata* / *Cupressus macrocarpa*	1 S	37 E	2600	1100	Mostly in III–VI	3.6 / 3.6+ / 3.6	c 450 / c 450 / c 450	c 300 / c 300 / c 300	965	1330[e]	E_T/E_o[i] : 0.86, 0.85, 0.86		Pereira and Hosegood (1962)
	SEVERE SOIL WATER DEFICIT													
	Estimated mean annual													
37	North Fork, California, U.S. Mixed chaparral species	37 N	119 W	850	1150	slight	Rock at 1.2		End July[h]	485	830[e]			Rowe (1948)
38	San Gabriel Mountains, California, U.S. *Rhus* species and *Cercocarpus montana* (1917–1924)	34 N	118 W	1200	780	slight				595	830			Hoyt and Troxell (1934)
39	California, U.S. Bass Lake *Pinus ponderosa*	37 N	119 W	1050	1260	slight			End July[h]	580	830[e]			Rowe and Colman (1951)
	North Fork Mixed chaparral species	37 N	119 W	850	1050	slight			End June[h]	415	830[e]			
	San Dimas Mixed chaparral species	34 N	118 W	850	1150	slight			End July[h]	560	830			

TABLE II—Continued

No.	Location and vegetation	Latitude	Longitude	Altitude (m)	Precipitation (mm) Annual	Precipitation Growing season[b]	Rooting depth[c] (m)	Available water (mm)	Maximum soil water deficit (mm)	Evaporation Annual (mm)	Evaporation Growing season (mm day⁻¹)[g]	Net radiation[d] Annual (mm)	Net radiation[d] Growing season (mm day⁻¹)	Author
40	San Dimas, California, U.S. Results for 1950	34 N	118 W	850	1230	slight								Patric (1961)
	Pinus coulteri						1.8	355		637		830		
	Adenostoma fasciculatum						1.8	355		648				
	Ceanothus crassifolius						1.8	355		599				
	Quercus dumosa						1.8	355		630				
41	Carmel Mountains, Israel, Maqui shrubs	33 N	35 E	c 450	650	c 100	> 7.0			509	III–X 1.90	1080	3.9	Shachori et al. (1967)
	EXTREME SOIL WATER DEFICIT													
42	San Dimas, California, U.S. Quercus dumosa and other chaparral spp.	34 N	118 W	c 1000	600	slight	3.5			535	V–X 1.87	830	3.2	Rowe and Reimann (1961)
43	San Dimas, California, U.S. Results for 5 dry years	34 N	118 W	850	525	slight			(June)[h]			830		Patric (1961)
	Pinus coulteri						1.8			392				
	Adenostoma fasciculatum						1.8			430				
	Ceanothus crassifolius						1.8			455				
	Quercus dumosa						1.8			458				

[a] Figures in brackets are estimates not directly derived from the original papers.
[b] Mean growing season precipitation given either as a depth or as an approximate percentage of annual precipitation.
[c] Rooting depth or depth of soil water depletion, whichever was observed.
[d] From Budyko (1955).
[e] Net radiation estimated by interpolation into unmapped regions of Budyko (1955).
[f] c = about.
[g] Months in which records made are shown by Roman numerals.
[h] Time when available water exhausted.
[i] Ratios of forest evaporation to open-water evaporation in seasons of variable length.

precipitation. In specifying conditions of supply in Table II, the first step was to give the annual precipitation and the second was to estimate (often only approximately) the proportion that falls in the growing season, or summer months. The extent to which available water in the soil is depleted in summer has then been estimated, and the results have been arranged in the following five classes:

1. Negligible soil water deficit: not exceeding 50 mm in an average year.
2. Small soil water deficit: maximum depletion not exceeding half the available water within the root range; where this is not calculable, then not exceeding 150 mm.
3. Moderate soil water deficit: maximum depletion exceeding half the available water (or exceeding 150 mm), but only just reaching permanent wilting percentage before restoration.
4. Severe soil water deficit: available water exhausted and remaining so for at least 1 month.
5. Extreme soil water deficit: available water exhausted and winter rains insufficient to restore the whole root range to field capacity.

Available water is taken to be that between field capacity and permanent wilting percentage, or similar specifications, within the rooting depth of the trees. The necessary data are given in a fair proportion of cases, especially when evaporation has been estimated as rainfall plus depletion of soil moisture. Where rooting depth and texture are given, storage capacity may be estimated approximately from the texture by comparison with the analyses of soils of a wide range of textures made by Salter and Williams (1965). Approximate monthly contributions to annual evaporation have frequently been calculated in watershed experiments by Thornthwaite's (1948) formula or by comparison with pans. By comparing these with monthly precipitation it is possible to estimate the development of soil water deficit during the season. Where no other guide is available, the maximum soil water deficit has been roughly estimated as the difference between annual evaporation and precipitation during the summer months.

Many rather subjective decisions have to be made in this classification. It would take too much space to justify the assignment of each example to its particular class, but considerable care has been given to each. An outline of the justification is contained in the data set out in Table II, and where numerical estimates are presented that are not given or readily calculable from the text of a particular paper, they are set in brackets. Some of the evidence on the soil water balance of experiments on watersheds is discussed more fully in Section III,D1.

There must be investigations that might have been included in the table but have been overlooked. In addition, there are deliberate exclusions. Where there is good reason to believe that watersheds were far from water-tight or that a significant part of the rooting depth was not sampled, the results have not been used. Apart from these considerations no attempt has been made to assess the reliability of results before inclusion in the table, though some are discussed further. Almost all are based on several years' observations. The translation of Molchanov (1960) is the main source of Soviet data. In particular, Molchanov's Table 218, which includes more than 100 estimates of evaporation from forests in the USSR in Europe, has been summarized as five forest zones in Table II. Most Japanese results have been obtained from Nakano (1967).

B. Effects of Variation in Forest Structure and Age

In considering the water consumption of a forest in a given environment it is necessary that characteristics of the forest that may affect its consumption should be described. A forest is normally defined by naming the species of which it is composed and specifying their relative numbers. However, given the botanical composition, consumption may vary with age, canopy density, stratification, and other aspects of structure. Fundamentally, the water consumption of forest is determined by its radiation balance, by the availability of water at evaporating surfaces (including leaves and soil), and by factors controlling exchanges of heat and water vapor between the forest and the air above it. This suggests that for analysis of water consumption a forest should be described in terms of (1) its ability to absorb, reflect, and emit radiation, (2) the distribution of roots and leaves, (3) stomatal characteristics, and (4) aerodynamic roughness. Such an analysis will be attempted in Section IV, but first evidence of changes in water consumption with age and density of tree and shrub layers, all of which can be controlled experimentally, will be discussed.

1. Effects of Reducing Stand Density

As might be expected, experimentally imposed reductions in stand density cause corresponding but not necessarily proportional, changes in interception. Wilm and Dunford (1948) compared five densities of *Pinus contorta*, having a sparse understory of *Populus tremuloides* and *Shepherdia canadensis*, at 3000 m in the Rocky Mountains. Initially there were 300–400 trees per acre exceeding 3.5 inches diameter at breast height and an average volume of 12,000 board feet per acre. This stocking was reduced in treatments with fourfold replication, as shown in Table III. The most extreme treatment removed all large timber and left only 147 trees per acre exceeding 3.5 inches diameter at breast height. Between the control and the most extreme treatment, interception of snow in spring and rain in summer was decreased from 104 mm to 28 mm, a reduction of over 70%. Interception from rain alone was reduced from 36 mm to 8 mm.

TABLE III

EFFECTS OF THINNING *Pinus contorta* STANDS ON WATER BALANCE
IN SUMMER (JUNE THROUGH SEPTEMBER)—MEANS OF 1942 AND 1943[a]

	Water balance (mm) at residual stocking (board feet acre^{-1}) of:				
	12,000	6,000	4,000	2,000	0
Precipitation	112	112	112	112	112
Soil water depletion	53	61	42	51	45
Total evaporation	165	173	154	163	157
Interception	36	22	17	15	8
Transpiration + evaporation from soil (by difference)	129	151	137	148	149

[a] From Wilm and Dunford (1948).

Rogerson (1967) measured throughfall in plots of pole-sized *Pinus taeda* thinned to various densities. The least dense plots had about 30% of the crown area and 20% of the basal area of the most dense plots. Estimated interception from mean storm size of 16 mm was 3.75 mm in the most dense and 1.5 mm in the least dense plots, a reduction of 60%.

Wilm and Dunford (1948) found, however, that although interception was proportionally greatly reduced by thinning, the change in total evaporation was slight (Table III). In the summer (June through September) total evaporation can be calculated as the sum of rainfall and soil water depletion. The differences in soil water depletion, and hence in estimated total evaporation, were not statistically significant.

Goodell (1958) removed half the trees in a forest of *Pinus contorta, Picea engelmannii,* and *Abies lasiocarpa* on a calibrated watershed in Colorado. The trees were clear-felled in strips varying in width from 22 to 132 yards. The increases in stream flow in the 2 following years were 274 and 219 mm. However, when Rich (1959) removed 36% of the basal area of a mixed conifer stand in Arizona, using a method of selection that did not create the large gaps of Goodell's study, stream flow was not increased significantly.

Reinhart and Eschner (1962) applied differential thinning to 5 watersheds carrying mixed hardwoods in West Virginia, having first compared them for 6 years. Changes in stream flow during the year following treatment are shown in Table IV. Mean annual evaporation from the control watershed, estimated from the difference between rainfall and run-off, was 865 mm. Therefore, large reductions in density apparently reduced evaporation by less than 10%. Most of the reduction occurred during the growing season. The effects diminished in subsequent years.

TABLE IV

Effects of Cutting on Stream Flow from Mixed Hardwood Forest, in the Year Following Cutting at Fernow, West Virginia[a]

Method of treatment	Basal area square feet $acre^{-1}$		Increase in stream flow (mm)
	Original	Residual	
Commercial clear-cut	98.2	17.2	86
Diameter limit	95.0	60.7	64
Extensive selection	114.4	87.4	18
Intensive selection	104.6	84.8	8
Control	105.7	105.7	

[a] From Reinhart and Eschner (1962).

Knoerr (1965) studied the depletion of soil water in stands of *Abies magnifica* varying in density of cover from 50 to 100%. The maximum rate of water uptake was found in stands with 70–80% cover, and the rate averaged about 15–20% lower at both 50 and 100% cover. Knoerr considered that the reduction in the rate of soil water depletion at high cover might be the result of restriction of transport of sensible heat from the surface of the canopy to its shaded parts.

Finally, Hibbert (1967) brought together the results of 30 experiments in various parts of the world on the increase in stream flow following forest thinning. Maximum increases in stream flow in the first year were 45 mm for each 10% of forest cover removed, but in 20 experiments the increase was less than half of this, and in 9 or 10 experiments it was negligible.

2. Evaporation from the Understory and Field Layer

In considering the relation between the structure of a forest and its water consumption, the existence of different strata of vegetation and the contribution of each stratum to total consumption must be taken into account. By frequent measurements under conifer and hardwood forests, with and without understories, Maran (1947), as reported in Penman (1963), found no major differences in soil water content that could be ascribed to effects of the understory. Similarly, Romanov (1951), cited in Penman (1963), found that the soil water content under oak was independent of whether the understory was complete, half removed, or completely removed. In a probably better-controlled experiment, Johnson and Kovner (1956) cut the dense understory of *Kalmia latifolia* and *Rhododendron maximum* that between them covered 54 of the 70 acres of a watershed under mixed hardwoods at Coweeta. The increase in stream flow was 100 mm in the first year, declining after 6 years to 16 mm as the understory was reestablished. When annual evaporation is estimated from the difference between precipitation and streamflow, it appears that the understory accounted for about 13 % of the total evaporation from the watershed. It should be noted that this was an evergreen understory beneath deciduous trees. Zahner (1958) eradicated a hardwood, presumably mainly deciduous, understory that ranged from small shrubs to trees 6–8 m tall and had a density of up to 20,000 stems per acre, beneath 50-year old stands of mixed *Pinus taeda* and *P. echinata* in Arkansas. While the soil was moist the rate of depletion of soil water was similar on control and treated plots, but by midsummer depletion rates were 25 % greater on the control plots. Molchanov (1960) investigated evaporation from the field layer of a number of pine, spruce, birch, and oak forests by using weighable "evaporators" (sic). In most cases the field layer accounted for 20–25 % of the total evaporation from the forest.

Barrett and Youngberg (1965) compared thinning treatments, with or without eradication of the dense understory of *Purshia tridentata, Ceanothus velutinus*, and *Arctostaphylos pinetorum*, in *Pinus ponderosa* in Oregon. Unfortunately, precipitation was not recorded accurately, but it appears to have been about 100 mm during the growing season. Reducing density from 1000 to 62 stems acre^{-1}, with understory present, reduced soil water depletion from 119 to 102 mm (and evaporation by about 8 %). Eradicating the understory from the most dense stand reduced soil water depletion from 119 to 90 mm (and evaporation by about 13 %). There was a suggestion of interaction between the treatments, as would be expected, but it was not significant statistically.

The investigations that have been quoted show that the effects of reducing stem density either by cutting the dominants or the understory are variable. Goodell (1967) and Douglass (1967) discussed the processes that relate evaporation from forests to changes in their density and structure. If a forest is thinned, the lower branches of the remaining trees, and any shrub and herbaceous layers present, are exposed to greater solar radiation and air movement. Conditions can be imagined in which considerable thinning can be made with only small reductions in the total radiation absorbed at leaf surfaces. Where larger gaps are created, and in the absence of understory or herbaceous vegetation, water should be conserved in the areas from which trees are removed, and Douglass (1960) found this to be so in a thinned *Pinus taeda* stand. If gaps created by thinning are small, e.g., where one to four trees are removed, then the empty crown and soil space is likely to be filled again in a few years by the growth of surrounding trees. If the gaps are much larger they may not be invaded by trees for many years. Nevertheless, development of dense ground vegetation may begin to consume water made available by thinning. The extent to which herbaceous communities in gaps will consume water may be reduced by their sheltered positions in relation to wind and, when precipitation is less than potential transpiration, by their possibly smaller rooting depth. Goodell (1967) summed up the situation by stating that "evapotranspiration depends on the spatial and temporal confluence of water and thermal energy at watershed surfaces," including transpiring surfaces of all the plants and the soil and snow pack, and he examined the effects of watershed treatments from this point of view.

3. Variation with Age of Stand

Evidence on the correlation of evaporation with stand age can be obtained either from adjacent stands of different age in comparable conditions of climate and soil or by observing changes with time in conditions in which a suitable control is available for comparison.

Holstener-Jorgensen (1961) compared 99 stands, mainly of *Fagus sylvatica*, *Quercus* species, and *Picea abies*, having a wide range of ages, at Bregentved, Denmark. The stands were on a low-lying, heavy-textured soil, and he considered that there was no drainage from them except when the water table rose above a certain level in winter. The fall of the water table in summer was therefore brought about only by evaporation. The extent of the fall was positively correlated with stand age. It is difficult to assess the corresponding change in evaporation rate, but consideration of Holstener-Jorgensen's data suggests that in most species annual evaporation increased between the ages 10 and 60 years by 50–80 mm (about 1 mm yr^{-1}) on a mean annual evaporation of about 400 mm. The effect on the water table was most marked in the first 20 years.

Molchanov's (1960) results, for ages up to 160 years of pine stands in the Moscow region, oak stands in the steppe forest zone, and spruce stands in the zone of mixed forest, show that in the first 40–60 years annual evaporation increased by 2–3 mm $year^{-1}$ and thereafter declined for an additional 100 years by about 1 mm $year^{-1}$. Schneider and Ayer (1961) reported changes in stream flow from Shackham Brook in New York State. In 1931 this watershed carried mixed deciduous woodland on 27% of its area, and an additional 57% was planted with *Pinus* and *Picea* species. By 1958 most of the forested area had attained a crown cover of over 90%. Between 1934 and 1957 stream flow declined by the equivalent of 5.3 mm $year^{-1}$. Douglass (1967) and Hibbert (1967) discuss the results obtained on Watershed 13 at Coweeta, North Carolina, which originally carried a mixed hardwood forest. This was clear-cut at the beginning of 1940 and then allowed to grow again from the cut stumps. In the year following cutting, stream flow increased by 373 mm. It then fell by about 300 mm over the next 24 years, more rapidly at first, and at a rate of 5–6 mm year^{-1} after the first 10 years.

On the Jonkershoek multiple watershed experiment in South Africa, run-off from two watersheds afforested with *Pinus radiata* declined during 16 years by about 22 mm year^{-1} compared with controls under native "fynbos" or sclerophyllous scrub (Wicht, 1967b).

Douglass (1967) argued that the apparent increase of annual evaporation with time on Coweeta Watershed 13 could not be the result of changes in ground cover, root development, total leaf surface, or soil moisture but was probably the result of a factor closely related to the height of the trees. Holstener-Jorgensen (1967) concluded that the greater seasonal fall of the water table he observed in older stands was the result of an increase of rooting depth with the age of stand. Molchanov (1960) separated the contributions of the soil and herbaceous vegetation to total evaporation, apparently by using weighed evaporators. Much of the increase in the first 50 years appeared to be the result of increased evaporation from the herbaceous layer, at least in pine and oak, but this was offset by falling interception and transpiration of the trees in the later life of the stands. Knoerr (1965), comparing young and mature stands of *Abies magnifica* in California, found that depletion rates from a 5-foot profile were similar when the profile contained adequate water but that at reduced water contents the depletion rate was less under a young stand than under a mature one.

C. Differences Among Species

1. Transpiration

Transpiration rates, per unit leaf weight or area, of trees growing in the same conditions often vary greatly among species (see Table I). However, extrapolation of such data to the transpiration of forest stands depends on assumptions that, except for estimating orders of magnitude, cannot be justified (Section II,B,1), and such calculations are disregarded here.

Ladefoged (1963) compared the transpiration of forest trees in closed stands in Denmark by the heat-pulse method, and some of his results are given in Table V for 1954, a wet summer when it was believed that little soil water deficit developed. The trees were all about 30 years old and measurements were made on 7–10 trees of each species. Ladefoged considered that the *Fagus* stand was typical, the *Picea* stand and possibly the *Quercus* stand were more dense than normal, and the *Fraxinus* stand was less dense than normal for Danish forests. Transpiration averages for June–August were based on only 3–5 days for each species, and observations on different species were not made on the same day. Adding to this the uncertainty of whether calibration of the method was satisfactory (Section II,B,2), these comparisons among species must be accepted with reservations.

TABLE V

DAILY TRANSPIRATION (mm) OF FOUR SPECIES IN
CLOSED STANDS IN DENMARK[a]

	Cloudless summer day, no morning dew	Mean, June–August
Fagus sylvatica	4.1	2.9
Quercus species	4.3	2.7
Fraxinus excelsior	3.3	1.7
Picea abies	3.6	2.4

[a] After Ladefoged (1963). Two methods of calculating transpiration on an areal basis were used, and the values given are the means of these.

2. Interception

The forest canopy has a capacity to retain precipitation that may be defined in different ways, i.e., maximum surface detention, detention after drip, and detention after shake (Grah and Wilson, 1944). Neglecting differences between these definitions, the interception process may be viewed as a filling of a capacity by precipitation and its emptying by evaporation. If 2 periods of precipitation are separated by a rainless period too short to evaporate all the intercepted water, then interception from the second fall will be less than from the first. The total amount of water intercepted and evaporated in a given period of time, therefore, is affected by the distribution of precipitation in time, by the capacity of the canopy to retain precipitation, and by the evaporation rate, which may also be a function of the canopy insofar as there are differences among canopies in density and aerodynamic roughness.

Curves relating interception to storm size are often fairly sharply inflected since a large proportion of incident precipitation is intercepted from storms less than canopy saturation capacity, but once saturation is reached there can be no further interception unless there is evaporation during the storm. Extrapolation of the regression of throughfall on storm size gives, at zero through fall, a statistical estimate of the precipitation required to saturate the canopy. However, this will be an underestimate unless a large proportion of the storms exceed the saturation capacity of the canopy. Zinke's (1967) review of forest interception studies in the United States showed that most estimates of canopy saturation capacity are in the range of 0.5–2.0 mm, with little evidence of consistent differences between deciduous trees (in leaf?) and conifers. A discussion of the evaporation of intercepted water is postponed to Section IV,D, but comparisons between species of total interception in a year or summer season are made now.

Eidmann (1959) compared interception in 30 stands of *Fagus sylvatica* and 26 stands of *Picea abies* in Sauerland, West Germany, for 5 years (Table VI).

TABLE VI

INTERCEPTION BY *Picea abies* AND *Fagus sylvatica* IN
SAUERLAND, WEST GERMANY—MEANS OF 5 YEARS[a]

	Precipitation (mm)	Interception (mm)	
		P. abies	*F. sylvatica*
Winter	587	118	25
Summer	629	196	68
Year	1216	314	93

[a] From Eidmann (1959).

Picea intercepted annually more than 3 times as much as *Fagus*, and the contrast between summer and winter was proportionately greater in *Fagus*, as might be expected from its deciduous habit. Molchanov (1960) quotes other studies made in Europe that also show high interception by *Picea abies* but no consistent difference between *Pinus* and deciduous hardwoods in summer. Ovington's (1954) comparison of adjacent one-fourth acre plots of 12 species in Kent, England, showed interception to vary from 21 to 34% of precipitation in deciduous trees and 36 to 54% in coniferous trees, but "rain gauges were never exposed if the canopy was wet from a previous shower." Therefore, these estimates were obtained in such a way as to reflect differences in canopy saturation capacity but to exclude, at least partially, differences between species in evaporation rates between showers.

Helvey and Patric (1965a) reviewed interception by mature mixed hardwoods in the eastern United States and concluded that there was little difference among hardwood forest types in the regressions of throughfall and stem flow on precipitation. Within the same region Helvey (1967) measured interception in three adjacent *Pinus strobus* stands of different age. Annual interception increased with the age of stand from 325 mm in a 10-year-old stand to 535 mm in a 60-year-old stand. Using statistical relationships established by Helvey and Patric, Helvey calculated that annual interception by mature mixed hardwood forest in the same precipitation climate as the *Pinus strobus* plots would be only 254 mm. Although lower interception by hardwoods was partly the result of their leafless condition in winter, the calculations showed that they would intercept less than *Pinus strobus* in summer also. Other controlled comparisons of broad-leaved and coniferous stands have not shown important differences while both stands were in leaf (Bodeux, 1954; Rogerson, 1960; Leyton *et al.*, 1967).

3. Total Evaporation

A number of controlled comparisons have been made between conifers and deciduous hardwoods in temperate climates and conditions of small or moderate soil water deficit.

On lysimeters 25 m square, 2.5 m deep, and with a water table at 2.25 m, so long as there is percolation to maintain it, a plantation of *Pinus nigra* and a mixed plantation of *Quercus*, *Betula*, and *Alnus* have been compared in the coastal dunes at Castricum in Holland since they were planted in 1941 (Deij, 1948). Each plantation is continued in the surroundings of the lysimeter to a distance of about 30 m. In the early years growth of trees on the lysimeters was poorer than on the surroundings. This was corrected by manuring, but the only published results, for 1948–1953 (Deij, 1954), may have been affected by the initial check. The results shown in Table II, No. 17 (kindly made available by Dr. Deij of the Royal Netherlands Meteorological Institute) are for the decade 1955–1964. During this period apparent evaporation (precipitation minus drainage) from the lysimeter with *P. nigra* was 655 mm yr^{-1}, and from mixed hardwoods, 500 mm yr^{-1}. Analyzing the data for the same decade, Penman (1967) concluded that evaporation from *P. nigra* exceeded that from the hardwoods by about 55 mm, or 13%, from May through October, and by about 100 mm, ot 110%, from November through April. It seems reasonable to conclude that a major cause of the annual difference is the evergreen habit of *P. nigra*, enabling it to intercept and transpire more water than the deciduous trees in the 6 winter months.

At a slightly higher latitude, in Denmark, Holstener-Jorgensen (1959, 1961) was able to estimate evaporation from the fluctuation of the water table in stands of various species between the middle of April and some time in the following winter. In 1956 and 1957, average evaporation from a 45-year-old stand of *Picea abies* was 470 mm and from a 65-year-old stand of *Fagus sylvatica*, 449 mm (Holstener-Jorgensen, 1959; Table II, No. 14). In 1959 Holstenger-Jorgensen compared a much larger number of stands of different species (Holstener-Jorgensen, 1961; Table II, No. 15). Apparent evaporation from *Fagus sylvatica*, *Quercus* species and other deciduous species varied significantly with age, but not among species, and averaged about 400 mm. In stands of *Picea abies* it was about 50 mm higher. The method excluded part of the winter, including March and early April, when net radiation is sufficient to allow a further 50–75-mm evaporation, at least from the evergreen *Picea*.

Both Deij's and Holstener-Jorgensen's studies suggest that annual evaporation from conifers exceeds that from deciduous trees, at least in a climate mild enough to allow appreciable evaporation in winter. Several more comparisons between temperate conifers and hardwoods that have been made through studies of soil water depletion give estimates of evaporation for the summer months only.

Urie (1959) found no difference between total evaporation for the period April to October from *Pinus resinosa* and *Quercus* species ("scrub oak") in lower Michigan (Table II, No. 19) although depletion of soil water was more rapid under pine in April and May and under oak in July. Moyle and Zahner (1954) compared soil water depletion to a depth of 1.2 m under hardwood and coniferous stands in Arkansas. The largest differences in water consumption during June through September were between a mixed stand of *Quercus falcata* and *Quercus stellata* (30–35 years old) and an all-aged stand of *Pinus echinata* and *P. taeda* for which the values were 550 mm and 475 mm respectively (Table II, No. 35). In view of differences in age, structure, and previous treatment between these stands and the incomplete sampling of the whole rooting depth, Zahner (1955) then compared a 20-year-old stand of *P. echinata* and *P. taeda* with a 35-year-old stand of *Q. falcata*, *Q. stellata* and *Q. marilandica* of similar basal area and previous treatment. The course of soil water depletion with time was almost identical in the two stands. In other comparisons involving *P. taeda*, *P. echinata*, and hardwoods, no significant differences were found by Metz and Douglass (1959) in South Carolina and by Marston (1962) in Ohio (Table II, No. 33 and No. 32). Marston found that evaporation from "brush" (*Rhus glabra*, *Crataegus* species, *Smilax glauca*, and *Sassafras albidum*) was about 65 mm less than from forest in April through September. Lull and Axley (1958) also compared soil water depletion under various mixtures of *P. echinata* and *Quercus* species and found no significant differences in the first 1.5 m of soil. However, the depth to which depletion was observed varied from 1.8 to 3.6 m between plots, and in only one case (Table II, No. 30) was the whole rooting depth sampled adequately.

In a comparison of *Populus tremuloides* with *Picea engelmannii* in the Gunnison National Forest, Colorado, Brown and Thompson (1965) found that *Populus* withdrew 350 mm from the soil as an average in three growing seasons, and *Picea* only 240 mm (Table II, No. 22). However, soil water storage at the beginning of the growing season was 730 mm under *Populus* and only 350 mm under *Picea*. Although in neither case was more than half the available water utilized, the observed difference in depletion may have depended on different soil conditions. Rooting depth was a little greater in *Picea* than *Populus*.

Another comparison of species was made in an entirely different climate in the Aberdare Mountains of Kenya. This location is almost on the equator, has an annual precipitation of about 1100 mm, and there is no sharp division of the year into dormant and growing seasons. Pereira and Hosegood (1962) compared virgin bamboo forest (*Arundinaria alpina*), about 12 m high, with plantations of *Pinus radiata* (36 m) and *Cupressus macrocarpa* (15 m) (Table II, No. 36). Evaporation was estimated from rainfall plus soil water depletion in periods of 6 to more than 12 months when soil moisture was below field capacity. Observations on the first two species were for 8 years and on the third for 3 years, and in order to eliminate differences in time, evaporation (E) in each stand was expressed relative to calculated evaporation (E_0) from open water. The ratio was 0.85 to 0.86 for all three species.

All the foregoing comparisons, which suggest that differences between evaporation from stands of different species are negligible during summer, were made in conditions where the soil was not dried to the permanent wilting percentage. In some there is room for doubt whether full rooting depth was sampled, although there is not much evidence of significantly different rooting depths between species on the same soil. Where the soil is dried to permanent wilting percentage well before the end of the season, any differences in rooting depth or density might be significant, but no comparisons of species in these conditions were found except those made in the San Dimas lysimeters, California (Patric, 1961), and in this case the limited depth of the lysimeters (1.8 mm) probably restricted normal root development. On these lysimeters *Pinus coulteri*, *Quercus dumosa*, *Ceanothus crassifolius*, *Adenostoma fasciculatum*, and other species of the Californian chaparral have been compared since 1946. When the plantings were about 10 years old, a comparison extending over the next 5 years showed that annual evaporation from *P. coulteri* was about 390 mm and from the other species was about 450 mm (Table II, No. 43). Water was withdrawn when available from the whole depth of the lysimeters, but in most winters only a limited depth was rewetted by rain. An appreciable amount of rain was lost to the soil by surface run-off. In these conditions the woody species used all the water that entered the soil, and consumption by *P. coulteri* was lower simply because surface run-off was greater under this species. In a wet year (1200 mm of rain instead of about 500 mm) the difference among species virtually disappeared (Table II, No. 40).

D. Differences Between Forests and Grassland

1. Negligible Soil Water Deficit (Table VII)

The oldest watershed experiment in the world, started in 1903 in Emmental, Switzerland (Engler, 1919; Burger, 1943, 1954), consists of a comparison of the Sperbelgraben, almost entirely forested, with the Rappengraben, which is about 70% pasture and meadow and 30% forest. In 25 years, 1927–1952, annual rainfall was about 1650 mm, and both rainfall and stream flow were greatest in June, July, and August, so that it is unlikely there was an appreciable soil water deficit in summer. Annual losses from the forested Sperbelgraben were on average 24% higher than from the mainly grass-covered Rappengraben. Bochkov (1959a) pointed out, however, that the difference between the losses from the two watersheds was similar in all seasons of the year, whereas, if the difference were the result of greater evaporation from the forest, it would be expected to arise mainly in summer. He suggested that some of the loss could be the result of the percolation of water beneath the stream gauge on the forested watershed.

TABLE VII

COMPARISON OF ANNUAL EVAPORATION FROM FOREST AND ADJACENT GRASSLAND IN
CONDITIONS OF NEGLIGIBLE SOIL WATER DEFICIT

Reference in Table II	Forest type	Evaporation (mm)				Author
		Forest	Grass	Difference	Grass/forest	
3	*Picea sitchensis*	800	416	384	0.52	Law (1958, and unpublished)
4	*Picea abies*	579	521	58	0.90	Delfs *et al.* (1958)
6	Mixed conifer/ deciduous	861 (or less)	696	165 (or less)	0.81 (or more)	Burger (1943, 1954
8	Mixed conifer/ deciduous	684	603	81	0.88	Hirata (1929)
as 11	Mixed deciduous	(850)		0–150	(0.82–1.00)	Douglass (1967)
10[a]	*Eucalyptus niphophila*			0	1.0	Costin *et al.* (1964)

[a] Data for November through March (Australian summer).

Penman's (1959) analysis of the internal evidence of this experiment suggests that evaporation from the two watersheds was almost identical and that there was a further loss from the forested Sperbelgraben amounting to 30% of the run-off (or 14% of the precipitation) that could not be ascribed to evaporation. It appears that the 24% difference in apparent evaporation may owe something to unmeasured percolation.

Hirata (1929), cited by Penman (1963), described results from a watershed near Tokyo, which was covered initially with 30% coniferous trees and the remainder with broadleaved trees, and was cleared in 1914–1915 after 6 years of recording. Grassland developed on the area and records were continued until 1919. Hirata compared results in the 3 years 1911–1913 with the 3 years 1916–1918. The mean values of rainfall minus run-off were 684 and 603 mm, respectively, suggesting that losses were 13% more when the watershed was forested than when it was under grass. If all the available years, viz,

1908–1913 and 1915–1919, are used, the corresponding figures are 658 and 646 mm.

Delfs *et al.* (1958) compared two adjacent watersheds in the Harz Mountains, Germany, for 6 years. One bore *Picea abies* forest and the other, which had been clear felled, had a cover of grass (*Aira flexuosa*) and the dwarf shrub *Vaccinium*. The possibility of some flow from adjacent watersheds into the forested watershed was noted by the authors. Annual estimates of evaporation, from precipitation minus run-off, were divided into monthly totals in proportion to evaporation observed in lysimeters, and from these monthly totals it can be deduced that the soil water deficit in summer did not exceed 50 mm. The annual values of precipitation minus run-off were 521 mm from the grassed watershed and 579 mm, or 11 % more, from the forested watershed. Results on the grassed watershed agree fairly well with an estimate from Penman's formula (Penman, 1963).

The 3 foregoing investigations were all obtained on uncalibrated and uncontrolled watersheds. An experiment on a watershed, No. 6 at Coweeta, which was calibrated by comparison with a control, is described by Douglass (1967). In the calibration period the watershed was under hardwood forest that had been cut on 12 % of its area, adjacent to the stream, in 1942 and had not fully recovered. Another 5 % of the area was in roads and rain-gauge openings (Douglass, personal communication). The watershed was cleared, fertilized and planted to grass (Kentucky 31 fescue) during 1958–1960. In the first full year under grass, run-off was unchanged. In succeeding years the density of the grass declined and run-off increased until in the fifth year it was 150 mm higher than could have been expected from the original forest. In April, 1965, the watershed was fertilized and the density of the grass was restored, and in the following hydrologic year water yield fell again to a level not exceeding that expected from the original forest (Douglass, personal communication). Rainfall and stream flow are not published but, assuming their difference to be similar to Watershed 18 at Coweeta (both watersheds face north) and about 850 mm year^{-1} (Table II, No. 11), the change in apparent annual evaporation varied from nil when grass density was highest to a reduction of about 17 % at the lowest density.

Costin *et al.* (1964) estimated evaporation from various vegetation types in the region of Mount Kosciusko, New South Wales, Australia, from the sum of rainfall and soil water depletion less surface run-off, in summer. At lower altitudes they considered that there was an effect of drought, but at higher altitudes soil water deficits could hardly have exceeded 25 mm, and their results provide replicated comparisons of snow gum (*Eucalyptus niphophila*) and snow grass (*Poa caespitosa*). From September 24 to April 30, evaporation rates in these two communities were almost identical. With so low a water deficit it may be suspected that the authors could not justifiably neglect losses by percolation, but in the shorter period of November 20–March 21, when their graphs show that soil water content was without doubt below field capacity, evaporation rates fron snow gum and snow grass were still virtually the same.

In contrast to all the above investigations that show relatively small differences between evaporation from forest and grassland, Law (1958 and later data supplied in a personal communication) found that annual evaporation over 10 years from a lysimeter bearing *Picea sitchensis* in Yorkshire, England, averaged 800 mm, whereas that from nearby grass-covered lysimeters was only 400–450 mm. The experimental site is a plantation of only 0.6 acre in "a relatively exposed situation," and within this the lysimeter occupies 0.11 acre. Interception constituted about half the total evaporation, and in view of the sensitivity of the evaporation rate of intercepted water to wind, it seems likely that the high value of annual evaporation is partly because of the exposed nature of this small and relatively isolated plantation. Excluding this result, it appears that the ratio of evaporation from grassland, unlimited by water supply, to that from forest falls between 0.8 and 1.0.

2. Grass and Forest Compared in Conditions of Water Deficit

Table VIII compares grass and forest with respect to rooting depth, soil water depletion, and total evaporation in conditions of soil water deficit. In almost every case, considerable differences in rooting depth (or maximum depth of water withdrawal) were found, with grass roots always penetrating soil less deeply. Within any one example the amounts of water withdrawn from the soil were usually in similar proportion to rooting depth.

In discussing the differences in total water consumption, a comparison may be made with the results obtained under negligible soil water deficit, which are summarized in Table VII. It should be noted, however, that most of the results in Table VII were obtained from watershed experiments and are for complete years, whereas some of those in Table VIII were obtained through studies of soil water depletion and are for summer seasons only.

In conditions of small soil water deficit, Brown and Thompson (1965) found the ratio of evaporation from grass to that from forest (G/F) to be 0.6. Compared with most of the results of Table VII and with the next two examples this ratio seems very low. It is accepted as an example of an effect of water deficiency with some hesitation because in a wet summer when rainfall, 198 mm, was sufficient to supply all but 18 mm of the water consumption of the grass, the difference between grass and forest was as marked as in two dry summers when rainfall accounted for less than half the consumption, the remainder coming from the soil. There was no important difference between consumption in wet and dry summers.

Under moderate soil water deficit, Marston (1962) and Croft and Monninger (1953) found both absolute and relative differences that do not differ

TABLE VIII

Comparison of Soil Water Depletion and Total Evaporation from Forest and Grassland in Varying Conditions of Soil Water Deficit

Reference in Table II	Vegetation	Rooting depth (m)	Available water (mm)	Soil water depletion Period days	(mm)	Total evaporation Period months	(mm)	Grass: forest	Author
Small soil water deficit									
22	Picea engelmannii	2.1	480	111	221	VI–IX	361		Brown and Thompson (1965)
	Festuca idahoensis	1.2		114	94	VI–IX	218	0.60	
Moderate soil water deficit									
32	Quercus species	Bed rock at 0.9 m	160	c 100	136	V–X	634		Marston (1962)
	Pinus echinata		160	c 100	125	V–X	637		
	Andropogon species		160	c 100	94	V–X	555	0.87	
31	Populus tremuloides	1.8+	320	c 120	293	year	567		Croft and Monninger (1953)
	Native grass and forbs	1.2		c 120	211	year	466	0.82	
33	Pinus taeda	1.5	175	40	149				Metz and Douglas (1959)
	Andropogon species and forbs	1.0		40	102				
Severe soil water deficit									
41	Maqui	7.0–8.0		c 200	291	III–X	384		Shachori et al. (1967)
	Sown pasture	0.9–1.8		c 200	157	III–X	250	0.65	
40	Chaparral shrubs	1.8	355			year	638		Patric (1961), data for wet year
	Native grass	1.2				year	419	0.66	
Extreme soil water deficit									
42	Chaparral	3.5		c 200	375	year	560		Rowe and Reimann (1961)
	Lolium multiflorum	2.0		c 200	176	year	419	0.75	
43	Chaparral shrubs	1.8	355			year	434		Patric (1961), data for 5 dry years
	Native grass	1.8				year	399	0.92	

noticeably from results obtained under negligible soil water deficit. The ratios G/F were 0.87 and 0.82, respectively. It could be held reasonably in these two cases that, although there are differences in rooting depth, the differences in the quantities of water made available are hardly greater than the difference in potential consumption of the two types of vegetation.

Two examples with severe soil water deficit (Shachori et al., 1967; Patric, 1961) show marked reductions of G/F, 0.65 and 0.66, respectively. In this class of water deficit, Patric's results for the wet year 1958 have been used. In that year rainfall was sufficient to bring the whole depth of soil in the lysimeter to "field capacity" at the beginning of the growing season.

In conditions of extreme water deficiency, when winter rainfall is insufficient to rewet the whole profile, the advantage of a greater rooting depth should be reduced, and further results from the San Dimas lysimeters (Patric, 1961) appear to demonstrate this. In a run of 5 consecutive years of subaverage rainfall when the lower horizons of the profile were never completely rewetted, the chaparral species used all the water that infiltrated into the soil, 430–460 mm yr^{-1}, except in *Pinus coulteri* where run-off was greater and infiltration and evaporation were correspondingly less. Under grass, infiltration was the same as under the shrubs (excluding *P. coulteri*) and evaporation, 400 mm, was only slightly less, leaving 45 mm that drained from the lysimeters; the ratio G/F was about 0.92. These results may have been affected by a restriction on rooting depth imposed by the lysimeters which were only 1.8 m deep. This suggestion finds support in the investigation of Rowe and Reimann (1961), who found that a mixture of *Quercus dumosa* and other chaparral species in the San Dimas region removed water from a depth of 3.5 m. Annual evaporation from chaparral during 1952–1956 was about 560 mm and lay within the range of values observed by Patric. In the natural situations observed by Rowe and Reimann, rainfall rewetted the whole rooting depth of the chaparral only in some years. The ratio of consumption of G/F was 0.75.

The discussion of the data of this section has suggested that the proportionate difference between evaporation from forest and grassland reaches a maximum in conditions of severe water deficit, when soils are reduced to wilting point for a month or more in summer and that the difference between the two vegetation types is less marked in conditions of extreme water deficit when the whole rooting depth of the trees is not rewetted in winter. The data are insufficient in number, and there are too many exceptions to justify this conclusion as more than a working hypothesis.

3. Soviet Evidence on Evaporation from Forested and Nonforested Land

There is considerable evidence in Soviet literature, mostly from regions with annual precipitation of 600 mm or less, that cannot be reconciled with the comparisons of forested and nonforested land that have so far been cited.

Evidence will first be presented that in general the annual flow of rivers in the USSR in Europe increases with the degree of afforestation of the river catchment. Molchanov (1960) quotes data of the annual flow of rivers within some of the major river basins. Table IX, taken from these data, compares

TABLE IX

Run-Off from Rivers in the Dnieper Basin in Relation to Forest Cover[a]

River	Area (km²)	Forestation (%)	Period	Underground	Surface	Total
				Average annual run-off (mm)		
Drut	4650	41	1932–1940	116	114	230
Pronya	4650	16	1933–1940	72	122	194
Bobr	2150	35	1936–1939	105	75	180
Sozh	2600	6	1936–1939	59	199	158

[a] From Sokolovskii, in Molchanov (1960).

four rivers in the Dnieper basin. It is not stated how surface and underground run-off were defined or distinguished, but the conclusion is that surface run-off decreases and total run-off increases with increasing forest cover; the same is demonstrated for the Volga basin.

Bochkov (1959b) analyzed data from 200 river basins, exceeding 1000 km², of the USSR in Europe. His graphs show that total run-off increases with forest cover but that forest cover also increases with latitude. In order to eliminate climatic effects, he considers two groups of data, for basins with 625- and 550-mm annual precipitation, respectively. Within a rainfall region there is still a correlation of run-off with latitude and much of the apparent effect of forest cover results from this. Without correcting for latitude, an increase of forest cover of 10% appears to increase run-off by about 10 mm annually. If one compares basins at similar latitudes, the apparent effect of forestation is much less, but there still appears to be a residual relation in which run-off increases by about 5 mm for an increase in forest cover of 10%. Bochkov (1959a) criticized experiments on small watersheds on the grounds that in such watersheds the proportion of the total run-off that flows beneath rather than through the stream gauge is increased by afforestation. Although Molchanov (1960) described a large number of investigations of forest hydrology made by methods other than comparisons of watersheds, he did not cite any controlled comparisons of adjacent forested and nonforested land.

Results from analyses of river basins seem to be conclusive that run-off is greater from the more forested basins, and yet experience from many other parts of the world shows that they cannot be explained by the hypothesis that evaporation from forests is less than from nonforested land. It is clearly difficult to eliminate differences in climate from comparisons of watersheds of several thousand square kilometers. It is possible, too, that selection may have been to develop deeper and heavier textured soils for agriculture and leave the shallower and lighter ones under forest. In this way a greater opportunity for storage, and hence for evaporation, would be found on nonforested land. As will be shown below, however, a restriction of annual evaporation through limitation of soil water storage is likely to occur only in conditions where there is considerable water deficit in summer.

E. Annual or Seasonal Consumption in Relation to Net Radiation and Water Supply

Table X shows the proportion of net radiation accounted for by evaporation from forests in each of the moisture classes in Table II. In conditions of negligible and small soil water deficit, almost all net radiation is used in

TABLE X

The Proportion of Net Radiation Used in Evaporation from Forests During a Year or Growing Season in Varying Conditions of Soil Water Deficit

Soil water deficit	Number of examples	Mean ratio evaporation: net radiation	Standard deviation	Standard error of mean
Negligible	13	0.94	0.25	0.07
Small	17	0.99	0.16	0.045
Moderate	12	0.865	0.12	0.035
Severe	7	0.63	0.11	0.045
Extreme	2	0.58 (0.52–0.64)		

evaporation. In conditions of moderate soil water deficit there is an indication of a restriction of evaporation below the maximum that the net radiation could allow, and in severe conditions, where soil moisture is at or below permanent wilting percentage for 1 or more months in summer, there is a clear restriction of evaporation to, on the average, 60% of the equivalent net radiation. It should be pointed out that all the examples in the last group (severe and extreme water deficit) come from California, except for the study of Shachori et al. (1967) on the maqui in Israel. Although estimates of evaporation in California are independent of one another, and were obtained by a variety of methods, the net radiation estimated from Budyko's map is the same in every case. If it is wrong for one, it is wrong for all. It is unfortunate that examples from other parts of the world cannot be found. The experimental watersheds in South Africa (Wicht, 1967a), where forest plantations are established in native sclerophyllous scrub, would make an interesting comparison with the Mediterranean maqui and the California chaparral. Wicht does not consider that the watersheds are watertight, and it is to be hoped, therefore, that studies of soil water balance will be made to provide estimates of evaporation.

Since soil water conditions appear to have had so little effect on evaporation in the first three sections of Table II (negligible, small, and moderate soil water deficit), the relation of the individual estimates of evaporation (E mm) to net radiation (R_n mm) has been examined and is plotted in Fig. 1. A close linear relation is found with a mean value

$$E = 63 + 0.84 \ R_n \ \text{mm}$$

which hardly differs from

$$E = 0.92 \ R_n \ \text{mm}$$

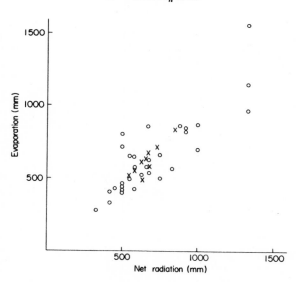

Fig. 1. The relation between evaporation from forest and equivalent net radiation (from Budyko, 1955), excluding conditions of severe or extreme soil water deficit (for definitions see Section IIA). O = annual evaporation: X = evaporation during growing season.

The conclusion that evaporation from forests is approximately equivalent to the net radiation when water supply is not limiting supports an hypothesis that already has been advanced by Penman (1967), who discussed in detail the data of Pereira *et al.* (1962b,c) from Kericho and Kimakia in Kenya and of Deij from Castricum in Holland.

In Table II there are two regions of high altitude where net radiation was not mapped by Budyko (1955). One is in Kenya (2000–2600 m), where net radiation obtained by rough interpolation in the map is equivalent to 1330 mm yr^{-1}. From meterological records at two sites in this region, Penman (1967) calculated net radiation to be between 1500 and 1600 mm yr^{-1}. The mean of three estimates of forest evaporation (Pereira *et al.*, 1962a,b; Pereira and Hosegood, 1962) is 1230 mm yr^{-1}. The other region is in Colorado, at heights of over 3000 m, where interpolated net radiation is equivalent to 950 mm yr^{-1}. Forest evaporation, estimated by Bates and Henry (1928) and Wilm and Dunford (1948), was only 383 and 360 mm yr^{-1}, respectively. Net radiation from June through September, again by rough interpolation, is equivalent to 3.9 mm day^{-1}, and estimates of forest evaporation in the growing season were 3.3 to 4.2 mm day^{-1} (Brown and Thompson, 1965) and 1.5 mm day^{-1} (Wilm and Dunford, 1948). It is unlikely that soil water deficit could account for the low evaporation recorded by Bates and Henry and Wilm and Dunford. Data from Colorado are excluded from the regression of evaporation on net radiation, because the estimation of net radiation by interpolation in Budyko's map seems unjustifiable.

IV. SOME THEORETICAL CONSIDERATIONS

A. AN EXPRESSION FOR EVAPORATION RATE

The aim of Section IV will be to examine the effect certain characteristics of forests have on their water consumption. For this purpose it will be assumed that the form of Penman's formula that includes the resistances to vapor diffusion and transport (Penman, 1952) is an adequate expression of evaporation from a continuous area of more or less uniform vegetation. This equation has been set out by Monteith (1965) in the following form:

$$\lambda E = \frac{sR_n + \rho c(e_s(T) - e)/r_a}{s + \gamma[(r_a + r_s)/r_a]} \tag{2}$$

where E is the evaporation rate in cm^3 cm^{-2} sec^{-1}, λ is the latent heat of vaporization of water, taken as 590 cal cm^{-3}, R_n is the net radiation intensity in cal cm^{-2} sec^{-1}, $e_s(T) - e$ is the vapor pressure deficit, mbar, of the air at height z, r_s is the internal resistance to diffusion of water vapor from leaves and soil, over unit land area in sec cm^{-1}, r_a is the external resistance to diffusion and transport of vapor from leaves and soil to the air at height z above the vegetation in sec cm^{-1}, ρc is the product of density in gm cm^{-3} and specific heat in cal gm^{-1} $°C^{-1}$ of the air, taken as 2.9×10^{-4} cal cm^{-3} $°C^{-1}$,

s is the change of saturation vapor pressure with temperature, mbar $°C^{-1}$, at air temperature, and γ is the psychrometric constant, 0.66 mbar $°C^{-1}$.

This equation has been critically examined by Businger (1956), Tanner and Pelton (1960), and Monteith (1965). When R_n is defined as above, the equation assumes that net radiation is equal to the sum of evaporation and heat exchange between the vegetation or soil surfaces and the air, and it therefore neglects changes in heat storage in the soil. It can be refined by defining R_n as the difference between net radiation and heat flux into the soil.

Some further expansions and derivations are needed. Net radiation may be written

$$R_n = R_i (1 - a) - R_b \tag{3}$$

where R_i is the incoming shortwave radiation, R_b is the net outgoing long-wave radiation, and a is the reflection coefficient of the surface for shortwave radiation (albedo).

The air resistance r_a may be derived from wind profiles as follows:

$$r_a = (\ln z - d/z_0)^2 / k^2 u \tag{4}$$

where u is the wind velocity in cm sec^{-1} at height z cm, z_0 is the roughness length in cm, d is the zero plane displacement in cm, and k is von Karman's constant.

The parameters z_0 and d are themselves derived from the analysis of wind profiles. Strictly, this derivation of r_a is valid only when the change of temperature with height is close to the adiabatic lapse rate, and Monteith (1963), Tanner (1963), and Monteith et al. (1964) discussed the corrections that must be applied in other conditions of temperature gradient.

By definition, the resistance to vapor diffusion or transport is the ratio of a vapor pressure gradient to an evaporation rate, i.e.,

$$r = \frac{\rho c [e_s(T_0) - e]}{\gamma \lambda E} \tag{5}$$

where $e_s(T_0)$ is the saturated vapor pressure of water at the evaporating surface. The diffusive resistances of single leaves frequently have been estimated by comparing the evaporation from leaves with either wet or dry surfaces and at the same time measuring leaf temperature and atmospheric humidity. With a wet surface, an estimate of r_a' is obtained, and with a dry surface, $r_a' + r_s'$, so that r_s' is obtained by difference, where r' signifies the resistance associated with unit leaf area—as distinct from r in Eq. (2), the resistance over unit land area. If the stomates are assumed to constitute the major part of r_s', this resistance can also be estimated from observations on stomatal geometry and number per unit area by the methods of Penman and Schofield (1951) and

Bange (1953); for a critical comparison of these two methods see Lee and Gates (1964). If a number of units of leaf area above unit ground area are regarded as resistances in parallel, then as a first approximation

$$r = r'/L*$$

where L is the ratio of leaf area to ground area, or leaf area index (Watson, 1947). Rutter (1967b; see Fig. 2[†] and Table XIII) measured r_a' and r_s' on needles in a canopy of *Pinus sylvestris* in different conditions throughout the year

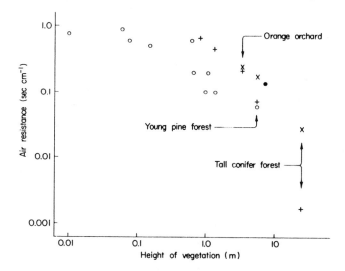

Fig. 2. The relation between height of vegetation and resistance of the air to transport of water vapor. O = data from Monteith (1965) adjusted to wind speed of 2 m sec^{-1} at 2 m above various crops. + = calculated from roughness lengths given by Kung (1961), assuming wind speed of 3 m sec^{-1} at 20 m above some crops and forests. X = calculated for wind speed of 3 m sec $^{-1}$ at 20 m, using regression of roughness length on height given by Kung (1961). ● calculated by Rutter (1967b) from observations on the evaporation of intercepted water for wind speed approximately 3 m sec^{-1} at 20 m.

and obtained corresponding values of r_a and r_s by dividing by L. Other values of r_a (Fig. 2) have been derived by Monteith (1965) from wind profile data in his own and other investigations. Monteith (1963) considers that r_s also may be estimated by analysis of the profiles of wind and other climatological elements above vegetation, and values so obtained (Monteith, 1965) are given below. Note, however, that Philip (1966) has criticized this procedure.

* For convenience in subsequent discussion, r'_s will be called stomatal resistance, and r_s, internal resistance.

† For main discussion of Fig. 2, see page 64.

B. Characteristics of the Forest Affecting Evaporation Rate

1. Reflection Coefficient

Appreciable differences exist between reflection coefficients [Eq. (3)] of different vegetation types. As shown in Table XI forests on the whole absorb more radiation than grass or crops. Budyko (1956) gave some low values for

TABLE XI

Coefficients of Reflection of Shortwave Radiation
in Various Vegetation Types

Vegetation type					
Coniferous forest	Broadleaved forest	Maqui	Grass	Crops	Author
0.10–0.14	0.175		0.22–0.33		Ångstrom (1925)
0.10–0.15	0.15–0.20		0.15–0.25	0.10–0.25	Budyko (1956)
				0.25–0.27	Monteith (1959)
0.12–0.13	0.16–0.23	0.13–0.17			Stanhill et al. (1966)
0.15–0.20	0.17–0.19		0.22–0.26	0.24–0.25	Barry and Chambers (1966)
0.11			0.23	0.20	Baumgartner (1967)

cropped fields that may have been obtained when the soil was not completely covered, because Monteith's (1959) experience was that incomplete cover reduced reflection from cropped fields. Monteith considered that the low values in forests are the result less of their dark color and more of the depth of distribution of the foliage which causes a proportion of the light reflected from the lower leaves to be absorbed by the undersurfaces of higher leaves.

2. Longwave Radiation

Vegetation emits longwave radiation in proportion to the fourth power of its absolute temperature and to its emission coefficient. According to Dennheim (1965), cited in Baumgartner (1967), the emission coefficients of forests, meadows, and arable crops are 0.97, 0.96, and 0.95, respectively, differences that are unimportant. Baumgartner points out, however, that forest canopies are commonly cooler than the foliage of crops or grassland in summer and may be warmer in winter. Assuming some reasonable temperature differences and reflection coefficients, he calculated the probable order of magnitude of the differences in shortwave and longwave radiation for conditions at Munich. These calculations suggest that, although not as important as reflection coefficients, differences in leaf temperature resulting from different microclimates within the same climate may not be negligible in their effects on radiation balance (see Table XII).

TABLE XII

CALCULATIONS OF PROBABLE DIFFERENCES BETWEEN VEGETATION TYPES
IN NET-SHORTWAVE AND LONGWAVE RADIATION NEAR MUNICH[a]

	Forest	Cereal	Grass	Bare
Net shortwave radiation	1370	1240	1190	1000
Temperature, relative to				
forest canopy, °C: summer	0	+2	+5	+10
winter	0	−4	−4	−4
Net longwave radiation	−370	−290	−440	−465
Net longwave and short-				
wave radiation	1000	950	750	595

[a] Annual totals in equivalent millimeters of water. From Baumgartner (1967).

Reifsnyder and Lull (1965) fully reviewed the radiant energy relations of forests, and detailed analyses of the radiation balance of branches have been made by Tibbals *et al.* (1964), Gates *et al.* (1965), and Knoerr and Gay (1965).

3. Roughness Length and Air Resistance

Kung (1961) brought together the few available estimates of roughness length and zero plane displacement over orchard and forest (Table XIII).

TABLE XIII

ROUGHNESS LENGTH (z_0) AND ZERO PLANE DISPLACEMENT (d)
ABOVE BUSHES AND TREES[a]

Cover	Height (m)	Windspeed (m sec^{-1})	z_0 (cm)	d (cm)	Author
Brush-field	0.7–0.9	3.95	6.4	−88.4	Fons (1940)
Orange	3.3	1.36	24.1	−332	Poppendiek (1959)
orchard		2.29	122	−244	
Pine forest	5.5	2.82	190	−190	Baumgartner
		1.15	399	−399	(1956, 1957)
Idaho forest	24.4	4.11	1847	−1847	Poppendiek (1959)
(spruce,					
hemlock, and		1.60	1347	−1637	
Douglas fir)					

[a] From Kung (1961).

Where a sufficient number of anemometers was used to measure the wind profile, Kung made duplicate calculations by using either the odd or even numbers of a stack of anemometers. Agreement between duplicate determinations was poor, and Kung suggested that at least seven anemometers were required for reliable measurements of the roughness lengths of forests. Roughness length increased with height in all the data for grass, crops, and forest that he examined, and their relation was fitted by the curve

$$\log z_0 = -1.24 + 1.19 \log h \text{ cm} \tag{6}$$

However, the points for two forests of about 5 and 25 m height were well above this line.

Calculations of air resistance r_a over vegetation types of different height are shown in Fig. 2. In part they are attributed to Monteith (1965), who calculated them from his own and other data by Eq. (4). Wind speeds, presumably those measured about 2 m above the vegetation, were variable, but in Fig. 2 Monteith's data have been reduced to a common wind speed of 2 m sec^{-1}, ignoring the possibility that roughness may change with wind speed. Other calculations have been made from the data assembled by Kung using alternatively the values of z_0 in Table XIII or those derived from Eq. (6). In order to accommodate the long roughness length of the "Idaho forest," a wind speed of 3 m sec^{-1}, 20 m above the zero plane, has been assumed in calculations from Kung's data. Both Monteith's and Kung's data include the pine forest, 5.5 m high, of Baumgartner (1956). The two sets of data are reasonably consistent and suggest that with an increase of height from 0.1 to 10 m, the air resistance in these wind conditions falls by one order of magnitude, from about 1.0 to around 0.1 sec cm^{-1}. Rutter (1967b) determined air resistances for a canopy of *Pinus sylvestris*, 7 m high, by observations on the evaporation rate and the humidity gradient above the wet canopy. With a wind speed of 2 m sec^{-1} a few meters above the canopy, the resistance was 0.14 sec cm^{-1}.

4. Stomatal Resistances

Table XIV shows measurements of the diffusive resistance of stomates per unit leaf area, the leaf area being that circumscribed by its outline. In some cases the resistance shown represents that of the stomatal array and cuticle in parallel, but since the latter is usually high, e.g., 15–150 sec cm^{-1} (Holmgren *et al.*, 1965), the stomatal resistance will not differ much from that of the whole epidermis. Most of the measurements have been made by dividing a vapor pressure gradient by a transpiration rate. Lee and Gates (1964) estimated the stomatal resistance of *Pinus resinosa* from the geometry and number per unit area of the stomates. Using Penman and Schofield's (1951) method they obtained a value of 9.34 sec cm^{-1}, but using Bange's (1953) method, which they considered more reliable on theoretical grounds, they obtained 2.05 sec cm^{-1}. Using Penman and Schofield's method on some crudely fixed leaves of *Pinus sylvestris*, Rutter (unpublished) obtained a value

TABLE XIV

COMPARISON OF MINIMAL STOMATAL RESISTANCES OF SINGLE LEAVES
IN HERBACEOUS PLANTS AND TREES

	Minimal stomatal resistance (sec cm^{-1})
By geometric measurement	
Herbaceous species	
7 species quoted by Holmgren *et al.* (1965)	0.2–0.9
Tree species	
Pinus resinosa (Lee and Gates, 1964) by Penman and Schofield's formula	9.34
by Bange's formula	2.05
Pinus sylvestris (Rutter, unpublished) by Penman and Schofield's formula	3.0
By the ratio vapor pressure gradient: transpiration rate	
Herbaceous species	
8 species quoted by Holmgren *et al.* (1965) or Monteith (1965)	0.4–3.4
Lamium galeobdolon (Holmgren *et al.*, 1965)	9.4
Circaea lutetiana (Holmgren *et al.*, 1965)	12.4
Tree species	
Populus tremula (Holmgren *et al.*, 1965)	2.30
Betula verrucosa (Holmgren *et al.*, 1965)	1.23
Acer platanoides (Holmgren *et al.*, 1965)	7.9
Quercus robur (Holmgren *et al.*, 1965)	11.6
Populus sargentii (Parkhurst and Gates, 1966)[a]	c[b]4.0
Liquidambar styraciflua (Knoerr, 1967)[a]	4.02
Liriodendron tulipifera (Knoerr, 1967)[a]	5.40
Magnolia grandiflora (Knoerr, 1967)[a]	c6.0
Quercus velutina (Knoerr, 1967)	c6.0
Pinus sylvestris (Rutter, 1967b)	c2.0

[a] These include some external resistance.
[b] c = about.

of 3.0 sec cm^{-1}, which agreed quite well with the mean of many determinations made by the transpiration method in the field. Stomatal resistances of *Populus tremula*, *Betula verrucosa* (Holmgren *et al.*, 1965), *Quercus velutina* (Knoerr, 1967), and *Pinus sylvestris* (Rutter, 1967b) lie between 1 and 3 sec

cm^{-1}, but the remaining six tree species are between 4 and 12 sec cm^{-1}. By contrast, seven herbaceous species cited from other authors by Holmgren *et al.* and investigated geometrically gave values between 0.2 and 0.9 sec cm^{-1}, and eight herbaceous species cited by either Holmgren *et al.* or Monteith (1965) and investigated by the transpiration method gave values of 0.4–3.4 sec cm^{-1}. Only the shade plants *Circaea lutetiana* and *Lamium galeobdolon* (Holmgren *et al.*, 1965) gave values of 12.4 and 9.4 sec cm^{-1}, respectively. So far as is known none of the plants, either trees or herbaceous, was seriously limited by light or water supply. Although the results are variable in each group, and the groups overlap, the tree species tend to have higher stomatal resistances.

As an approximation, the stomatal resistances of single leaves may be divided by the leaf area index to estimate the internal resistance of unit area of vegetation. For various crops, Watson (1947, 1956) found the leaf area index at the time of maximum development to be between 2.5 and 5.0. Etherington and Rutter (1964) obtained values of just under 4.0 in grass about 10 cm high, and Donald and Black (1958) quote maximum values between 6.0 and 9.0 in three swards of *Trifolium subterranium* and one of mixed *T. repens* and *Lolium perenne*. About 25 determinations of leaf area index in deciduous forest stands in Europe are cited by Carlisle *et al.* (1966). They range from 2.5 to 8.5 and are mostly about 5–6. Ovington (1957) obtained a value of 3.3 for *Pinus sylvestris* (corrected to projected area of leaf), and Rutter (1966) found a seasonal variation between 2.5 and 4.5 in a plantation of the same species. Ovington (1956] found, however, that the dry weight of the canopy (leaves and branches per hectare) was 2–3 times larger in, e.g., *Pseudotsuga taxifolia*, *Abies grandis*, and *Picea omorika* than in *P. sylvestris* and *P. nigra*, and their leaf area indices may be correspondingly higher. Table I also shows that *Pseudotsuga* may have 3 times as much foliage per hectare as *Pinus*.

Table XIV shows that the leaves of trees tend to have higher stomatal resistances than those of herbaceous plants. Since the leaf area indices of many forests are similar to those of crops, the higher stomatal resistances of some tree leaves may produce correspondingly higher values of r_s in forest canopies than in crops. Most of Monteith's (1965) estimates of internal resistance of unit area of grass or crop are in the range 0.3–0.7 sec cm^{-1}, with some higher values where a restriction of water supply was likely. Direct information on internal resistance of unit area of forest is almost nonexistent. From Baumgartner's (1956) meteorological profiles above a plantation of *Pinus sylvestris*, Monteith calculated the internal resistance to be 0.9 sec cm^{-1} in early July. By combining the stomatal resistance of leaves with the leaf area index, Rutter (1967b) calculated the internal resistance of a *Pinus sylvestris* plantation to vary between about 0.5 sec cm^{-1} in midsummer to 2.0 sec cm^{-1} in winter.

Monteith points out that when r_s is obtained from meteorological measurements over an area of vegetation, it includes a contribution from the resistance to diffusion from the soil. A number of rough estimates of evaporation from forest soil have been made. In what appears to have been a fairly open forest of *Picea abies*, Stålfelt (1963) found interception (and reevaporation) of water by the surface soil to make up nearly 20% of total annual evaporation. Rutter (1966) found evaporation from the surface soil in a *Pinus sylvestris* plantation to be about 8% of annual evaporation, and Baumgartner (1967) calculates from considerations of energy balance that evaporation from soil in forests, meadows, and cultivated crops accounts for 10%, 25%, and 45%, respectively, of total evaporation. Helvey and Patric (1965a) criticize the use of boxes and trays (i.e., small lysimeters) such as were used by Stålfelt and by Rutter, and they cite evidence that these lead to overestimates of interception by, and evaporation from, forest litter and surface soil. They consider that the most reliable evidence is that annual interception by forest litter in the southern Appalachian mountains is about 5 mm. This is probably not much more than 1% of the annual evaporation, but there may be further evaporation from soil below the litter. It appears that no methodologically satisfactory investigations of evaporation from forest soil have been made.

C. The Interaction of Climatic and Vegetational Characteristics on Evaporation Rate

In order to assess the relative importance of reflection factors, aerodynamic roughness, and internal resistance, some representative values have been assumed for grass, 10 cm high, a crop 100 cm high, and for coniferous forests 5 and 25 m high. Evaporation from each of these has been calculated in conditions where net radiation on the forest is equivalent to 0.5, 1.5, 3.0, and 5.0 mm day^{-1}, and over a range of humidities. In order to assess the effect of a difference in the reflection coefficient, values of total incoming radiation must also be assumed, and Table XV shows characteristic combinations of solar and net radiation derived from inspection of Budyko's maps (1955). Somewhat arbitrarily chosen air temperatures for each radiation climate are given (Table XV), and the range of vapour pressure deficits, corresponding to relative humidities between approximately 60 and over 90% is shown in Fig. 4. In the lower part of Table XV are given air resistances for an assumed wind speed of 3 m sec^{-1}, appropriate to the various heights and consistent with Fig. 2, and some assumed internal resistances. The resistances are very similar to those assumed by Monteith (1965) in calculations of annual transpiration from grass, crops, and forest in the climates of the Thames Valley and the Sacramento Valley. The main difference is that in the present calculations it is assumed that internal resistances at the low levels of radiation are double those at the high levels.

TABLE XV

Values of Climatic Factors and Vegetational Characters
Assumed for Model Calculations of Evaporation

Radiation and temperature

Solar radiation, equivalent mm day^{-1}	3.0	4.5	7.5	10.0
Net radiation, forest, for $a = 0.15$	0.50	1.50	3.00	5.00
Net radiation, grass and crop, for $a = 0.25$	0.20	1.06	2.38	4.00
Mean air temperature, °C	2	7.5	13	26

Air resistance for wind speed 3m sec^{-1}

	Grass	Crop	Forest	Forest
Height, m	0.1	1.0	7	25
r_a, sec cm^{-1}	0.5	0.2	0.1	0.02

Internal resistance r_s sec cm^{-1}

	Grass and crop	Forest
Low radiation (3.0 and 4.5 mm day^{-1})	1.0	2.0
High radiation (7.5 and 10.0 mm day^{-1})	0.5	1.0

Figure 3 shows the change in transpiration rate with air resistance for net radiation equivalent to 3.00 or 2.38 mm day^{-1} (appropriate to $a = 0.15$ in forest and 0.25 in grass or crop) and for two values of internal resistance. It is clear that the increased transpiration that might occur as a result of the greater height and lower air resistance of forests would be more than offset if their internal resistance were in fact double that of grass and short crops. In succeeding calculations, the assumption of an internal resistance in the forest in summer of 1.0 sec cm^{-1} has been maintained, but this factor clearly needs much more investigation.

A feature of Fig. 3 is that increasing net radiation by the equivalent of 0.62 mm day^{-1} appears to increase transpiration by only 0.1 to 0.2 mm day^{-1}. This serves to draw attention to a disadvantage of using Eq. (2) for this kind of hypothetical exploration, namely, that the vapor pressure deficit is not independent of net radiation. If an increase of net radiation of 0.6 mm day^{-1} produced an increase in transpiration of only 0.2 mm day^{-1}, the additional energy must be transferred mostly to the air. However, this transfer would increase the vapor pressure deficit and the change in transpiration would be greater than that shown in Fig. 3, where vapor pressure deficit is held constant. Eq. (2) should really be used only with observations of temperature and humidity made above the crop or forest under consideration.

In constructing Fig. 4, the hypothetical forest 25 m high has been

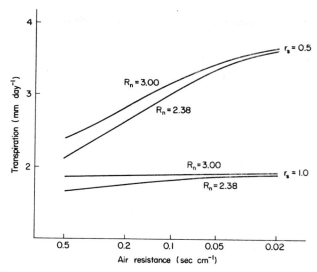

Fig. 3. Effects of variation in height, internal resistance r_s, and reflection coefficient a on calculated transpiration when solar radiation is equivalent to 7.5 mm day^{-1}, net radiation for $a = 0.15$ is 3.00 mm day^{-1}, wind speed at 20 m is 3 m sec^{-1}, air temperature is 13°C, and vapor pressure deficit is 3 mbar. Reflection coefficients 0.15 and 0.25 are transformed to differences in net radiation ($R_n = 3.00$ and 2.38), and height is transformed to air resistance consistently with Fig. 2.

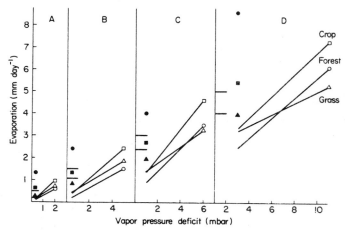

Fig. 4. Effects of varying conditions of solar radiation and vapor pressure deficit on calculated evaporation from vegetation types having the characteristics specified in Table XV. Lines with open symbols at end = transpiration rates. Closed symbols = rates of evaporation of intercepted water at high air humidity. Solar radiation equivalent in A to 3.0 mm day^{-1}; B: 4.5; C: 7.5; and D: 10.0 mm day^{-1}. Horizontal bars on right of vertical lines show net radiation; the higher on the forest, the lower on grass and crop.

abandoned for lack of any direct measurements of r_a and r_s in such a forest, and calculations have been restricted to parameters of the forest consistent with those derived from Baumgartner's (1956) and Rutter's (1967b) observations on *Pinus sylvestris*. In many of the combinations of conditions shown in Fig., 4 the calculated transpiration rates exceed the energy of net radiation, i.e., some energy is also obtained from the air. From Eq. (2) Monteith (1965) derived a parameter

$$r_i = \rho c \frac{[e_s(T) - e]}{\gamma R_n}$$

a property of the climate that has the units of resistance (sec cm^{-1}) and that he calls an isothermal resistance. Transpiration will exceed the energy of net radiation when $r_i > (r_s + r_a)$, and the evaporation of intercepted water will do this when $r_i > r_a$. At the two lower levels of radiation, transpiration exceeds net radiation in all but the most humid conditions. Transpiration from grass slightly exceeds that from the forest, and transpiration from the tall crop exceeds both grass and forest. At the two higher levels of radiation, transpiration is greater from grass than from the forest when humidity is fairly high, but the position is reversed at lower humidities when heat from the air is making a substantial contribution. The result of all these calculations is very similar to that obtained in Monteith's calculations of potential transpiration from grass and forest in the climates of the humid Thames Valley and the hot and dry Sacramento Valley. In the former case, annual transpiration from the two vegetation types was the same; in the latter, transpiration from the forest was 1820 mm year^{-1} and from grass was 1480 mm year^{-1}. This calculation of course neglects the effects of drought in California.

Results in Table XVI (Section IV, D) derived from actual observations in southeast England also show annual potential transpiration of forest to differ little from that of grass. However, total evaporation from the forest includes interception as a large component. Since the evaporation of intercepted water usually occurs in conditions of fairly high humidity, the rate of this process is shown in Fig. 4 only for relative humidities exceeding 90%. Assuming grass and forest to be capable of intercepting the same amount of water, say 1.5 mm, from a storm of sufficient size, it may be seen that, under the two lower levels of radiation, evaporation of this quantity could occur in a day or less from the forest but would take several days in grass. Thus, interception by grass in several successive rainy days would be much less than 1.5 mm per day. With net radiation equivalent to 2–3 mm day^{-1}, evaporation of 1.5 mm of water intercepted by grass would take nearly a day but would be accomplished in a few hours in the forest, leaving a substantial part of the day available for transpiration. In southeast England it can be observed that water intercepted by conifer canopies rarely persists for more than a day in winter, although grass remains wet for long periods after rain, and that intercepted water disappears from forest canopies in a few hours in summer. It can also be seen in Fig. 4 that at the higher levels of radiation the evaporation rate of intercepted water in grass only slightly exceeds the transpiration rate. With a lower wind speed and higher air resistance than those assumed, these rates would approach each other even more closely. McMillan and Burgy (1958) found no difference between the ratio of loss of water from grass with either wet or dry leaves in weighed lysimeters in a field of perennial rye grass at Davis, California, but Rutter (1967b) found that the rate of evaporation of intercepted water from a pine canopy was about 4 times as great as the transpiration rate.

D. THE SIGNIFICANCE OF INTERCEPTION IN THE WATER CONSUMPTION OF FORESTS

Rijtema (1965) derived a formula for evaporation from vegetation that takes account of the separate contributions of intercepted and transpired water. Rutter (1967b) developed another formula that is only an alternative expression of the same concepts. In order to express total evaporation in a form consistent with Eq. (2) and also appraise the significance of interception as a potential "loss," a new approach will be made to these formulas.

Equation (2), containing an internal diffusion resistance, is strictly an expression of a transpiration rate E_T. If r_s is set equal to zero, the equation becomes an expression for the rate of evaporation of intercepted water E_I. In a forest of *Pinus sylvestris*, Rutter (1967b) found that on the average E_I was 4 times as great as E_T. If it is assumed that there is no transpiration while intercepted water is being evaporated—and in any case Rutter's estimates of E_I would have included any transpiration that occurred at the same time—then one quarter of the intercepted water was equivalent to transpiration that would otherwise have occurred in the same atmospheric conditions, while three quarters was evaporation that would not have occurred in the absence of precipitation and interception. Generalizing, the additional evaporation consequent on interception, which is the net interception loss of Burgy and Pomeroy (1958), is

$$I(E_I - E_T)/E_I \qquad \text{or} \qquad (1 - E_T/E_I)I$$

where I is the depth of water intercepted.

Remembering that Eq. (2) gives E_T when r_s is included, and E_I when $r_s = 0$, then

$$E_T/E_I = \frac{s + \gamma}{s + \gamma[(r_a + r_s)/r_a]}$$

and

$$(1 - E_T/E_I)I = \frac{\gamma r_s I/r_a}{s + \gamma[(r_a + r_s)/r_a]} \tag{7}$$

The total evaporation E' during a time t that includes interception I is therefore given by

$$\lambda E' = \frac{sR_n t + (1/r_a)\{\rho c[e(T) - e]t + \gamma r_s \lambda I\}}{s + \gamma[(r_s + r_a)/r_a]} \tag{8}$$

Returning to Eq. (7), the net interception loss may be rewritten

$$\frac{\gamma r_s I}{(s + \gamma)r_a + \gamma r_s}$$

The fraction of I is less than 1, but it approaches 1 when r_s is large relative to r_a and when s is small. If $s = 0.5$ (air temperature = 2°C) and $r_s \simeq 10\, r_a$, as it may be in conifer forest, then the fraction of interception lost is 0.85, whereas with $r_s \simeq r_a$, as it may be in grass, the fraction is 0.36. With $s = 2$ (air temperature = 26°), the corresponding fractions are 0.71 and 0.2.

Therefore it appears that a high proportion of the water intercepted by forest is evaporated without any saving of stored water in the soil. The simplest demonstrations of high evaporation rates of intercepted water are found in Rutter (1963), Patric (1966), and Helvey (1967), who in coniferous forest in England, the Pacific coast of Alaska, and North Carolina, respectively, have all found that interception in winter months exceeds by many times the potential transpiration for these months calculated by Penman's or Thornthwaite's formulas.

Calculations of the significance of interception throughout the year in a *Pinus sylvestris* forest are shown in Table XVI. Potential transpiration for each month was calculated from Eq. (2), using values of r_a appropriate to mean monthly wind speeds and mean values of r_s appropriate to the month, both derived from observations in the canopy (Rutter, 1967b). Comparisons with transpiration measured by the cut-leaf method on 44 randomly chosen days suggested that estimates made by the formula were 10% too high. Interception was directly measured, and actual transpiration in the month was calculated by a method (Rutter, 1967b) equivalent to Eq. (8) and then corrected by the 10% reduction factor. The sum of interception and calculated actual transpiration agreed closely with the sum of precipitation and soil water depletion, and the calculations in Table XVI are therefore firmly based in field observations. Net radiation (assuming a reflection factor of 0.15 in the forest) and the potential transpiration of grass were calculated by the formulas of Penman (1956).

Annual interception averaged 229 mm, and the calculated actual transpiration was 48 mm less than the potential. The net increase in evaporation resulting from interception therefore was 181 mm. Rijtema (1966) also calculated that interception by *Pinus nigra* at Castricum, Holland, increased

TABLE XVI

Evaporation and Radiation Balance in a *Pinus sylvestris* Forest in Southeast England, Compared with the Potential Transpiration of Grass[a]

	May–August (mm)	September–December (mm)	January–April (mm)	Total (mm)
Forest				
Calculated potential transpiration	320	93	62	475
Interception	75	91	63	229
Calculated actual transpiration	300	72	55	427
Total evaporation	375	163	118	656
Net radiation (equivalent)	412	52	95	559
Grass				
Calculated potential transpiration	330	61	99	490

[a] Transpiration in the forest was not limited by soil water conditions.

total evaporation over potential transpiration by an amount that appears from his Fig. 7 to be equivalent to about 150 mm year^{-1}. In his Fig. 11 he shows a high correlation between calculated evaporation and actual evaporation obtained from studies of the water balance of the lysimeter.

This assessment of the significance of interception, however, is oversimplified if it does not take account of the extent of the forest. Evaporation of intercepted water is achieved mainly at the expense of heat in the air (Fig. 4 and Table XVI), and in a sufficiently large forest the cool, moist air that is formed may be expected to depress transpiration in the area downwind. In an infinitely large forest, or mainly forested area, removed from sources of advective heat, evaporation over any but short times must be limited by net radiation, and rapid evaporation of intercepted water can only be achieved at the expense of a subsequent depression of transpiration. With a wind speed of, e.g., 200 miles day^{-1} the area in which this might occur would appear to be one of several hundred miles across. Calculations like those in Table XVI, which are based on observations, or like those of Monteith (1965), which are based on reasonable assumptions about the size of r_a and r_s, suggest that the potential transpiration rate of forests in some circumstances may be no greater than that of grass. Table VII shows that there are circumstances, notably in deciduous forests, where only small differences between total annual evaporation of forest and grassland have been found. Where larger differences in total evaporation are found without appreciable soil water deficit, it appears likely that this is the result of the greater potential for evaporation of intercepted water in the forest and that the forest is taking heat that is either derived from large sources of advective heat, e.g., a dry area or a warm ocean upwind, or is simply obtained at the expense of the surrounding land if the forests form a relatively small proportion of the total cover. Equations (2) and (8) may be adequate to express the effect of environment on evaporation, but they do not show the reaction of evaporation on environment.

E. Response of Uptake to Soil Water Deficit

In Section III,E evidence was presented that suggested that where more than half the available water is utilized during the growing season there is some restriction of total seasonal consumption relative to net radiation and that when soil water is reduced to the wilting point there is a clear restriction. This is not unexpected and is only a crude indication of the response of transpiration to a reduction in soil moisture.

This response has been examined more closely by taking eight studies in which the progress of soil water depletion under forest has been followed during periods with little or no rain. The uptake of water, in millimeters per day, has been calculated from the published data for successive stages of depletion, and the resulting curves are shown in Fig. 5. Where curvilinear regressions have been fitted (Zahner, 1955; Zinke, 1959), these have been utilized; otherwise, the data have been read from graphs or calculated from tables. These curves are likely to be affected by climatic variables. Some of them, e.g., those of Rowe and Colman (1951) for Bass Lake, Croft and Monninger (1953), and Moyle and Zahner (1954), show increases in rate of uptake during the initial stages of drying that are no doubt the result of an increase in potential transpiration up to midsummer. If this is so, then the curves are probably affected also by reduced potential transpiration in September and October, because the reduction of soil moisture to near wilting point may not occur until then. The data would be better expressed if rates of uptake could be calculated as a proportion of potential transpiration, and this has been done by Zahner (1967) for 3 examples, viz, Metz and Douglass (1959), Moyle and Zahner (1954), and Lull and Axley (1958). The last of these has not been included in Fig. 5 because soil moisture was frequently recharged during the growing season, but Lull and Axley's data suggest that the depletion rate was not markedly reduced until 75% of available water had been utilized.

Zahner (1967) proposed empirical expressions for the response of uptake, as a fraction of potential transpiration, to falling soil water content in soils of different texture. The main feature of these expressions is that they assume uptake to equal potential transpiration until 33% of the available water has

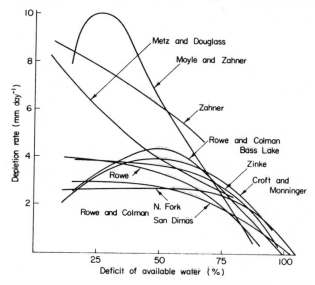

Fig. 5. The relation between the rate of soil water depletion and the deficit of available water under forests. Soil textures, depths of rooting, and available water capacities as follows: Moyle and Zahner (1954), silt loam, 1.5 m, 300 mm, Zahner (1955), silt loam, 1.8 m, 400 mm; Metz and Douglass (1959), clay, 1.7 m, 175 mm; Rowe and Colman (1951), Bass Lake, clay loam, 1.8 m, 270 mm; North Fork, Sandy clay loam, 1.0 m, 155 mm; San Dimas, sandy clay loam, 1.5 m, 250 mm; Croft and Monninger (1953), texture not given, 1.8 m, 300 mm; Zinke (1959), sandy clay loam, 1.8 m, 190 mm; Rowe (1948), clay loam, 1.0 m, 210 mm.

been withdrawn from a clay soil, 50% from a loam, and 75% from a sandy soil. The expressions satisfactorily fitted the observed course of soil moisture depletion in the three examples mentioned above, which represented, in the order in which they are given, a clay, a loam, and a sandy soil. In the other examples in Fig. 5, which are mostly described as clay loams or sandy clay loams, uptake is not markedly restricted until between 50 and 75% of the available water has been withdrawn. On a sandy clay loam in southeast England transpiration of *Pinus sylvestris* was not restricted until about 60% of the available water had been utilized (Rutter, 1967a).

Hallaire (1961), Makkink and van Heemst (1962), and Denmead and

A. J. Rutter

Shaw (1962) showed that the relation of actual to potential transpiration remained constant over a wider range of available water when the potential transpiration rate was low than when it was high. A suggestion of this effect is seen in Fig. 5, in which the curves with the highest initial rates show the earliest decline. Two of Zahner's calculated examples are from localities where the initial transpiration rates were of the order of 5–7 mm day^{-1}, while the third, which was also the sandy soil (Lull and Axley, 1958), had a maximum rate of uptake of only 2–2.5 mm day^{-1}. It would be interesting to fit Zahner's functions to the remaining data of Fig. 5 to see whether the shape of the curves is affected by potential transpiration rate as well as the soil texture.

Mathematical analyses of the relations between soil moisture conditions and the water balance of plants have been made by Gardner (1960, 1965), Visser (1964), and Cowan (1965). They have resulted in quantitative explanations of the effects of both soil water and climatic conditions on water uptake and of the interaction between these factors of supply and demand. They also predict the variation in uptake with density and depth in the soil of absorbing roots. Cowan's analysis suggests that the relation between actual and potential transpiration will be constant over a range of soil moisture that increases with the density of roots per unit volume of soil. The practical application of these formulas is limited by a lack of knowledge about root systems, about resistances to water movement through plants, and about the relation between leaf water potential and stomatal behavior.

The generalizations drawn from Table X may have some practical value in fields in which estimates only of annual or growing season consumption are needed, but in investigations on the ecology and growth of trees a more detailed analysis of changing water balance during drought is required. Workable empirical functions like Zahner's are of value in these circumstances. The more fundamental mathematical treatments referred to define the characteristics of plants that are significant in the analysis of the effects of water stress on transpiration, internal water potential, and other aspects of water balance, and they point the way to further research.

ACKNOWLEDGMENTS

I am grateful to Dr L. J. L. Deij of the Royal Netherlands Meteorological Institute and to Mr. F. Law of the Fylde Water Board for permission to use unpublished data and to Professor J. L. Monteith for helpful criticism of Section IV.

REFERENCES

Andersson, N. E., Hertz, C. H., and Rufelt, H. (1954). A new fast recording hygrometer for plant transpiration measurements. *Physiol. Plantarum* **7**, 753
Ångstrom, A. (1925). The albedo of various surfaces of ground. *Geograf. Ann.* **7**, 323

Bange, G. G. J. (1953). On the quantitative explanation of stomatal resistance. *Acta Botan. Neerl.* **2**, 255.

Barrett, J. W., and Youngberg, P. T. (1965). Effect of tree spacing and understory vegetation on water use in a pumice soil. *Soil Sci. Soc. Am. Proc.* **29**, 427

Barry, R. G., and Chambers, R. E. (1966). A preliminary map of summer albedo over England and Wales. *Quart. J. Roy. Meteorol. Soc.* **92**, 543.

Bates, C. G., and Henry, A. J. (1928). Forest and stream-flow experiment at Wagon Wheel Gap, Colorado. *Monthly Weather Rev. Suppl.* **30**, 79 pp.

Baumgartner, A. (1956). Untersuchungen über den Wärme-und Wasserhaushalt eines jungen Waldes. *Ber. Deut. Wetterdienstes* **5**, 53 pp.

Baumgartner, A. (1957). Beobachtungswerte und weitere Studien zum Wärme-und Wasserhaushalt eines jungen Waldes. *Wiss. Mitt. Univ. München Meteorol. Inst.* **1**.

Baumgartner, A. (1967). Energetic basis for differential vaporisation from forest and agricultural stands. *In* "Forest Hydrology" (W. E. Sopper and H. W. Lull, eds.), p.381. Pergamon, Oxford.

Bochkov, A. P. (1959a). The elements of water balance in the forest and on the field. *Publ. Intern. Assoc. Sci. Hydrol.* **48**, 164.

Bochkov, A. P. (1959b). The forest and river run-off. *Publ. Intern. Assoc. Sci. Hydrol.* **48**, 174.

Bodeux, A. (1954). Recherches écologiques sur le bilan d'eau sous la forêt et la lande de Haute-Campagne. *Agricultura, (Louvain)* **2**, 47.

Brown, H. E., and Thompson, J. R. (1965). Summer water use by aspen, spruce and grassland in western Colorado. *J. Forestry* **63**, 756.

Budyko, M. I. (1955). "Atlas of Heat Balance" (2nd ed., 1963), Leningrad. (Translation of text accompanying the maps by M. Dunbar, 1960, Dept. Natl. Defence, Canada, Arctic Meteorol. Res. Group, *Publ. Meteorol.* **19**, 35 pp.)

Budyko, M. I. (1956). The heat balance of the Earth's surface. (Translation by N. A. Stepanova, 1958.) *U.S. Dept. Comm., Office Tech. Serv. P. B. Rept.* 131692, 258 pp.

Burger, H. (1943). Einfluss des Waldes auf den Stand der Gewässer, *Mitt. Schweiz. Anstalt Forstl. Versuchswesen* **23**, 167.

Burger, H. (1935). Holz, Blattmenge und Zuwachs. 2. Die Douglasie. *Mitt. Schweiz. Anstalt Forstl. Versuchswesen* **19**, 21.

Burger, H. (1945). Holz, Blattmenge und Zuwachs. 7. Die Lärche. *Mitt. Schweiz. Anstalt Forstl. Versuchswesen* **24**, 7.

Burger, H. (1947). Holz, Blattmenge und Zuwachs. 8. Die Eiche. *Mitt. Schweiz. Anstalt Forstl. Versuchswesen* **25**, 211.

Burger, H. (1948). Holz, Blattmenge und Zuwachs. 9. Die Föhre. *Mitt. Schweiz. Anstalt Forstl. Versuchswesen* **25**, 435

Burger, H. (1950). Holz, Blattmenge, und Zuwachs. 10. Die Buche. *Mitt. Schweiz. Anstalt Forstl. Versuchswesen* **26**, 419

Burger, H. (1951). Holz, Blattmenge und Zuwachs. 11. Die Tanne. *Mitt. Schweiz. Anstalt Forstl. Versuchswesen* **27**, 247

Burger, H. (1953). Holz, Blattmenge und Zuwachs. 13. Fichten im bleichalterigem Hochwald. *Mitt. Schweiz. Anstalt Forstl. Versuchswesen* **29**, 38.

Burger, H. 1954. Einfluss des Waldes auf den Stand der Gewässer. *Mitt. Schweiz. Anstalt Forstl. Versuchswesen* **31**, 9.

Burgy, R. H., and Pomeroy, C. R. (1958). Interception losses in grassy vegetation. *Trans. Am. Geophys. Union* **39**, 1095.

Businger, J. A. (1956). Some remarks on Penman's equations for the evapotranspiration. *Neth. J. Agr. Sci.* **4**, 77.

Carlisle, A., Brown A. H. F., and White, E. J. (1966). Litter fall, leaf production and the effects of defoliation by *Tortrix viridana* in a sessile oak (*Quercus petraea*) woodland. *J. Ecology* **54**, 65.

Costin, A. B., Wimbush, D. J., and Cromer, R. N. (1964). Studies in catchment hydrology in the Australian Alps. 5. Soil moisture characteristics and evapotranspiration. *Australia, CSIRO, Div. Plant Ind. Tech. Paper* **20**, 20 pp.

Cowan, I. R. (1965). Transport of water in the soil-plant-atmosphere system. *J. Appl. Ecol.* **2**, 221.

Croft, A. R., and Monninger, L. V. (1953). Evapotranspiration and other water losses on some aspen forest types in relation to water available for stream flow. *Trans. Am. Geophys. Union* **34**, 563.

Decker, J. P., Gaylor, W. G., and Cole, F. D. (1962). Measuring transpiration of undisturbed tamarisk shrubs. *Plant. Physiol.* **37**, 393.

Deij, L. J. L. (1948). The lysimeter station at Castricum (Holland). *Intern. Assoc. Sci. Hydrol. Gen. Assembly Oslo* p. 11.

Deij, L. J. L. (1954). The lysimeter station at Castricum (Holland). *Intern. Assoc. Sci. Hydrol. Gen. Assembly Rome* **3**, 203.

Delfs, J. Friedrich, W., Kiesekamp, H., and Wagenhoff, A. (1958). Der Einfluss des Waldes und des Kahlschlages auf den Abflussvorgang, den Wasserhaushalt und den Bodenabtrag. *Mitt. Niedersächsischen Landesforstverwaltung* **3**, 223 pp.

Denmead, O. T. (1964). Evaporation sources and apparent diffusivities in a forest canopy. *J. Appl. Meteorol.* **3**, 383.

Denmead, O. T., and Shaw, R. T. (1962). Availability of soil water to plants as affected by soil moisture content and meteorological conditions. *Agron. J.* **54**, 385.

Doley, D., and Grieve, B. J. (1966). Measurements of sap-flow in a Eucalypt by thermoelectric methods. *Australian Forest Res.* **2**, 3.

Donald, C. M., and Black, J. N. (1958). The significance of leaf area in pasture growth. *Herbage Abstr.* **28**, 1.

Douglass, J. E. (1960). Soil moisture distribution between trees in a thinned Loblolly pine plantation. *Forestry* **58**, 221.

Douglass, J. E. (1967). Effects of species and arrangement of forests on evapotranspiration. *In* "Forest Hydrology" (W. E. Sopper and H. W. Lull, eds.), p. 451. Pergamon, Oxford.

Eidmann, F. E. (1959). Die Interception in Buchen-und Fichtenbeständen; Ergebnis mehrjähriger Untersuchungen im Rothaargebirge (Sauerland). *Publ. Inter. Assoc. Sci. Hydrol.* **48**, 5.

Engler, A. (1919). Einfluss, des Waldes auf den Stand der Gewässer. *Mitt. Schweiz. Anstalt Forstl. Versuchswesen* **12**, 626 pp.

Etherington, J. R., and Rutter, A. J. (1964). Soil water and the growth of grasses. 1. The interaction of water-table depth and irrigation amount on the growth of *Agrostis tenuis and Alopecurus pratensis. J. Ecol.* **52**, 677.

Fons, W. L. (1940). Influence of forest cover on wind velocity. *J. Forestry* **38**, 481.

Fritschen, L. J. (1965). Accuracy of evapotranspiration determinations by the Bowen ratio method. *Bull. Intern. Assoc. Sci. Hydrol.* **2**, 38.

Gardner, W. R. (1960). Dynamic aspects of water availability to plants. *Soil Sci.* **89**, 63.

Gardner, W. R. (1965). Dynamic aspects of water availability to plants. *Ann. Rev. Plant Physiol.* **16**, 323.

Gardner, W. R., and Nieman, R. H. (1964). Lower limit of water availability to plants. *Science* **143**, 1460.

Gates, D. M. (1964). Leaf temperature and transpiration. *Agron. J.* **56**, 273.

Gates, D. M., Tibbals, E. C., and Kreith, F. (1965). Radiation and convection for Ponderosa pine. *Am. J. Botany* **52**, 66.

Goodell, B. C. (1951). Comparison of stream flow in two experimental watersheds. *Trans. Am. Geophys. Union* **32**, 927

Goodell, B. C. (1958). A preliminary report on the first year's effects of timber harvesting on water yield from a Colorado watershed. *U.S. Dept. Agr. Forest Serv. Rocky Mt. Forest Range Expt. Sta. Sta. Paper* **36**, 12 pp.

Goodell, B. C. (1967). Watershed treatment effects on evapotranspiration. *In* "Forest Hydrology" (W. E. Sopper and H. W. Lull, eds.), p. 477. Pergamon, Oxford.

Grah, R. F., and Wilson, C. C. (1944). Some components of rainfall interception. *J. Forestry* **42**, 898 .

Halevy, A, (1956). Orange leaf transpiration under orchard conditions. 4. A contribution to the methodology of transpiration measurements in citrus leaves. *Bull. Res. Council Israel Sect.* **5D**, 155.

Hallaire, M. (1961) Irrigation et utilisation des réserves naturelles. *Ann. Agron.* **12**, 9.

Hamilton, E. L., and Rowe, P. B. (1949). Rainfall interception by chaparral in California. *Calif. Forestry Range Expt. Sta.* 46 pp.

Hartig, R. (1865). "Vergleichende Untersuchungen über den Wachstumgang und Ertrag der Rotbuche und Eiche im Spessart, der Rotbuche im Ostlichen Wesergebirge, der Kiefer in Pommern, und der Weisstanne im Schwarzwald." Stuttgart.

Helvey, J. D. (1967). Interception by eastern white pine. *Water Resources Res.* **3**, 723.

Helvey, J. D., and Patric, J. H. (1965a). Canopy and litter interception of rainfall by hardwoods of eastern United States. *Water Resources Res.* **1**, 193.

Helvey, J. D., and Patric, J. H. (1965b). Design criteria for interception studies. *Publ. Intern. Assoc. Sci. Hydrol.* **67**, 131.

Hibbert, A. R. (1967). Forest treatment effects on water yield. *In* "Forest Hydrology" (W. E. Sopper and H. W. Lull, eds.), p. 527. Pergamon, Oxford.

Hirata, T. (1929). Contributions to the problem of the relation between forest and water in Japan. *Imp. Forestry Expt. Sta.* (*Meguro, Tokyo*) 41 pp.

Holmgren, P., Jarvis, P. G., and Jarvis, M. S. (1965). Resistances to carbon dioxide and water vapour transfer in leaves of different plant species. *Physiol. Plantarum* **18**, 557.

Holstener-Jorgensen, H. (1959). A contribution to elucidation of evapotranspiration of forest stands on clayey soils with a high water-table. *Publ. Intern. Assoc. Sci. Hydrol.* **48**, 286.

Holstener-Jorgensen, H. (1961). An investigation of the influence of various tree species and the ages of the stands on the level of the ground-water table in forest tree stands at Bregentved. (Danish, with English summary.) *Forstlige Forsøksvaesen Danmark* **27**, 233.

Holstener-Jorgensen, H. (1967). Influences of forest management and cutting on ground water fluctuations. *In* "Forest Hydrology" (W. E. Sopper and H. W. Lull, eds.), p. 325. Pergamon, Oxford.

Hoover, M. D. (1944). Effect of removal of forest vegetation upon water yields. *Trans. Am. Geophys. Union* **25**, 969.

Hoover, M. D., Olson, D. F., and Greene, G. E. (1953). Soil moisture under a young Loblolly pine plantation. *Soil Sci. Soc. Am. Proc.* **17**, 147.

Hoyt, W. D., and Troxell, H. C. (1934). Forests and stream-flow. *Trans. Am. Soc. Civil Engrs.* **99**, 1.

Huber, B. (1932). Beobachtung und Messung pflanzlicher Saftströme. *Ber. Deut. Botan. Ges.* **50**, 89.

Ivanov, L. A., Silina, A. A., and Tsel'niker, J. L. (1950). Rapid weighing method for determining transpiration under natural conditions (Russian). *Botan. Zh.* **35**, 171; *Forestry Abst.* **12**, 1748.

Johnson, E. A., and Kovner, J. L. (1956). Effect on stream-flow of cutting a forest understory. *Forest Sci.* **2**, 82.

Johnston, R. D. (1964). Water relations of *Pinus radiata* under plantation conditions. *Australian J. Botany* **12**, 111.

Knoerr, K. R. (1965). Partitioning of the radiant heat load by forest stands. *Proc. Soc. Am. Foresters* **1964**, 105.

Knoerr, K. R. (1967). Contrasts in energy balances between individual leaves and vegetated surfaces. *In* "Forest Hydrology" (W. E. Sopper and H. W. Lull, eds.), p. 391. Pergamon, Oxford.

Knoerr, K. R., and Gay, L. W. (1965). Tree leaf energy balance. *Ecology* **46**, 17.

Kovner, J. L., and Evans, T. C. (1954). A method for determining the minimum duration of watershed experiments. *Trans. Am. Geophys. Union* **35**, 608.

Kung, E. (1961). Derivation of roughness parameters from wind profile data above tall vegetation. *Wisconsin Univ. Dept. Meteorol. Ann. Rept.* **1961**, 27.

Ladefoged, K. (1960). A method for measuring the water consumption of large intact trees. *Physiol. Plantarum* **13**, 648.

Ladefoged, K. (1963). Transpiration of forest trees in closed stands. *Physiol. Plantarum* **16**, 378.

Law, F. (1958). Measurement of rainfall, interception and evaporation losses in a plantation of Sitka spruce trees. *Intern. Assoc. Sci. Hydrol. Gen. Assembly Toronto* **2**, 397.

Lee, R., and Gates, D. M. (1964). Diffusion resistances in leaves as related to their stomatal anatomy and microstructure. *Am. J. Botany* **51**, 963.

Leyton, L., and Juniper, B. E. (1963). Cuticle structure and water relations of pine needles. *Nature* **198**, 770.

Leyton, L., Reynolds, E. R. C., and Thompson, F. B. (1967). Rainfall interception in forest and moorland. *In* "Forest Hydrology" (W. E. Sopper and K.H. Lull, eds.), p. 163. Pergamon, Oxford.

Lull, H. W., and Axley, J. H. (1958). Forest soil-moisture relations in the Coastal Plain sands of Southern New Jersey. *Forest Sci.* **4**, 2.

McMillan, W. D., and Burgy, R. H. (1958). Interception loss from grass. *J. Geophys. Res.* **65**, 2389.

Makkink, G. F., and van Heemst, H. J. D. (1962). The actual evapotranspiration as a function of the potential evapotranspiration and the soil moisture tension. *Neth. J. Agr. Sci.* **4**, 67.

Marshall, D. C. (1958). Measurement of sap flow in conifers by heat transport. *Plant Physiol.* **33**, 385.

Marston, R. B. (1962). Influence of vegetation cover on soil moisture in south-eastern Ohio. *Soil Sci. Soc. Am. Proc.* **26**, 605.

Metz L. J., and Douglass, J. E. (1959). Soil moisture depletion under several Piedmont cover types. *U.S. Dept. Agr. Tech. Bull.* **1207**, 23 pp.

Molchanov, A. A. (1960). "The Hydrological Role of Forests" (Translation by A. Gourevitch, 1963), 407 pp. Israel Program Sci. Transl., Jerusalem.

Monteith, J. L. (1959). The reflection of short-wave radiation by vegetation. *Quart. J. Roy. Meteorol. Soc.* **85**, 386.

Monteith, J. L. (1963). Gas exchange in plant communities. *In* "Environmental Control of Plant Growth" (L. T. Evans, ed.), p. 95. Academic Press, New York.

Monteith, J. L. (1965). Evaporation and environment. *Symp. Soc. Exptl. Biol.* **19**, 205.

Monteith, J. L., Szeicz, G., and Yabuki, K. (1964). Crop photosynthesis and the flux of carbon dioxide below the canopy. *J. Appl. Ecol.* **1**, 321.

Moyle, R. C., and Zahner, R. (1954). Soil moisture as affected by stand conditions. *U.S. Dept. Forest Serv. Southern Forest Expt. Sta. Occasional Paper* **137**, 14 pp.

Nakano, H. (1967). Effects of changes of forest conditions on water yield, peak flow and direct run-off of small watersheds in Japan. *In* " Forest Hydrology " (W. E. Sopper and H. W. Lull, eds.), p. 551. Pergamon, Oxford.

Ovington, J. D. (1954). A comparison of rainfall in different woodlands. *Forestry* **27**, 41.

Ovington, J. D. (1956). The form, weights and productivity of tree species grown in close stands. *New Phytologist* **55**, 289.

Ovington, J. D. (1957). Dry matter production by *Pinus sylvestris* L. *Ann. Botany (London)* [N.S.] **21**, 287.

Ovington, J. D. (1962). Quantitative ecology and the woodland ecosystem concept. *Advan. Ecology* **1**, 103.

Parker, J. (1957). The cut-leaf method and estimations of dirurnal trends in transpiration from different heights and sides of an oak and a pine. *Botan. Gaz.* **119**, 93.'

Parkhurst, D. F., and Gates, D. M. (1966). Transpiration resistance and energy budget of *Populus sargentii* leaves. *Nature* **210**, 172.

Patric, J. H. (1961). The San Dimas large lysimeters. *J. Soil Water Conserv.* **16**, 13.

Patric, J. H. (1966). Rainfall interception by mature coniferous forests of south-east Alaska. *J. Soil Water Conserv.* **21**, 229.

Penman, H. L. (1952). The physical bases of irrigation control. *Proc. 13th Intern. Hort. Congr., London* **2**, 913.

Penman, H. L. (1956). Evaporation—an introductory survey. *Neth. J. Agr. Sci.* **4**, 9.

Penman, H. L. (1959). Notes on the water balance of the Sperbelgraben and Rappengraben. *Mitt. Schweiz. Anstalt Forstl. Versuchswesen* **35**, 99.

Penman, H. L. (1963). " Vegetation and Hydrology." *Commonwealth Bur. Soil. Sci. (Gt. Brit.) Tech. Commun.* **53**, 124 pp.

Penman, H. L. (1967). Evaporation from forests: a comparison of theory and observation. *In* " Forest Hydrology " (W. E. Sopper and H. W. Lull, eds.), p. 373. Pergamon, Oxford.

Penman, H. L., and Long, I. F. (1960). Weather in wheat: an essay in micrometeorology. *Quart. J. Roy. Meteorol. Soc.* **86**, 16.

Penman, H. L., and Schofield, R. K. (1951). Some physical aspects of assimilation and transpiration. *Symp. Soc. Exptl. Biol.* **5**, 115.

Pereira, H. C., and Hosegood, P. H. (1962). Comparative water use of softwood plantations and bamboo forest. *J. Soil. Sci.* **13**, 299.

Pereira, H. C., McCulloch, J. S. G., Dagg, M., Hosegood, P. H., and Pratt, M. A. C. (1962a). A short term method for catchment basin studies. *E. African Agr. Forestry J.* **27**, 4.

Pereira, H. C., Dagg, M., and Hosegood, P. H. (1962b). The development of tea estates in tall rain forest: the water balance of both treated and control valleys. *E. African Agr. Forestry J.* **27**, 36.

Pereira, H. C. Dagg, M., and Hosegood, P. H. (1962c). The water balance of bamboo thicket and of newly planted pines. *E. African Agr. Forestry J.* **27**, 95.

Philip, J. R. (1966). Plant water relations: some physical aspects. *Ann Rev. Plant Physiol.* **17**, 245.

Pisek, A., and Tranquillini, W. (1951). Transpiration und Wasserhaushalt der Fichte (*Picea excelsa*) bei zunehmender Luft-und Bodentrockenheit. *Physiol. Plantarum* **4**, 1.

Polster, H. (1950). "Die physiologischen Grundlagen der Stofferzeugung im Walde."
    Bayerischer Landwirtschaftsverlag, Munich.
Poppendiek, H. F. (1959). Investigation of velocity and temperature profiles in air layers
    within and above trees and bush. *Univ. Calif. (Los Angeles) Dept. Eng. Contract* N6—
    ONT—75, *Task Order VI, NR*—082—036.
Rawitscher, F. (1955). Beobachtungen zur Methodik der Transpirations-messungen bei
    Pflanzen. *Ber. Deut. Botan. Ges.* **68**, 287.
Reifsnyder, W. E., and Lull, H. W. (1965). Radiant energy in relation to forests. *U.S.
    Dept. Ag. Tech. Bull.* **1344**, 111 pp.
Reigner, I. C. (1964). Evaluation of the trough type rain-gage. *U.S. Dept. Agr. Forest
    Serv. Northeast. Forest Expt. Sta. Res. Note* **20**, 4 pp.
Reinhart, K. G. (1967). Watershed calibration methods. *In* "Forest Hydrology" (W. E.
    Sopper and H. W. Lull, eds.), p. 715. Pergamon, Oxford.
Reinhart, K. G., and Eschner, A. R. (1962). Effect on stream-flow of four different forest
    practices in the Allegheny Mountains. *J. Geophys. Res.* **67**, 2433.
Reynolds, E. R. C. (1966). The internal water balance of trees. *Forestry (Suppl.)* p. 32.
Reynolds, E. R. C., and Leyton, L. (1963) *In* "The Water Relations of Plants" (A. J.
    Rutter and F. H. Whitehead, eds.), p. 127. Blackwell, Oxford.
Rich, L. R. (1959). Watershed management research in the mixed conifer type. *U.S. Dept.
    Agr. Forest Serv. Rocky Mt. Forest Range Expt. Sta. Prog. Rept.* **1959**.
Rijtema, P. E. (1965). An analysis of actual evapotranspiration. *Agr. Res. Repts., Wageningen*
    659, 107 pp.
Rijtema, P. E. (1966). Evapotranspiration. *Inst. Land Water Management Res., Wageningen,
    Tech. Bull.* **47**, 89 pp.
Ringoet, A. (1952). Recherches sur le transpiration et le bilan d'eau de quelques plantes
    tropicales. *Publ. Inst. Natl. Étude Agron. Congo Belge, Sér. Sci.* **56**.
Rogerson, T. L. (1960). Influence of natural hardwoods and plantation red pine on net
    precipitation reaching the forest floor. M.S. Thesis (unpubl.) Pennsylvania State Univ.,
    University Park, Pennsylvania.
Rogerson, T. L. (1967). Throughfall in pole size Loblolly pine as affected by stand density.
    In "Forest Hydrology" (W. E. Sopper and H. W. Lull, eds.), p. 187. Pergamon,
    Oxford.
Rowe, P. B. (1948). Influence of woodland chaparral on water and soil in Central Cali-
    foria. *Calif. Dept. Nat. Resources, Div. Forestry*, 70 pp.
Rowe, P. B., and Coleman, E. A. (1951). Disposition of rainfall in two mountain areas
    of California. *U.S. Dept. Agr. Tech. Bull.* **1048**.
Rowe, P. B., and Reimann, L. F. (1961). Water use by brush, grass and grassforb vegetation.
    J. Forestry **59**, 175.
Russell, E. W. (1961). "Soil Conditions and Plant Growth," 9th ed. Longman, London.
Rutter, A. J. (1963). Studies in the water relations of *Pinus sylvestris* in plantation conditions.
    1. Measurements of rainfall and interception. *J. Ecology* **51**, 191.
Rutter, A. J. (1964). Studies in the water relations of *Pinus sylvestris* in plantation con-
    ditions. 2. The annual cycle of soil moisture change and derived estimates of evapora-
    tion. *J. Appl. Ecology* **1**, 29.
Rutter, A. J. (1966). Studies in the water relations of *Pinus sylvestris* in plantation con-
    ditions. 4. Direct observations on the rates of transpiration, evaporation of inter-
    cepted water, and evaporation from the soil surface. *J. Appl. Ecol.* **3**, 393,
Rutter, A. J. (1967a). Studies in the water relations of *Pinus sylvestris* in plantation con-
    ditions. 5. Responses to variation in soil moisture. *J. Appl. Ecol.* **4**, 73.
Rutter, A. J. (1967b). An analysis of evaporation from a stand of Scots pine. *In* "Forest
    Hydrology" (W. E. Sopper and H. W. Lull, eds.), p. 403. Pergamon, Oxford.

Rutter, A. J., and Fourt, D. F. (1965). Studies in the water relations of *Pinus sylvestris* in plantation conditions. 3. A comparison of soil water changes and estimates of total evaporation on four afforested sites and one grass-covered site. *J. Appl. Ecol.* 2, 197.

Salter, P. J., and Williams, J. B. (1965). The influence of texture on the moisture characteristics of soils: 2. Available water capacity and moisture release characteristics. *J. Soil Sci.* 16, 310.

Schneider, W. J., and Ayer, G. R. (1961). Effect of reforestation on stream-flow in Central New York. *U.S. Geol. Surv. Water Supply Papers* 1602.

Shachori, A. Y., Stanhill, G., and Michaeli, A. (1965). The application of integrated research approach to the study of effects of different cover types on rainfall disposition in Carmel Mountains in Israel. *Arid Zone Res.* 25, 479.

Shachori, A., Rosenzweig, D., and Poljakoff-Mayber, A. (1967). Effect of Mediterranean vegetation on the moisture regime. *In* " Forest Hydrology " (W. E. Sopper and H. W. Lull, eds.), p. 291. Pergamon, Oxford.

Skau, C. M., and Swanson, R. H. (1963). An improved heat pulse velocity meter as an indicator of sap speeds for detecting sap movements in woody plants. *J. Geophys. Res.* 63, 4743.

Stålfelt, M. G. (1963). On the distribution of the precipitation in a spruce stand. *In* " The Water Relations of Plants. " (A. J. Rutter and F. H. Whitehead, eds.), p. 116. Blackwell, Oxford.

Stanhill, G., Hofstede, G. J., and Kalma, J. D. (1966). Radiation balance of natural and agricultural vegetations. *Quart. J. Roy. Meteorol. Soc.* 92, 128.

Stone, E. C. (1957). Dew as an ecological factor, 2. The effect of artificial dew on the survival of *Pinus ponderosa* and associated species. *Ecology* 38, 414.

Stone, E. C. (1958). Dew absorption by conifers. *In* " The Physiology of Forest Trees " (K. V. Thimann, ed.), p. 125. Ronald Press, New York.

Swanson, R. H., and Lee, R. (1966). Measurement of water movement from and through trees and shrubs. *J. Forestry* 64, 187.

Tanner, C. B. (1963). Energy relations in plant communities. *In* " Environmental Control of Plant Growth " (L. T. Evans, ed.), p. 141. Academic Press, New York.

Tanner, C. B., and Pelton, W. L. (1960). Potential evapotranspiration estimates by the approximate energy balance method of Penman, *J. Geophys. Res.* 65, 3391.

Thornthwaite, C. W. (1948). An approach toward a rational classification of climate. *Geograph. Rev.* 38, 85.

Tibbals, E. C., Carr, E. K., Gates, D. M., and Kreith, F. (1964). Radiation and convection in conifers, *Am. J. Botany* 51, 529

Urie, D. H. (1959). Pattern of soil moisture depletion varies between Red pine and oak stands in Michigan. *U.S. Dept. Agr. Forest Serv. Lake States Forest Expt. Sta. Tech. Note* 564, 1 pp.

Vaadia, Y., and Waisel, V. (1963). Water absorption by the aerial organs of plants. *Physiol. Plantarum* 16, 44.

Valek, Z. D. (1959). Beitrag zur hydrologischen und hydrotechnischen Vervendbarkeit der Holzarten. *Publ. Intern. Assoc. Sci. Hydrology* 48, 322.

van Bavel, C. H. M. (1961). Lysimetric measurements of evapo-transpiration rates in the eastern United States. *Soil Sci. Soc. Am. Proc.* 25, 138.

Visser, W. C. (1964). Moisture requirement of crops and rate of moisture depletion of the soil. *Inst. Land Water Management Res. Wageningen, Tech. Bull.* 32, 21 pp.

Watson, D. J. (1947). Comparative physiological studies on the growth of field crops. 1 Variation in net assimilation rate and leaf area between species and varieties, and within and between years. *Ann. Botany (London)* [N.S.] 11, 41.

Watson, D. J. (1956). Leaf growth in relation to crop yield. *In* "The Growth of Leaves" (F. L. Milthorpe, ed.), Proc. 3rd Easter School Agr. Sci. Univ. of Nottingham, p. 187. Butterworth, London.

Wicht, C. L. (1967a). Forest hydrological research in the South African Republic. *In* "Forest Hydrology" (W. E. Sopper and H. W. Lull, eds.), p. 75. Pergamon, Oxford.

Wicht, C. L. (1967b). The validity of conclusions from South African multiple watershed experiments. *In* "Forest Hydrology" (W. E. Sopper and H. W. Lull, eds.), p. 749. Pergamon, Oxford.

Wilm, H. G. (1943a) Determining net rainfall under a conifer forest. *J. Agr. Res.* **67**, 501.

Wilm, H. G. (1943b). Statistical control of hydrologic data from esperimental watersheds. *Trans. Am. Geophys. Union* **2**, 618.

Wilm, H. G., and Dunford, E. G. (1948). Effect of timber cutting on water available for stream-flow. *U.S. Dept. Agr. Tech. Bull.* **968**, 43 pp.

Zahner, R. (1955). Soil water depletion by pine and hardwood stands during a dry season. *Forest Sci.* **1**, 258.

Zahner, R. (1958). Hardwood understory depletes soil water in pine stands. *Forest Sci.* **4**, 178.

Zahner, R. (1967). Refinement in empirical functions for realistic soil-moisture regimes under forest cover. *In* "Forest Hydrology" (W. E. Sopper and H. W. Lull, eds.), p. 261. Pergamon, Oxford.

Zinke, P. J. (1959). The influence of a stand of *Pinus coulteri* on the soil moisture regime of a large San Dimas lysimeter in Southern California. *Publ. Intern. Assoc. Sci. Hydrology* **49**, 126.

Zinke, P. J. (1967). Forest interception studies in the United States. *In* "Forest Hydrology" (W. E. Sopper and H. W. Lull, eds.), p. 137. Pergamon, Oxford.

CHAPTER 3

WATER DEFICITS AND PHYSIOLOGICAL PROCESSES

A. S. Crafts

DEPARTMENT OF BOTANY, UNIVERSITY OF CALIFORNIA, DAVIS, CALIFORNIA

I. INTRODUCTION

It is common knowledge that water stress (the state of water when the supply is inadequate) inhibits plant growth. It is also well known that the potential of water in the xylem conduits of many plants is negative during a major portion of the plant's life (Crafts *et al.*, 1949; Scholander *et al.*, 1965). In fact, the water potential of a plant may pass through a series of states that are critical in its life. In an air-dry seed the water content may be around 5% of the dry weight and the water present may have a negative potential of hundreds of bars. When the seed is planted it absorbs water; the water potential is increased until it approaches that of the soil. As the embryo expands and grows into a seedling, water potential in the young xylem elements passes the zero value and becomes positive; the seedling exhibits guttation from its leaf tips and the excised root produces xylem exudate. Such positive pressure is usually only a fraction of a bar; in grape and maple it may be as high as 2 bars.

As leaves expand and transpiration increases, water potential is again lowered from a positive value to zero and then into the negative range; the actual values depend on three parameters: (1) excess of transpiration rate over absorption rate, (2) soil moisture potential, and (3) height of the plant. All plants, even when growing in water culture, may suffer temporary wilting when transpiration is excessive. All plants growing in soils having negative water potentials must endure similar or lower potentials. Tall plants growing even in moist environments will have negative potentials in their upper extremities; Scholander *et al.* (1965) show negative hydrostatic pressures for redwood approaching -20 bars and for Douglas fir around -25 bars. Crop plants growing in soils approaching the permanent wilting percentage have negative hydrostatic pressures in the -16 to -20 bar range at night; in the day the negative pressures may range from -10 to -20 bars lower (Scholander *et al.*, 1965). Desert plants that are well adapted to soils of low moisture and high solute content exhibit negative pressures around 80 bars. And plants growing in fresh water have a pressure from -5 to -15 bars; those in sea water such as mangrove, around -40 bars. Since the water in the plant body is a continuum, it seems evident that water in the total apoplast system must exist at approximately these potential levels.

To go a step farther, when a plant has reached the stage of permanent wilting and continues to lose water, its water potential must continue to lower until death occurs; water will then be released from dying cells, but under most conditions this will not compensate for rapid drying and the plant then goes into the air-dry state with a water potential comparable to that of the seed.

While the term stress is somewhat ambiguous, it might be defined as the state a plant enters when the water potential crosses the zero mark and becomes negative. From this point on water in the xylem is in a metastable state (Crafts, 1939).

In terms of flower and vegetable crops, stress might be considered as the condition entered when the potential ranges from 0 to -5 bars. Field and forage crops thrive until the potential approaches -16 bars, and desert plants live normally with potentials in the range of -20 to -80 bars. They grow, however, only when rainfall occurs and available soil moisture brings the potential above the -20 bar level.

In thinking of plant responses to these varied potential conditions it is well to remember that lowering pressure has a structuring effect on liquid water that increases viscosity and lowers fluidity and so, in a sense, serves to retard loss and conserve water.

II. WATER IN PHYSIOLOGICAL PROCESSES

In the living plant, water occurs in many states and is involved in all physiological processes. Water of hydration and imbibition in colloidal phases such as cell walls, osmotic water in vacuoles and in phloem conduits, and hydrostatic water in the xylem all enter into life processes. As shown by experiments with isotopic water (Ordin *et al.*, 1956; Biddulph and Cory, 1957; Biddulph *et al.*, 1961; Ordin and Gairon, 1961; Vartapetyan, 1965; Lebedev and Askochenskaya, 1965), liquid flow, diffusion across membranes, and exchange between liquid–liquid and liquid–solid phases all enter into plant water relations. Because of the physical structure of water involving hydrogen– and dipole–dipole bonding that result in high internal pressure (tensile strength) and high surface tension, water equilibria serve as important integrating forces in plant life. The first contact of a new plant with the water in its environment is through the seed.

A. SEED GERMINATION

As a seed absorbs water from the soil, hydration processes are initiated first in cell walls and then in protoplasm, metabolic activity is increased, swelling occurs, and the embryo pushes the radicle into the soil. Water content of the soil is a very important factor in seed germination; germination rate may be closely correlated with the rate of water absorption. Soaking in water is a common practice as a means of speeding up germination and assuring an adequate supply early in the germination process.

The amount of hydration required to bring about seed germination varies widely among plant species. Owen (1952) found that approximately 20% of a sample of wheat seeds germinated at a soil moisture content below the PWP. This is probably true of seeds of a number of crop plants that have been selected for high germination. It would not be true for a wide array of plants; germination of some seeds requires the presence of free water at atmospheric pressure.

Satoo and Goo (1954) as reported by Kozlowski (1964) found seeds of *Chamaecyparis obtusa*, *Pinus densiflora* and *P. thunbergii* to germinate in soil having a moisture tension up to 8 atm; *C. obtusa* was most sensitive to reduced soil moisture. Kramer and Kozlowski (1960) found *P. strobus* seed sensitive to immersion time in water: 12 hours soaking almost doubled germination percentage during 44 days; the percentage was essentially the same for soaking periods up to 48 hours; periods of 72 and 120 hours had a deleterious effect on germination and longer periods would have been even more harmful. In contrast, seeds of many weeds, particularly those of aquatic habitats, survive soaking for a year or more. Most seeds, however, have an optimum period of moistening before germination; moisture stress reduces and may even prevent germination; immersion or soil saturation will result in the death of most seeds within a few days or weeks.

Collis-George and Sands (1962) studied seed germination using equipment that enabled them to distinguish between matric potential (largely imbibitional forces) and osmotic potential. Under their conditions matric potential was more effective than osmotic potential in decreasing germination rate. Osmotic potentials reduced germination only when values were above 1000 cm of water when sodium chloride and glycerol were the solutes. Matric potentials up to 100 cm reduced germination rate but not total germination. A value of 15,000 cm, equivalent to the permanent wilting point, prevented germination of perennial ryegrass and oats; alfalfa germinated 38 %. Different solutes had different effects on germination for each seed species. Cadmium sulfate reduced germination more than sodium chloride or glycerol, a result they attributed to penetration of its ions into the seeds. Thus, the type of solute, particularly its toxicity, is more important than its osmotic potential in affecting seed germination.

Effects of moisture stress on seed germination have definite implications for agriculture. In regions of irrigation agriculture it is customary, when necessary, to preirrigate and then to prepare the seed bed and plant as soon as possible to ensure a good stand of the crop. In areas of dry-land agriculture, seeding usually follows a rainy period. When soil moisture is limiting, catch crops such as milo, Sudan grass, or safflower are often used, and under unfavorable soil moisture conditions even these may fail. Even if the seeds germinate and a crop gets underway, lack of soil moisture may stunt it and reduce yield.

B. WATER AND MINERAL ABSORPTION

When the radicle pushes out into moist soil, water absorption is increased and soon the embryo develops into a seedling. Mineral absorption by roots has long been recognized as an active process dependent on expenditure of metabolic energy (Hoagland, 1944; Steward, 1964; Laties and Budd, 1964). Nevertheless, roots cannot absorb ions that do not impinge upon their surfaces, and water, moving through the soil and into roots, carries nutrient ions into the rhizosphere in a form available for absorption. Without this transport the roots of most plants would have to be much more extensive and more finely divided in order to obtain the nutrients they do. Therefore, the supply of nutrients to a plant is directly related to water movement into roots, and when such movement ceases because of lowered soil moisture, roots are limited to those nutrient ions within the range of diffusion; this supply must become limiting within a very short time.

As the seedling starts growth an active process in the roots brings about absorption of salts, water follows, and these move into the xylem and undergo acropetal flow to the shoot under positive pressure (Crafts and Broyer, 1938; Laties and Budd, 1964).

With the expansion and growth of leaves and the development of the shoot, water loss by transpiration soon exceeds water uptake by roots and the pressure of sap in the xylem becomes subatmospheric. This may be studied by a simultaneous measure of transpiration and absorption (Kramer, 1937) by noting exudation lag upon detopping plants (Ulehla, 1965) or by determining the water potential of leaves while the transpiration rate is altered (Weatherley, 1965).

The status of the transpiration stream from this time on fluctuates, but under most conditions stress mounts. Although this deficit may be temporarily relieved by rainfall or irrigation, the xylem stream is usually under stress throughout the major growth period of the plant (Scholander *et al.*, 1965). Scholander *et al.* (1966) have shown that plants such as mangroves that grow in sea water have constant water potentials in the xylem ranging from -30 to -60 atm.

Water in the xylem, under these conditions, is in a metastable state made possible by the inherent tensile strength of the continuous water columns and by the absence of unwet surface upon which cavitation may take place (Crafts, 1939).

Glinka and Reinhold (1962) have shown that a high CO_2 treatment reversibly reduces both influx and efflux of water in the sunflower hypocotyl. This they attribute to an effect on the permeability of cell membranes to water. High CO_2 in the soil has been known to hinder water uptake by roots.

C. Translocation of Minerals

While the great use of water by plants has long been recognized, certain aspects of water function in plants were controverted for years (Dixon, 1914; Curtis, 1935). With the introduction of isotopes as tracers, studies on intact plants and increased knowledge of plant structure (Esau, 1953; Esau and Cheadle, 1959, 1961) have shown that water serves many useful functions and that transpiration is an essential process in plant growth and production. The critical experiments of Stout and Hoagland (1939) served to prove that minerals absorbed from the soil are transported to the foliar parts of plants in the xylem via the transpiration stream. Thus, one major function of water movement in plants is solute distribution (Crafts, 1961b).

During the height of the free-space controversy, doubt was cast on the essentiality of active solute absorption by roots (Hylmo, 1953; Broyer, 1956; Briggs and Robertson, 1957; Kramer, 1957; Levitt, 1957; Russell and Shorrocks, 1957). However, with more critical experiments (Levitt, 1957; Russell and Shorrocks, 1957; Bernstein and Gardner, 1961), it seems evident that the older view of Hoagland is valid, i.e., that salts are actively absorbed whereas, under conditions of low water potential in the xylem, water is pulled through the plant passively to satisfy the demands of transpirational loss.

There also was controversy over the state of water in xylem conduits

(Greenidge, 1957; Loomis *et al.*, 1960). While methods of study have varied and interpretations have not been in agreement, it seems obvious that intact, continuous columns of water must extend from the soil through the root cortex, endodermis, pericycle, root xylem, stem xylem, leaf xylem, and leaf parenchyma cell walls, in short, throughout the apoplast from the soil to the menisci in the walls bordering the stomatal chambers. Although many xylem conduits contain air or vapor through a large part of the year, there must exist vessels or tracheids in the most recently differentiated xylem that maintain their liquid contents. The continuous transpiration of leaves up to and during permanent wilting could not go on if all of the columns were broken (Scholander *et al.*, 1961, 1965).

Under extreme conditions, as when the continuous path of water in xylem conduits is broken by horizontal saw cuts (work of Greenidge, 1954, 1955a,b) or air embolism (Scholander *et al.*, 1955, 1957, 1961), some water must continue to move through the molecular interstices of xylem cell walls. Additional evidence of such movement is given by Hygen (1965), who demonstrated dye movement in frozen spruce trees in Norway. That continuous liquid columns do break under severe stress is documented by Milburn and Johnson (1966), who used an electronic amplifier to make the "clicks" caused by cavitation audible and capable of being recorded on tape.

The path of water movement in the leaf has been clarified by modern work. Whereas the classical picture of water movement and salt distribution postulated a cell-to-cell movement involving passage through cytoplasmic membranes, work by Strugger (1938) indicated that the principal flow was along the cell walls from tracheids to the evaporating surfaces of stomatal chambers. Subsequent work with labeled tracers (Leonard, 1958; Crafts and Yamaguchi, 1964) has shown clearly that apoplastic movement predominates. Evidently, salt uptake in leaves occurs from apoplast to symplast just as it does in roots in the soil.

Obviously, most plants absorb, translocate, and transpire water under conditions of continuous though fluctuating stress, and inorganic nutrients absorbed from the soil by roots are distributed throughout the plant by the transpiration stream. From our knowledge of water structure, what effects might one expect from such stress?

The water potential of the xylem stream is reduced by two means: (1) reduced pressure and (2) the presence of salts. Both of these are structuring influences. As stress builds up under mounting water deficit, the increase in ice-likeness of water throughout the system must add appreciably to resistance to flow and thus slow down movement. Gardner (1965) reports work by Brouwer and by Jensen in which there was increased conductivity of water under conditions of stress; Ordin and Gairon (1961) found tritium diffusion into roots to increase under stress. However, rates of water movement in xylem are recognized to be much higher than necessary for mineral distribution, and seldom is the distribution function rendered inadequate. Factors much more likely to interfere with normal metabolism and growth through mineral availability are deficiency or unbalance of minerals in the soil, root diseases, pests, etc.

D. TRANSLOCATION OF FOODS

The major function of phloem tissue in plants is long-distance transport of foods (Curtis, 1935; Crafts 1961b; Kursanov, 1963; Esau, 1965). One notable advance in our understanding of translocation physiology in recent years has been the demonstration by Duloy *et al.* (1961), Esau and Cheadle (1961), Esau (1965), Evert and Murmanis (1965), Engleman (1965), and others that the protoplasmic strands that traverse the sieve plates of functional sieve tubes are open tubules and hence are able to accommodate mass flow of the assimilate stream. Work by Leonard and Crafts (1956), Petersen (1958), Yamaguchi and Crafts (1959), Mason (1960), De Stigter (1961), van der Zweep (1961), Quinlan and Sagar (1962), Forde (1963), Yamaguchi (1961, 1965), Hartt (1967), and others with isotopic tracers in intact plants has demonstrated that the distribution patterns of many compounds in plants can be explained only on the basis of source-to-sink flow as along a stream. These several investigations would seem to eliminate any possibility of long-distance transport by diffusion.

One change from the original Münch (1930) hypothesis is the evidence that the total assimilate stream is apparently absorbed by growing cells in meristematic sinks, and hence the energy-consuming process of separating solutes from the solvent water may be a minor factor in the driving force of food movement (Crafts, 1961b). Since most or all of the water moving in the assimilate stream is absorbed into living cells of meristems it is apparent that it serves two purposes: (1) of solvent and matrix for the differential osmotic–pressure–flow mechanism, and (2) for the provision of water under positive pressure to growing cells even at times of high water stress in the apoplast system. Every leaf primordium at the shoot meristem is a sink for food movement until, sometime during expansion, it passes the compensation point and becomes a source. Considering the complex growth patterns of most plants and the diurnal cycle of food synthesis and use, it is obvious that source–sink relations are intricate and variable (Forde, 1963; Quinlan, 1966). Since the phloem operates under positive hydrostatic pressure and, in fact, is an inflated, elastic system with no mechanism to prevent reversal of flow, the direction of food movement in different organs of the shoot is variable, complex, and at times difficult to analyze. Bidirectional movement of foods may take place in certain plants, as demonstrated by Biddulph and Cory (1960, 1965) and illustrated by Crafts (1966).

Water stress in plants is reflected in reduced water movement into the phloem at the source; volume of the assimilate stream and velocity of flow are reduced, but concentration is increased in compensation. Transport should continue so long as a gradient in pressure between source and sink is maintained. Just as the xylem system seems able to maintain its function even under high stress and removal of at least a portion of its supply of conduits (Greenidge, 1955a,b) so also is the phloem capable of transporting foods under severe stress and when portions are eliminated (Mason and Maskell, 1928).

A number of workers have reported that water deficits decrease translocation. Roberts (1964) noted that moisture stress inhibited movement of ^{14}C-photosynthate from leaves of potted yellow poplar seedlings; distance and velocity as well as amount of transport were lowered. Weibe and Wihrheim (1962) found that translocation of ^{14}C-photosynthate from intact leaves of sunflower plants was decreased with increasing DPD. In wilted plants having DPD's of 10–12 atm, translocation was about one third that of controls having deficits of 1–2 atm. Zholkevich (1954) found that irrigation increased movement of ^{14}C from leaves to heads of wheat. Plaut and Reinhold (1965) using ^{14}C-sucrose applied to bean leaves reported less transport in water-deficient plants. After 15 hours, metabolism of the labeled sucrose resulted in a higher content of ethanol-insoluble ^{14}C in normal than in water stressed plants.

In a comprehensive study of food movement in sugarcane, Hartt (1965a,b, 1967) found that low moisture supply decreases the velocity and mass rate of translocation of ^{14}C-photosynthate. Surplus sucrose not utilized in growth moved more slowly in the phloem and was stored in the stalk.

Limiting moisture slowed transport of ^{14}C-sucrose more severely than it curtailed photosynthesis in the same leaf. From this result Hartt concludes that moisture supply has a primary effect upon transport; that is, transport is not secondary to photosynthesis, as has been suggested (Nelson's discussion of Robert's paper, 1964).

Low moisture supply retarded profile development in the stem and loss of moisture gradient caused a steepened slope of the profile. Hartt concludes that these results indicate a mass-flow mechanism of translocation. During the ripening of cane storage of sucrose in the stalk may be increased by withholding irrigation; less sucrose is hydrolyzed in transit; less is used in growth; the slowly moving stream allows more time for internal distribution to storage tissues. Figure 1 shows the results of a 90-minute experiment at three different levels of deficit (Hartt, 1967).

In their studies on translocation of labeled tracers in mistletoe-infected and noninfected trees, Leonard and Hull (1965) found movement to come to a halt in the autumn, even in conifers with healthy active foliage. Leonard attributes this development to moisture deficiency since soil moisture is at a minimum at this time of year (Leonard, 1967).

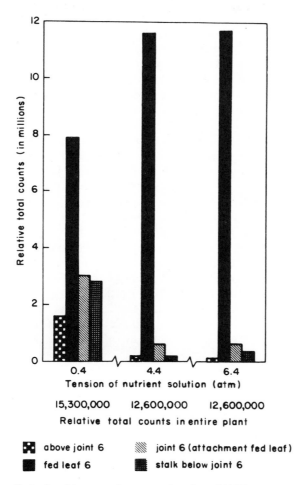

Fig. 1. The effect of moisture supply on translocation of ^{14}C in sugarcane. Plants were placed in salt solutions for 48 hours; they had 5 minutes exposure to $^{14}CO_2$, 90 minutes before harvest. From Hartt (1967).

In studies on the effects of water stress on translocation in relation to photosynthesis and growth in wheat, Wardlaw (1967) determined that from the onset of wilting there was a progressive reduction in the rate of photosynthesis. Even in limiting light, photosynthesis was lower in wilted than in turgid leaves, and the difference was not reduced by increasing CO_2 concentration. Grain growth was not affected by several days of leaf wilting; assimilates moved into the grains from the lower parts of wilted plants. Movement of assimilates into the conducting tissues was prolonged in wilted leaves, but the rate of translocation was little affected by stress. Wardlaw suggests that water stress acts directly on the leaf rather than indirectly through effects on growth or translocation. Evidently, wheat differs from sugarcane in its response to water deficiency. This may be related to the fact that wheat is transforming sugars to starch and so maintaining pressure gradients within the phloem and would seem to support rather than to controvert the mass-flow mechanism.

Smith *et al.* (1959) found that the rate of uptake and movement of maleic hydrazide decreased as water potential was lowered, even before wilting occurred. Penetration and translocation of 2,4-D was found to be reduced in soybean plants subjected to a drying cycle as compared with plants maintained in soil at field capacity (Hauser and Young, 1952; Hauser, 1955). Pallas (1960) concluded that translocation of 2,4-D and benzoic acid was appreciably reduced in plants growing in soil approaching the PWP.

Pallas and Williams (1962), using bean plants in soil culture in a growth chamber, found more ^{32}P to be absorbed and about 8 times as much to be translocated into epicotyls in plants at about one-third atm moisture tension as in plants at 3 atm tension. No effect of soil moisture stress was found on 2,4-D uptake but twice as much of the herbicide was translocated into epicotyls in plants at 1/3 atm as in plants at 4 atm tension. Differences in absorption in this case probably reflects the fact that 2,4-D penetrates the cuticle through the lipoid route whereas ^{32}P enters via the aqueous route (Crafts, 1961a).

In studies on the effect of moisture stress on absorption and movement of picloram and 2,4,5-T in bean plants, Merkle and Davis (1967) found that mannitol at 0.1 and 0.2 M concentrations reduced leaf disc turgidity about 7 and 12% and apparent sap velocity 28 and 91%. This seems to indicate that, percentagewise, sap movement is much more severely affected than is turgidity. When moisture stress was induced by withholding water for 96 hours, the average leaf turgidity was 90% of normal and the sap velocity was 26%. Moisture stress had no apparent effect upon absorption of either picloram or 2,4,5-T; it affected translocation of the latter more than it did picloram. When plants were stressed until wilting occurred, picloram movement was 57% of normal, 2,4,5-T movement was reduced below the limits of the method of detection. Under all conditions, picloram was more mobile than 2,4,5-T, probably because the latter is absorbed in transit more actively than is picloram.

Davis *et al.* (1967) found that moisture stress reduced transport of the above two herbicides in mesquite and winged elm. Stress reduced transport in proportion to growth reduction; stress sufficient to slow growth markedly reduced transport of both herbicides. From their work there is no way of determining whether inhibition of transport is primary or secondary to the effect on photosynthesis. Since growth reduction paralleled slowing of transport, the latter is probably the primary relation.

Badiei *et al.* (1967) in studies on movement of 2,4,5-T in blackjack oak found an inverse relation between moisture stress and both absorption and translocation. A decrease in soil moisture level from 16.0 to 7.5 to 2.8% reduced translocation from the treated leaf from 33.2 to 17.9 to 13.2%. The writers suggest that the poor kill of blackjack oak during the summer months may in part be the result of moisture stress in the plants.

In contrast to the above reports, Magalhaes and Foy (1967) found no effect of moisture stress on translocation of dicamba in nutsedge.

E. Transpiration

Because water movement in the xylem is essential to nutrient distribution and because water loss in the vapor form is a necessary corollary of CO_2 absorption for photosynthesis and to O_2 escape, transpiration is an extremely important physiological process. As explained above, water movement in the xylem is greatly in excess of that required for adequate nutrient transport. Therefore, if transpiration could be reduced without limiting CO_2 uptake, much water could be conserved, and hence water deficits in plants could be reduced. Transpiration has been the subject for an immense amount of research and, hence, only literature pertaining to the effects of water stress on water loss can be treated here.

As a plant transpires a lowering of water potential in the mesophyll cell walls is transmitted progressively throughout the apoplast system to the cell walls of the root hairs. While water may be drawn from living cells along the route, these develop their own potential deficits that are normally subject to diurnal fluctuations but reach an overall steady state. Thus, water lost from leaves is mostly replaced by water absorbed from the soil as a catenary process (Van den Honert, 1948), and the rate of movement is governed by the step involving the greatest resistance. Van den Honert concludes that evaporation of moisture from the mesophyll walls and its diffusion into the outer atmosphere are the rate-limiting steps under most conditions. These are the steps over which the plant exercises control through stomatal movements.

Root killing and removal prove that resistance to the movement of water through living cells of the root may constitute an appreciable portion of the total resistance (Kramer, 1933, 1938, 1949; Ordin and Kramer, 1956). Low soil moisture, high osmotic pressure of the soil solution, and low soil temperatures are additional factors. Philip (1957, 1958) and Bonner (1959) have invoked a " vapor gap " around roots to rationalize water uptake without an attendant uptake of salt. Work by Bernstein *et al.* (1959) and Gardner and Ehlig (1962) would seem to refute this concept. The latter workers present convincing evidence that the major resistance to water movement into roots derives from the low water conductivity of unsaturated soil. Weatherly (1965) provides additional evidence from experiments on potted and water culture plants conducted in a climatological wind tunnel. Jenny and Grossenbacher (1963) show by electron microscopy that the roots of barley plants are coated with a mucilaginous layer that extends from the cell walls into the micropores of the soil, for, in some instances, distances of up to 10 μ. This mucigel is pectic in nature and, hence, has cation binding sites throughout its mass. Under low water potential conditions this coating would shrink, water binding sites would be vacated, and ions possibly would be hindered in their movement into the root. Thus, transpiration under conditions of stress might bring about selective exclusion of salts, giving the impression that water was entering in the vapor phase. An alternative explanation might be that in the experiments cited the plants had low-salt roots and hence that, for a period, salts were stored in the vacuoles of root cells and not transported to tops. Broyer (1956) and Russell and Shorrocks (1957) have described such effects.

Physiologists have recognized for years that water vapor may escape from leaves by two routes: through the stomates and through the cuticles. Cuticular transpiration has been evaluated, and values of from 10 to 90 % of total transpiration are quoted. Pallas and Bertrand (1966) used high CO_2 concentrations in the atmosphere to bring about stomatal closing, and they give values ranging from 18 to 90 % for transpiration " after stomatal closure " for some five crops (Pallas and Bertrand, 1966, Table 3). Like Ting and Loomis (1963), they recognize that the degree of stomatal opening as reflected in size of the stomatal pore above several microns may be of minor importance in evaluating transpirational responses. Thus, the values of Pallas and Bertrand may be high because of inaccurate determinations of stomatal closure. Nevertheless, much modern work indicates that cuticles of leaves of many crop plants are appreciably permeable and cuticular transpiration continues after hydro-active stomatal closure has occurred (Szabo and Buchholtz, 1961; Crafts and Foy, 1962; Darlington and Cirulis, 1963; Yamada *et al.*, 1964; Bukovac and Norris, 1967). Such closure of stomates is the common response to low water potential in leaves.

Vaadia *et al.* (1961) stress the difficulty in determining accurately the role of stomatal closure in controlling transpiration, because stomatal reactions are usually associated with a lowering of leaf water content. Thus, the common method of estimating transpiration by measuring the change in total weight of a plant culture may be in error not only by the gross increase in fresh weight of the plant but by changes in the water content of the foliage. Hygen (1953) has pointed out three distinct phases in water loss from detached leaves: (1) a constant rate phase when open stomates exercise little control over water loss, (2) a decreasing rate phase when stomatal closure progressively reduces transpiration, and (3) a phase when closed stomates limit water loss to the cuticular route. The stomatal mechanism is discussed in more detail in Chapter 6, Volume I. It seems evident from many studies with detached leaves, attached leaves, and whole plants that the relation of water content to transpiration is very complex; water potential in the leaf is a better, yet still imperfect, index of the relation of stress to transpiration.

Vaadia et al. (1961) give an excellent analysis of control of nonstomatal waterloss. While students of leaf anatomy and stomatal mechanism have been led to assume that stomates exercise the main or even the sole regulation of transpiration, modern work on cell-wall structure (Frey-Wyssling, 1959; Roelofsen, 1965), cuticle permeability (Orgell, 1955; Crafts and Foy, 1962; Szabo and Buchholtz, 1961; Darlington and Cirulis, 1963; Yamada et al., 1964; Bukovac and Norris, 1967), and foliar uptake of nutrients and pesticides (Crafts and Foy, 1962; Yamada et al., 1964) certainly proves that changes in leaf water potential can be rapidly reflected, not only in changes in water vapor pressure in stomatal chambers but also in rates of cuticular water loss. While studies on stomatal control of transpiration may provide accurate values for plants having a copious water supply, as soon as the plant enters a period of stress nonstomatal control through lowered water-vapor pressure of stomatal chambers and maintained cuticular loss enter the picture and inaccuracies occur. Under these conditions there is internal redistribution of water and lowering of total plant water content so that absorption and transpiration become increasingly out of phase.

Jarvis and Jarvis (1963) studied transpiration in relation to osmotic potential of culture solutions using seedlings of aspen, birch, pine, and spruce. They found that sensitivity to water stress in both soil and solution culture varied in the following order: pine, birch, aspen, and spruce. They reasoned that these differences might be the result of differences in root length per unit volume of medium or of differences in water potential in the roots. A transient increase in transpiration in aspen and birch might result from an increase in permeability of roots.

Slavik (1958) carried out experiments on transpiration of excised sugar beet leaves in an attempt to solve the problem of how a water deficit regulates transpiration when acting through internal factors not associated with stomatal closure. Using Hygen's (1951, 1953) method, Slavik calculated the stomatal and cuticular transpiration intensities reduced to zero water deficit and the correlation and regression coefficients between cuticular transpiration intensity and initial water deficit and rate of increase of initial water deficit.

By such calculations Slavik found that the intensity of stomatal transpiration was reduced by about 3% and cuticular transpiration by 6% for every 1% of water deficit. Both stomatal and cuticular transpiration decreased by an average of 7% for every rise in the rate of increase of water deficit of 1% per hour.

Fry and Walker (1964) measured rates of transpiration and net photosynthesis of Douglas fir seedlings undergoing progressive water stress. Under stress, stomatal opening was about 1μ; during optimum hydration, 3 μ. Transpiration and net photosynthesis decreased as water stress in the needles increased. The stomata showed no increase in resistance to ethanol injection until both transpiration and photosynthesis had greatly diminished.

Boyer (1965) used sodium chloride in the culture solution to lower water potential around roots of cotton plants; osmotic potential values used were -0.5, -3.5, -6.5, -9.5, and -12.5 bars. Fresh and dry weights were not altered in the first three cultures but decreased rapidly between -6.5 and -9.5 bars and were slightly lower at -12.5 bars, at which level they were approximately 50% of the values in the other cultures.

Resistance to diffusion of CO_2 was not altered through the range of solute concentrations; transpiration likewise remained rather constant. Photosynthesis and respiration were reduced in proportion to the increase in salt concentration in the culture medium. At -8.5 bars reduction was 25% on the basis of leaf area, fresh or dry weight, or chlorophyll content. Stomata remained open through all salt concentrations; reduction in photosynthesis may have resulted from effects of salt on enzyme systems (Boyer, 1965). Gardner and Ehlig (1965) used cotton, sunflower, pepper, and trefoil plants in studies on the physical aspects of water stress in leaves. While it has generally been assumed that the permanent wilting point corresponds to zero TP in leaves, Gardner and Ehlig found visible wilting to occur at TP's of 2–3 bars. They consider that the symptoms of permanent wilting result from a change in the elastic properties of the cells when TP drops below a critical value somewhat above zero. While cotton leaves have more internal rigidity than pepper leaves, results of their tests were such that Gardner and Ehlig concluded that leaves wilt at a critical turgor, denoting a change in elasticity corresponding to a water potential level of -11 to -13 bars, a value that corresponds fairly well to the traditionally accepted permanent wilting point correlated with a soil–water potential of -15 bars.

Since it is obvious that transpiration under most conditions is wasteful of water, because much more water is lost than is required to bring about mineral distribution, improvement in transpiration efficiency would be desirable. Pallas and Bertrand (1966) illustrated differences in transpiration efficiency among corn varieties; under stress, Dixie 82 (a double-cross hybrid) required 343 lb of water to produce a pound of grain; Hastings, an open-pollinated variety, required 814 lb. Under irrigated conditions, Dixie 82 used 654 lb of water; Hastings, 863 lb. Thus, selection and breeding are methods for obtaining higher water use efficiency in plants (Briggs and Shantz, 1913; Maximov, 1923). Tumanow (1927) showed that sunflower plants, subject to occasional but not excessive wilting, increased in transpiration efficiency over unwilted plants, and Eidmann (1962) found pine and spruce seedlings to have increased transpiration efficiency in pots in which the soil was under stress.

To get at the basic reasons for changes in transpiration efficiency one must go the the process of gaseous exchange. Koch (1957) measured the photosynthesis: transpiration ratio, P/T, where P = photosynthesis (milligrams of CO_2 uptake) per unit area per unit time and T = transpiration (grams of H_2O lost) per unit area per unit time. Loustalot (1945) with *Carya* seedlings and Polster *et al.* (1960) and Neuwirth and Polster (1960) with tree species found not only that efficiency of transpiration was increased with decreasing available soil moisture over an entire growing season but that during early stages of hydroactive stomatal closing stomatal transpiration was more strongly restricted than CO_2 uptake.

Figure 2 shows results on P/T ratios for *Quercus ilex* and *Q. pubescens*

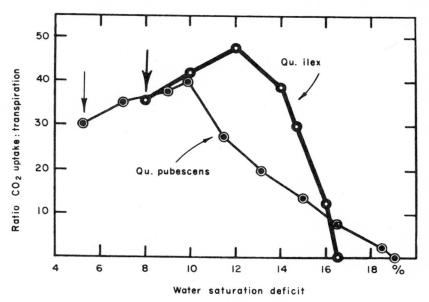

Fig. 2. The course of the P/T quotient (ratio of photosynthesis to transpiration expressed in milligrams of CO_2 per gram of H_2O) for drying, excised twigs of *Quercus ilex* and *Q. pubescens* under standard conditions. At the water-deficit values marked by the vertical arrows, transpiration and photosynthesis reached their highest absolute values. From Larcher (1965).

twigs found by Larcher (1965). Similar but less marked effects of water stress were found with potted olive trees.

Bierhuizen and Slatyer (1965) have shown that under constant adequate water supply the resistance to CO_2 diffusion into the cotton leaf, per unit resistance to water vapor movement, lowers as light intensity increases from 1000 to 6000 footcandles. It also decreases with decreasing air movement from around 4.6 at an air velocity of 3.1 cm second^{-1} to 2.2 at an air velocity of 0.6 cm second^{-1} when light intensity is 6000 footcandles. This difference of 2.4 at 6000 footcandles becomes about 2.8 at 4000 and 3.2 at 2000 footcandles. This means that P/T does change with environmental changes, becoming greater at light saturation and in still air. That P/T is also determined by plant species is shown in Fig. 3 from Bierhuizen and Slatyer, showing the relation between transpiration ratio E/A (grams of H_2O transpired per

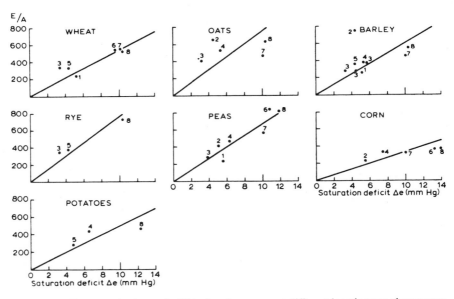

Fig. 3. The transpiration ratio E/A of various crops at different locations vs. the average saturation deficit. For details, see Bierhuizen and Slatyer (1965).

gram of carbohydrate produced) and water-saturation deficit of the atmosphere for 7 crops. The low slope for corn indicates a high water use efficiency as compared with oats and rye. Corn is a crop that matures in late summer and well may be exposed to water stress; oats and rye are winter annuals that usually ripen before drought occurs. Thus, the differences shown by these crops may well relate to the soil moisture conditions to which they have been exposed during evolution.

Ehrler *et al.* (1966), working with cotton plants in controlled environment chambers and measuring transpiration under different conditions of saturation deficit of water vapor in the air, found that the recovery from the initial loss of turgidity brought about by sudden illumination was faster and more complete at high than at lower values of saturation deficit. Recovery occurred even in the light at the higher deficit values, but it was enhanced by return to darkness and lowering the saturation deficit. Ehrler *et al.* concluded that recovery depended on increased absorption, a result that implies an adaptive response.

Another approach to increasing the efficiency of water utilization by plants is through chemical control of stomatal opening. Ferri and Lex (1948) showed that an aqueous solution of β-napthoxyacetic acid applied to soil closes the stomates of nasturtium leaves. Zelitch (1961) found that 8-hydroxyquinoline sulphate prevents closed stomates of tobacco leaves from opening under conditions favorable to opening. Stoddard and Miller (1962) tested the effects of this chemical on strawberry plants and found that it closes the stomates and reduces water loss, enabling the plants to withstand prolonged drought.

Smith and Buchholtz (1962, 1964) and Wills and Davis (1962) reported reduced transpiration of plants treated with 2-chloro-4-ethylamino-6-iso-propylamino-1,3,5-triazine (atrazine), and they attributed this to closure of stomates. Pallas and Bertrand (1966) found no effect of atrazine on corn in their field trials. A report from Australia (Freney, 1965) indicates that superior growth of plants in the presence of low amounts of triazine herbicides results from the increased uptake of minerals, not from an effect on water utilization.

Shimshi (1963a,b) tested the effect of phenylmercuric acetate, a chemical that brings about stomatal closure, on the transpiration of tobacco and sunflower, on stomatal closure, on transpiration, and on photosynthesis of corn. These tests proved that this chemical reduced photosynthesis less than it did transpiration. Decreased soil moisture, even within the available range, caused a further reduction in both transpiration and photosynthesis that could not be attributed to stomatal closure. Analysis of resistance to water movement indicated that as soil dries an appreciable resistance to water movement develops at the evaporating surfaces of the mesophyll cells. The water tension at these surfaces may reach 80 bars without the plants wilting.

Gale and Hagan (1966) have written an excellent review on antitranspirants. They discuss the cooling effects of transpiration but point out that in hot, arid regions, when soil moisture becomes limiting, stomates usually close during the hottest hours of the day and thus the cooling effect is lost. Antitranspirants that close stomata cause the leaves to warm up. Because of heat conduction to the atmosphere, it is doubtful if the cooling by transpiration or warming because of antitranspirants are important factors in plant growth except under the most severe conditions. Slatyer and Bierhuizen (1964) found leaf air temperature differences up to 9°C, when transpiration was completely inhibited, in cotton plants in Australia.

Gale and Hagan (1966) point out that whereas diffusion of water vapor is subject to two resistances, stomatal aperture resistance and boundary layer resistance, CO_2 diffusion to the chloroplasts is subject to three separate resistances: the two just mentioned plus the liquid phase diffusion resistance to movement from the mesophyll walls to the chloroplasts. Using the calculations of Gaastra (1959), they conclude that under certain conditions an increase in stomatal resistance will reduce transpiration more than photosynthesis, resulting in a favorable P/T ratio. Work by Zelitch (1961), Zelitch and Waggoner (1962), and Slatyer and Bierhuizen (1964) shows that this may be accomplished by use of certain antitranspirants, e.g., phenylmercuric acetate.

A second approach to transpiration reduction is by use of materials that form thin films. Higher alcohols, waxes, silicones, and plastics such as polyethylene, vinyl-acrylate, and similar polymers have been tried with varying results. Gale and Hagan conclude that the main practical problem is the development of new, improved antitranspirants of the stomata-closing type having greater specificity for crop species and longer functional life and of film-forming materials with greater selectivity to gases and vapors. Finally, these materials will need careful study by physiologists and agriculturalists to find their effects on mineral nutrition, photosynthesis, etc., and to avoid injury to crops.

F. PHOTOSYNTHESIS

Decreasing water potential in leaves is known to reduce photosynthesis. As Slavik (1965) points out, this is the result of three effects: (1) hydroactive closing of stomata bringing about a reduced CO_2 supply, (2) water stress in cytoplasmic ultrastructure affecting enzyme activity, and (3) dehydration of cuticle, epidermal walls, and cell membranes reducing their avidity for and permeability to CO_2. Attempts to correlate photosynthetic activity with tissue water potential have been complicated by the fact that stomates are photoactive as well as hydroactive, that actually water potentials of tissues are difficult to measure, and that when, as is often the case, limiting soil moisture is used to lower water potentials, results are influenced by stored water, internal resistance to water absorption, and other disturbing factors. Nevertheless, many carefully made studies have given an excellent picture of the relation of photosynthesis to water supply. Figure 4 from Ashton (1956) presents the rates of photosynthesis of sugarcane and attendant soil moisture percentages through five drying cycles. While photosynthesis decreases between field capacity and permanent wilting, in this study the rate of decrease is not uniform; the initial rate is maintained until a significant moisture stress develops and finally drops rapidly, approaching zero as permanent wilting is reached. Several days pass after an irrigation before the original photosynthesis rate is regained. With the repeated drying cycles, Ashton found that the

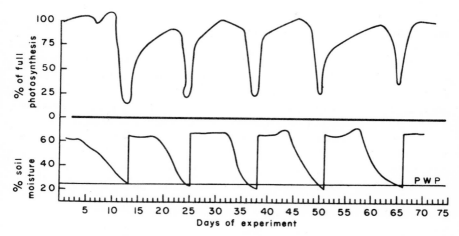

Fig. 4. Rates of photosynthesis of sugarcane and percentages of soil moisture during 5 drying cycles over a period of 75 days. From Ashton (1965).

plants tended to regain their photosynthetic capacity more rapidly; also, the level of photosynthesis gained with repeated drying cycles as if the plant were adapting to repeated drought treatments. During the course of a single day, photosynthetic rate was found to fluctuate, decreasing as stress mounted under bright weather, but often increasing temporarily during cloudy periods when transpiration rate fell off and water potential increased.

Heinicke and Hoffman (1933) and Heinicke and Childers (1935) proved the dependence of photosynthetic rate upon air flow over leaves, implying that the rate of CO_2 diffusion into leaves may be limiting under still conditions. Moss et al. (1961) showed that relative responses to fluctuating CO_2 supply are greater in photosynthesis than in transpiration. In diffuse sunlight net assimilation increased as CO_2 supply was raised, and in full sunlight turbulence was required for corn to maintain normal photosynthetic rates. Evidently, CO_2 movement into the leaf is one of the critical limiting factors for reaching the photosynthetic potential that the leaf is capable of attaining (Vaadia et al., 1961). Kozlowski (1958) reported that an unfavorable water balance in trees may reduce growth, cause early leaf fall, dieback, transplanting failure, sunscorch, and death. Water deficits in leaves check growth by influencing carbohydrate supply through effects on photosynthesis, largely because of increased resistance to CO_2 diffusion (Kozlowski, 1964).

Larcher (1965), studying the relation of photosynthesis to transpiration, found under laboratory conditions that as water begins to become limiting, water vapor output is more severely restricted than CO_2 uptake. Figure 2 shows his results with excised twigs of two *Quercus* species. The P/T quotient was not highest when the *Quercus* twigs were assimilating and transpiring most highly, that is, at a relatively high soil moisture. As shown in Fig. 2, the high points on the P/T curves were attained only when the water saturation deficit had reached values between 10 and 12 %; the falloff in CO_2 uptake overtakes and exceeds transpiration only at still higher saturation deficits. Working with potted olive trees, Larcher made similar tests. He found P/T ratios in the region of 10 mg CO_2 per gram H_2O on the first day; as the soil moisture was lowered on succeeding days by the transpiring plants, the average ratios were approximately 8.5 on the second day, 6.8 on the third day, 9.0 on the fourth day, and 1.0 on the fifth day as soil moisture became depleted. The high value of 9.0 on the fourth day represents a high efficiency of water utilization, but CO_2 uptake and hence photosynthesis was down to about one third of its initial value.

Tranquillini (1963) has pointed out that the effect of relative humidity through its influence on transpiration is important as a determinant of water use efficiency. High relative humidity tends to counteract low soil moisture, and low relative humidity tends to exaggerate soil moisture deficiency. He used young gymnosperm seedlings growing in soils having 100, 50, and 28 % FC (field capacity). In soil at FC all plants showed decreased photosynthesis at low relative humidities. In shade-tolerant spruce, photosynthesis dropped to 15 % of maximal; in pine, to 33 %; and in larch (light tolerant), to 70 %. With soil at 50 % of FC and in air having a relative humidity of 85 %, photosynthesis was greater than in soil at FC. When soil moisture dropped to 28 % of FC and relative humidity was 85 %, photosynthesis dropped far below values attained in soil at FC.

Different results in studies on P/T ratios may be caused by rooting characteristics of species. Yurina (1957) noted that drought affected photosynthesis of various species differently. Oaks were influenced less by drought than species having less extensive root systems. It has been noted many times where deep-rooted trees are removed in range improvement programs that springs long dry will flow again, and on large water sheds stream flow will increase.

In forestry, the spacing of trees for most efficient utilization of light, CO_2, and water is an important matter. In Monterey pine plantings in Australia, observed by the writer, 8 × 8 foot spacing had proved excellent during the early years as the young trees were becoming established. Such close planting had as its objective shading of the lower branches to produce straight, clear boles. However, as the trees closed in and covered the total soil area, water supply by the soil became limiting and, particularly in drought years, growth was severely inhibited. Experiments by foresters in the Canberra area proved that supplemental water would greatly increase the annual growth increment and, based on this observation, thinning practices were being initiated to fit the tree cover more nearly to the annual water supply.

Gaastra (1959) attempted to evaluate the relative effects of light, CO_2, temperature, and stomatal conductivity on photosynthesis of crop plants. He found 0.1 % the CO_2 saturation concentration for photosynthesis. Under low limiting light, considerable stomatal closure occurred without affecting the rate of photosynthesis. Since stomatal opening is more dependent on water balance than on light, photosynthesis may be limited severely during periods of water stress despite the fact that light and CO_2 supply are optimum. A fact stressed by Hartt (1967) is that translocation of photosynthetic products may also be limited by water stress and, hence, photosynthesis may be limited by assimilate saturation while other factors are not limiting. El-Sharkawy and Hesketh (1964) found net photosynthesis of 4 plant species to be depressed by high water deficits.

In studies with cabbage, Čatsky (1965) found that leaves of different age wilt unevenly; old leaves wilt before young ones when water is withheld. Measuring 5 leaves he found water stress to develop most rapidly in the older leaves. Photosynthesis decreased in the same order, and there was good correlation between photosynthesis rate and water potential between saturation deficits of 10 to 70%. Brix (1962) found photosynthesis to decrease as DPD increased in leaves of loblolly pine seedlings and tomato plants (Fig. 5 and 6). Carbon dioxide uptake declined when leaf DPD attained a value of 4.0 atm, and when it reached 11 atm, photosynthesis stopped. The rate of recovery of photosynthesis from water stress depended upon the condition of the root system; when permanent wilting has occurred, recovery may be very slow. When root hairs are destroyed and growing points of roots are retarded in growth, the water-absorbing surface of the root system is greatly reduced, and normal water uptake, following irrigation, depends upon restablishment of this surface. This requires root growth, and it may take days or weeks depending upon the age and vigor of the plants. In the case of annual or biennial herbaceous plants, the original form of the root system may never be restored.

Todd and Webster (1965) studied the affects of drying cycles on photosynthesis and survival of wheat and oat seedlings. All plants carried on a higher photosynthetic rate at a lower turgor after subjection to a single drought period. Photosynthetic rate paralleled moisture stress during successive drying cycles. In survival studies there was a continuing loss of plants

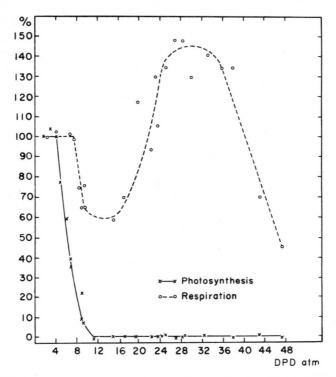

Fig. 5. The effect of water stress on the rates of photosynthesis and respiration of loblolly pine seedlings. Water stress in leaves as DPD in atmospheres; photosynthesis and respiration as percentage of rates in controls at field capacity. From Brix (1962).

with each successive cycle. The most drought-hardy varieties, based on field experience, could be separated from the least hardy by significant differences in survival.

To eliminate the stomatal factor and thus simplify interpretation of results, Slavik (1965) conducted photosynthesis trials using the liverwort *Conocephallum conicum*. Measuring CO_2 uptake of the thallus in plexiglass containers with an infrared CO_2 analyzer at $25° \pm 0.5°C$, $55 \pm 5\%$ relative humidity, 300 ppm CO_2, illumination at 6.8×10^4 erg cm^{-2} sec^{-1}, and in an airstream of 5 liters per hour per 300 mg thallus fresh weight, and starting with full water-saturated thalli, Slavik made hourly measurements on weight and CO_2. Figure 7 gives Slavik's results expressed as relative photosynthetic rate against water content in percent dry weight. This curve shows that in the case of this material there was a steady decrease in photosynthetic rate for each decrease in water content until at complete dryness photosynthesis ceased. Having determined the relation between osmotic pressure of the cell sap and water

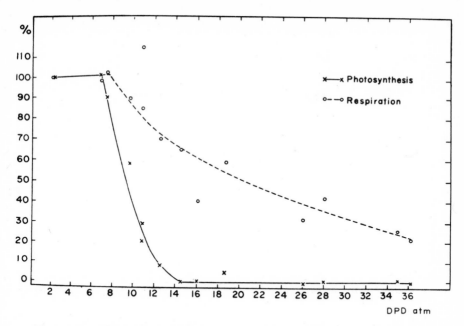

Fig. 6. The effect of water stress on the rates of photosynthesis and respiration of tomato plants. Results are expressed as in Fig. 5. From Brix (1962).

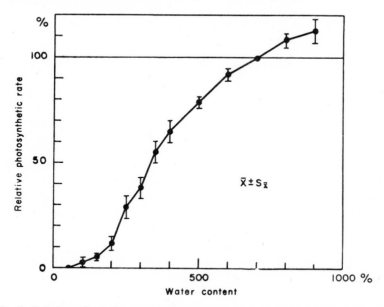

Fig. 7. Relation of relative photosynthetic rate to water content of liverwort thalli. Photosynthetic rate is expressed in percentage of the standard rate of each sample at 700% water content. From Slavik (1965).

content for this material, Slavik converted water content to osmotic potential, and Fig. 8 shows this relation. At full saturation these thalli had an osmotic potential of -5 atm; the osmotic potential of the air dry material was -12.6 atm, which corresponded to a water saturation deficit of about 90%.

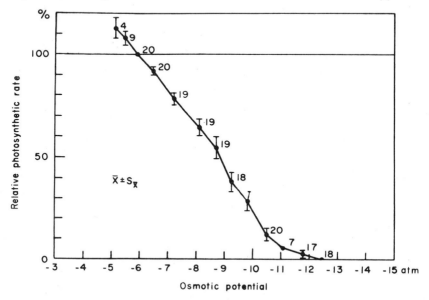

Fig. 8. Relation of relative photosynthetic rate to osmotic potential of the cell sap of liverwort thalli. Photosynthetic rate expressed as in Fig. 7; osmotic potential in atmospheres. From Slavik (1965).

It is difficult to generalize from data such as these because most studies on the relation of water availability to photosynthetic rate are complicated by factors involved in water absorption and transport, by photoactivity as well as hydroactivity of stomates, and by diurnal variation in light, temperature, transpiration rate, etc. If Slavik's results can be interpreted as being of general significance, then it is obvious that photosynthesis is highly dependent upon water supply, and as soon as the water potential of a leaf is lowered, assimilation slows down.

The above observations may have profound implications for certain agricultural practices. At one time it was common for sugar beet growers to withhold irrigation some days before harvest in order to raise the sugar percentage in the beet roots at the time of delivery to the factory. While this reduction in water content lowered somewhat the hauling costs of the beets, it probably cut the per acre production of sugar enough to more than compensate for this saving.

Currently there is interest in chemical transpiration depressants that act by closing stomates and thus conserving moisture. While this may achieve plant survival in times of critical water shortage, it certainly cannot add to dry weight production since photosynthesis is reduced when stomata are closed. What is needed is increased efficiency of water utilization. The most promising way to achieve this is through plant breeding and selection.

G. RESPIRATION

Since most if not all of the multitude of chemical reactions that make up the process of respiration are enzymatic, and since the function as well as the structure of enzymes are affected by the water status of plants, it seems that respiration should be subject to regulation by the hydration of tissues. Research has proved that this is true (Scholander *et al.*, 1952; Woodhams and Kozlowski, 1954; Ragai and Loomis, 1954; Takaoki, 1957; Stanley, 1958; Kozlowski and Gentile, 1958; Yarosh, 1959; Kramer and Kozlowski, 1960; Stocker, 1960; Brix, 1962; Jarvis and Jarvis, 1965).

Woodhams and Kozlowski (1954) grew bean and tomato plants in soil at the moisture equivalent, soil undergoing one drying cycle, and soil subject to 4 drying cycles. In all cases carbohydrates were higher in controls grown in frequently irrigated soil than in those undergoing drought. Differences were small between plants analyzed at the end of a single drought period and plants irrigated and held 24 hours after a single drought period. In plants subjected to 4 drying cycles carbohydrates were very low in leaves, somewhat higher in stems, and not greatly different from the partial and total one-cycle treatments. Changes in starch–sugar ratio in response to drought were also greatest in leaves. The greater amounts of starch in low-moisture-deficit plants compared with leaves from unstressed plants was not compensated by sugars in the high-moisture-tension treatments. Woodhams and Kozlowski took this to mean that respiration rates were increased in the moisture-stressed plants.

Ragai and Loomis (1954) found the respiration rate of corn seeds to increase exponentially with increasing moisture content. Takaoki (1957), using sucrose and salts as osmotic substrates, found increasing respiration with decreasing osmotic potential, except in the case of certain halophytes. Increased respiration of wilted leaves has been observed (Iljin, 1957), but here one must consider the time factor in relation to possible injury.

Yarosh (1959) observed increases in ascorbic acid and catalase activity in cotton plants without irrigation as compared with irrigated plants. Taking these changes to be indices of respiration he concluded that water stress in the nonirrigated plants led to increased respiration. He also found that the ratio of nonprotein to protein nitrogen was higher in seeds of unirrigated than irrigated cotton plants. He attributed this to slowing down of protein synthesis and to partial hydrolysis.

Stanley (1958), studying germination processes of stratified and unstratified pine seed, found oxygen uptake to follow water absorption. The respiratory quotient, in both cases, decreased during the first 72 hours of germination; these changes were attributed to activation of enzymes of the embryo; mitochondrial study indicated that the particulate oxidative enzymes were functioning.

Kozlowski and Gentile (1958) studied respiration of terminal buds of white pine. Removal of scales speeded respiration before bud break; the difference between buds with and without scales disappeared after bud break. There was a strong correlation of respiration rates with hydration of tissues until the moisture content reached a value of more than 500% of the oven-dry weight.

Although respiration in seeds is usually affected much more than in leaves by changes in moisture content, severe dehydration of leaves often leads to a burst in respiration as a result of hydrolysis of starch to sugar (Kramer and Kozlowski, 1960). Stocker (1960) has reported that leaves that become flaccid from rapid transpiration on warm days may exhibit high rates of respiration. Schneider and Childers (1941) made similar observations.

Brix (1962) studied photosynthesis, respiration, and transpiration in loblolly pine seedlings and tomato plants. Figure 5 shows the results of his photosynthesis and respiration studies in loblolly pine. It is obvious that the respiration response is complicated; photosynthesis was decreased before there was an effect on respiration. Brix proposes that reduction in concentration of respiratory substrates resulting from lowered photosynthesis might be responsible for the initial decrease in respiration. The subsequent increase may result from increase in sugars produced by hydrolysis of starch. The final drop could well represent permanent wilting and possibly induction of senescense. Tomato plants under comparable conditions showed a steady lowering of respiration with increasing stress (Fig. 6). Both pine and tomato showed close correlations between photosynthesis and transpiration under increasing water stress.

Jarvis and Jarvis (1965) measured respiration of roots in relation to the osmotic potential of the culture medium using root tips of pine, spruce, birch, and aspen. Figure 9 presents the results. It is evident that root respiration of all four species is negatively correlated with potential. These results differ from those of Takaoki (1957) and Iljin (1957). The problem of injury is often involved in treatments using strong osmotic agents. Jarvis and Jarvis used Carbowax 1540, a polyethylene glycol that evidently was not absorbed to any appreciable extent, nor did it cause injury as indicated by rapid fluctuations in O_2 release.

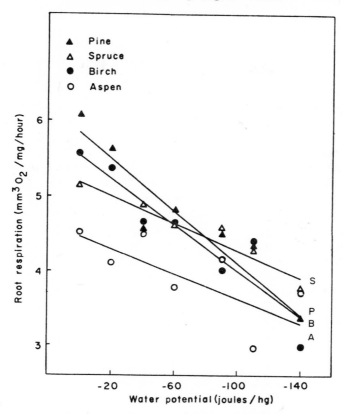

Fig. 9. The relation between respiration, measured as O_2 uptake, and osmotic potential of the medium, for apical root sections 1–2 cm long of pine, spruce, birch, and aspen. From Jarvis and Jarvis (1965).

H. Growth

It is widely recognized that plants grow only when they have available water. Irrigation agriculture is based on this fundamental premise, and while specialized desert-loving plants are adapted to long periods of drought and may survive such periods when mesophytes would die, it remains a fact that these plants grow only during the brief periods when there is available moisture in the soil.

The effects of stress on plant growth have been described in several reviews (Kramer, 1949; Slatyer, 1957, 1960; Kozlowski, 1958, 1962, 1964; Kramer and Kozlowski, 1960; Vaadia *et al.*, 1961; Slavik, 1965; Kozlowski and Keller, 1966). From this array of literature it is apparent that the relationship between water stress and growth is very complex, involving to some extent all of the physiological processes described above. Probably the process most directly and severely affected is photosynthesis; hydroactive closure of stomates may rapidly reduce photosynthesis to values near zero (Kramer, 1962). Figure 10 from Brix (1962) shows this relationship for pine seedlings and tomato plants. While the relation of respiration to water status is more complex, under conditions of permanent wilting, continued respiration, after photosynthesis has ceased, uses up stored foods, reduces substrate supply, and hinders growth.

Gingrich and Russell (1956) studied the effect of soil moisture tension and oxygen concentration on the growth of roots of corn seedlings. Germinating seeds having radicles 11–14 mm in length were placed in soils of varying moisture contents, sealed into closed containers, and subjected to atmospheres of differing oxygen compositions. The experiments lasted 24 hours, after which the seedlings were removed and measured. In the absence of other limiting factors, increases in radicle elongation, fresh weight, dry weight, and degree of hydration of the seedlings became progressively less as the soil moisture tension was increased from 1 through 12 atm. The growth properties measured in these experiments were more sensitive to changes in moisture tension in the range between 1 and 3 atm than for any other range when oxygen was not limiting. The growth responses to oxygen depended on the moisture tension of the soil. Under conditions of low moisture stress the oxygen concentration of the soil atmosphere had to be above 10.5% for maximum growth. The inverse relation between oxygen and water in soils near FC is apparently critical for root growth of seedlings.

Carolus et al. (1965) found tomato growth and development to be related to relative soil moisture levels but also to be influenced by variation in atmospheric stress. Fruit productivity was increased 120% in a season of high atmospheric stress with soil moisture held at 85% of FC. Productivity decreased 15% when atmospheric stress was low and soil moisture was high. As available soil moisture increased from low (20%) to high (85%), under high atmospheric stress fruit number increased 47% and fruit size 49%; under low atmospheric stress fruit did not vary significantly in size and 23% fewer fruits were produced. High soil moisture tension coupled with high atmospheric stress caused blossom-end rot; low DPD in the plants resulting from low soil moisture stress and/or low atmospheric stress resulted in blotchy ripening of fruits. This work brings out the fact that maximum growth and productivity depend upon a balance of stresses or supplies such that the plant is at no time subject to severe deficiency of necessary factors.

Vaadia et al. (1961) discuss the complex effects of water stress on growth. They mention the different abilities of plants to use water efficiently. One example is the result of Pallas and Bertrand (1966) mentioned above.

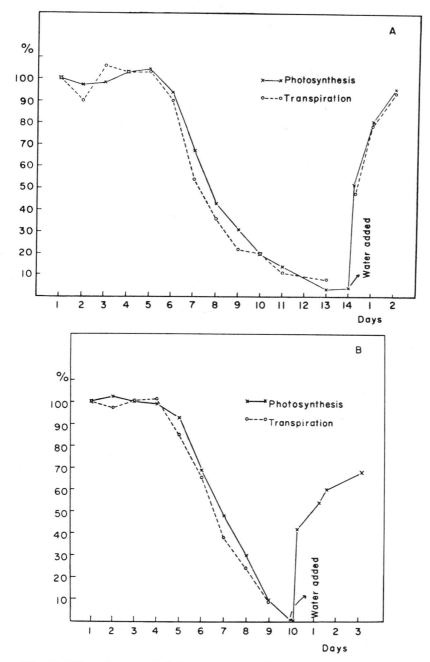

Fig. 10. Effect of stomatal closing on photosynthesis as shown by comparisons of photosynthesis and transpiration rates; these rates decreased steadily as the drying cycles progressed; recovery was rapid when water was added. A: Loblolly pine seedlings; B: tomato plants. From Brix (1962).

De Wit (1959) has calculated yields of potential photosynthesis for Netherlands conditions.

Vaadia *et al.* (1961) also discuss differences in rooting habit of different plants. One aspect of this that has not been emphasized is the fact that because the feeder roots of many crops are shallow, if irrigation is delayed until the approach of permanent wilting, roots in the upper foot of soil may become desiccated to the point of death, and absorption is displaced to lower soil horizons where nutrients are not only less plentiful but of different proportions. For example, sugar beets growing in a deep alluvial soil may root to a depth of 6 feet or more. Following irrigation if the crop is left until the whole occupied depth of soil reaches permanent wilting, roots in the top 6–12 inches may die; uptake of minerals is displaced downward into soil layers having less nitrogen and more salts than upper layers. Under these conditions growth may be permanently inhibited, alkali salts may become damaging to the vascular tissues, the plants may become chlorotic, and growth may be reduced despite the fact that irrigation is repeated.

Another factor affecting growth in stressed soils is soil texture. In heavy soils diffusion of water to feeder roots is slow, and under conditions of high transpiration cylinders of dry soil may develop around roots while an overall soil sample may indicate an abundance of moisture. This localized deficit can lead to a serious loss in fruit crops that require a plentiful water supply at the time of rapid fruit development (Aldrich *et al.*, 1940).

Quality of certain vegetable crops may be adversely affected by water stress. Brassica crops may develop an objectionably high fiber content during early growth if the water supply runs low. Succulent crops such as lettuce, radishes, and celery require adequate water supplies right up to the time of harvest; beans are sensitive to water stress during pollination and fertilization. If they are to be used green, beans resemble the succulents mentioned above; if they are to be raised for seed, irrigation may be terminated as soon as seed are fully grown. Most seed crops and fiber crops such as cotton may be benefited by withholding water during maturation (Vaadia *et al.*, 1961).

Fraser (1962) described the role of water in tree growth. From a 5-year study involving some 16 species he developed a convincing picture of the role of available soil moisture in tree establishment. He found radial growth to be closely correlated with moisture supply. Moisture influences initiation of cambial activity indirectly through the effects of soil-profile development and soil aeration on roots. On sites where there is a thick organic soil cover, growth initiation was delayed because the soil remains cool. Roots of trees in this organic layer make poor contact with the underlying mineral soil. In dry seasons these roots become dry and radial growth stops in early July whereas in normal years growth may continue until September.

Doss *et al.* (1962) made a 3-year study of 5 forage species under 3 moisture regimes involving irrigation after 30, 65, and 85% of the available moisture in the top 2 feet of soil had been removed. Average yield for all species except *Lespedeza cuneata* increased as available moisture increased. The yield of *L. cuneata* was highest at the intermediate moisture regime. Average annual yields of all species totaled 8631 lb for the low moisture (30%) regime, 9436 for the intermediate (65%) regime, and 9954 for the high (85%) regime. Average daily evapotranspiration rates were 0.09, 0.16, and 0.17 inch, respectively, for the 3 regimes.

Jarvis and Jarvis (1965) studied the relation of water potential to the growth of pine and spruce seedlings in culture solutions using polyethylene glycol as an osmotic substrate. Figure 11 shows the results of this study. They found the sensitivity of relative growth rate to decreasing water potential of the culture medium to increase in the following order: aspen < pine < birch < spruce. In soil they found the order of pine and birch to be reversed. All the differences in growth could not be explained in terms of leaf water potentials; sensitivity differences may involve stomatal mechanism or differences in osmotic or matrix potentials related to metabolism rather than to water status.

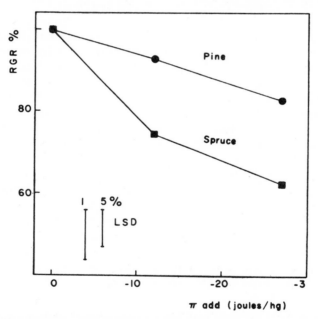

Fig. 11. The relation between relative growth rate of seedlings of pine and spruce and the osmotic potential of the root medium. The osmotic potential of the control solution (O on the abscissa) was -5 joules/hg. π add is additional osmotic potential resulting from added polyethylene glycol. The period of growth was 5 weeks. From Jarvis and Jarvis (1965).

I. Plant Composition

Water stress may have both qualitative and quantitative effects on plant constitutents. Probably the most direct effects are on carbohydrates through the inhibition of photosynthesis. Woodhams and Kozlowski (1954) noted the rapid conversion of starch to sugars in tomato and bean plants. The high percentage of starch in their control plants compared with plants that were subjected to drying cycles was not compensated by sugars in the latter, a fact that Woodhams and Kozlowski attributed to high respiration in the stressed plants.

Ordin (1960) found a parallel effect of turgor pressure upon elongation and cellulose synthesis in the growth of cell walls; stressed cells had limited cellulose synthesis. In further studies on the effects of stress on cell wall metabolism Plaut and Ordin (1964) found a small increase in the incorporation of ^{14}C-glucose into ^{14}C-cell wall constituents caused by nitrogen fertilization, a large decrease resulting from increasing the DPD of leaves to the 12–14 atm level by adding 0.25 M mannitol to the culture solution. Current research on auxin-induced cell growth implicates cell wall plasticity and cellulose synthesis.

Moisture deficits are reported to bring about increases in nitrate nitrogen in bean with less effect on soluble organic nitrogen; nitrogen contents of apple, tomato, and tobacco are increased in plants under stress (Kozlowski, 1964). Mothes (1956) reports a relationship between water content and proteolysis. Proteolytic enzymes activated by reduced sulfhydryl groups have increased activity in stressed plants with closed stomata because of the lowered oxygen tension. Petrie and Wood (1938a,b) found that formation of proteins from amino acids decreased with increasing moisture stress. Ivanov (1959) noted decreased amino acid synthesis with moisture deficit.

Zholkevich and Koretskaya (1959) grew pumpkin plants under normal and stressed soil moisture conditions and analyzed the roots for differences in constituents. They found that interruption of phosphorylation is the primary response to drought. This leads to a decrease in ATP, phosphated sugar esters, RNA, and DNA and thus to suppression of protein synthesis. The protein component of the total nitrogen was maintained, falling to 77 % of normal only at the end of the drought period. During the early part of the drought period, sugars in the roots increased; later there was loss under severe stress.

During drought the specific content of succinic and fumaric acids decreased abruptly; citric and malic acids decreased, but to a lesser degree. Later in the stress period citric and malic acids accumulated in roots, citric to 2½ times that in controls, malic to 6 times. When soil moisture was lowered to 15 % of field capacity all organic acids decreased in the roots. Since succinic and fumaric acids are characteristic of roots of young plants whereas malic acid is high in roots of mature plants, the authors suggest that drought and aging bring about similar shifts in root metabolism. Shah and Loomis (1965) found lowered amounts of RNA and protein in old leaves of sugar beet and suggested that moisture stress and age had similar effects on RNA metabolism. Pyruvic and α-ketoglutaric acids were much reduced in whole root systems of stressed pumpkin plants. Eighteen amino acids plus the amides glutamine and asparagine were identified; alanine and glutamic acid predominated. As drought itensified, the total amount of amino acids decreased, presumably because their synthesis was blocked. Amides increased with increasing stress, possibly because amidation exceeded amination. Changes in carbohydrates, phosphorylated compounds, organic acids, and amino acids suggest that soil drought restricts conversion of sugars by blocking glycolysis, the Krebs cycle, and amination; a recycling of sugars to the foliar organs sets in. The presence of increasing amounts of organic compounds in sap collected from roots during initial stages of drought reflects a decrease in retention of these in roots resulting from a weakening of synthetic activity. Evidently, the energy metabolism of the root is severely disturbed (Zholkevich and Koretskaya, 1959).

Gates and Bonner (1959) found a decrease in both DNA and RNA on a per-leaf basis in young tomato plants subjected to water stress. Since ^{32}P was readily incorporated into leaves of moisture-stressed plants, they reasoned that the failure of RNA accumulation was caused by increased breakdown that superceded synthesis under conditions of water stress.

Kessler (1959) reported that water stress is inversely related to nucleic acid level in sunflower seedlings and that drought tolerance in a number of plants is increased by the application of adenine to germinating seeds, young seedlings, and very young leaves. In a more detailed report Kessler (1961) states that the net rate of protein formation from amino acids decreases regularly with developing water stress. With tomato plants he found that RNA as a percentage of dry matter decreased to about one third its highest value as stress mounted for 20 days. Adenine treatment prevented this loss. Conversely, RNase activity about doubled during the same period in control plants but was unchanged in adenine-treated plants. Since ^{14}C-uracil incorporation into RNA was not affected by water stress, Kessler assumed that the change in RNA resulted from hydrolysis. When ^{14}C-thymine incorporation into DNA was similarly tested, less incorporation occurred, suggesting in this case that DNA synthesis and, hence, cell division may be impaired by water stress.

The rise in RNase activity reported above coupled with the failure of water stress to affect uracil incorporation suggest that the lowering of RNA during drought results from hydrolysis. Apparently, drought causes a freeing of RNase from a protoplasmatic complex, and Kessler found by fractionation studies that this enzyme is associated with the proteinaceous fraction of the ribonucleoprotein, possibly nucleoprotein particles that are surrounded by and bonded to membranes of lipid nature. In addition to adenine, Kessler reported that zinc, caffein, uracil, xanthine, and uridine triphosphate are effective in counteracting the effects of drought.

Kessler and Frank-Tishel (1962), working with olive and ligustrum, found that water stress induced synthesis of DNA in mature leaves, and this is paralleled by the accumulation of RNA. The new DNA differs from the original in that the ratio of quanidine plus cytosine (G + C) to adenine plus uridine (A + U) increases from 1.07 to 1.15 to 1.25 to 1.38 as the percentage of water loss went from 0 to 5 to 11 to 39 % in olive. In ligustrum, a drought-sensitive plant, the ratio did not change. If RNA in the olive leaf is double-stranded, then dehydration-induced synthesis of (G + C)-rich RNA may be involved, because the G + C pairs are more H-bonded, denser, and have higher thermal stability. The writers postulate that drought resistance may be related to a higher G + C content in the RNA or a heightened capacity to synthesize (G + C)-rich RNA under the stimulus of stress.

In continued studies Kessler et al. (1964) report a causal relationship between proteins and their ability to bind water. Straight relationships exist between the degree of hydrolysis and the rate and level of water loss. Ribosomal proteins and RNA are more stable than related nuclear and soluble cytoplasmic constituents. Osmotic pressures induce conformational changes in proteins, as evidenced by spectral and chromatographic properties. These changes paralleled changes in enzymatic activities. Osmotic pressures developed by KCl and sucrose stimulate formation of polyA from ADP. They give rise per se to typical changes in base composition of copolyribonucleotides synthesized from equimolar concentrations of ADP, GDP, CDP, and UDP. Increasing concentrations of hydrophilic and H-bonding compounds in salinized cells may alter water structures and affect integrity of the water–protein system essential to maintainance of secondary and tertiary protein structures. Kessler et al. suggest that changes in polyribonucleotide composition may arise mainly from activity of DNA-dependent RNA polymerase. Environmental stress then, through the effects of osmotic media on the activity of polynucleotide phosphorylase, may be coded into an altered composition of polynucleotides, leading to adaptive reactions by serving as primers for RNA-dependant polymerase. The possible role of water structure in this mechanism is stressed.

Shah and Loomis (1965) subjected sugar beet plants to wilting cycles and determined changes in ribonucleic acids and proteins resulting from water stress. Both stress and aging reduced adenylic and increased uridylic nucleotides. During moisture stress, soluble RNA increased and the amount of RNA in ribosomal and other cell fractions decreased. RNA and protein were affected before wilting was visibly evident. In the opinion of Shah and Loomis, changes in growth caused by water stress may be related to changes in RNA and protein metabolism.

Vaadia *et al.* (1961) report both lower and higher sugar contents in water-stressed sugar beets, increased nitrogen in stressed wheat, increased nicotine in tobacco, and increased rubber content in guayule. Kozlowski (1964) reports increased P, K, and Mg with increasing levels of moisture in corn; N and P in tomato were reduced during wilting; N, Mg, and Mn were highest in water-stressed apricot-tree leaves, while K, P, and Zn were more plentiful in trees having readily available water. (The reader is referred to these reviews for details.)

Chen *et al.* (1964), working with 6-month-old seedlings of rough lemon and sweet lime, found the total nitrogen level of stems and leaves to increase with increasing water deficit, whereas that in the roots decreased. The decrease of total nitrogen in roots did not balance the 40% higher total nitrogen increase of aerial parts; this was attributed to continued nitrogen absorption from the soil coupled with translocation of that lost from the roots. The changes mentioned were greatest in rough lemon, which has the lower drought resistance of the two species. The protein level increased at the beginning of dehydration, decreased in the median range, and then increased slightly after inception of permanent wilting.

No common pattern of changes in amino acid content was found during the entire dehydration process; the soluble nitrogen fraction went through a peak at about the time the cultures ran out of available water. Protein level and water content of the tissue on a dry-weight basis were positively correlated; protein level and the rate of water loss showed an inverse correlation. The greater drought tolerance of sweet lime was apparently related to some internal factor and not to stomatal closure.

In experiments on seedlings of Dixie 18 double-cross hybrid corn, West (1962) found that water stress, developed by the addition of mannitol to the culture media, reduced fresh and dry weight, protein and nucleotides; ribonucleic acid accumulated in the water-stressed plants. Chromatography indicated that reduced growth was a result of a shift in ATP production to guanosine and uridine triphosphates and formation of an RNA of altered nucleotide composition; this RNA contained a higher ratio of guanosine and uridine monophosphates to cytodine and adenosine monophosphates than obtained in the controls.

The effects of water stress on amino acid and protein metabolism in two strains of Bermuda grass were studied by Barnett and Naylor (1966). Water stress is known to induce changes in free amino acid levels, particularly a great increase in free proline and in amides; proteolysis caused by water stress releases amino acids that upon deamination yield free ammonia that is synthesized into amides (Barnett and Naylor, 1966). Using $^{14}CO_2$, Barnett and Naylor ran labeling experiments; they found that amino acids were continually synthesized during drying treatment but protein synthesis was inhibited and protein level was decreased. Water stress induced a 10–100-fold accumulation of free proline and a 2–6-fold accumulation of free asparagine in Bermuda grass shoots, responses that are characteristic of water-stressed plants. Valine levels increased; alanine and glutamic acid levels decreased. Free proline turns over more slowly than other free amino acids during water stress; this proline is synthesized from glutamic acid and accumulated. Barnett and Naylor suggest that it may function as a storage compound during water stress.

Itai and Vaadia (1965) found that water-stressed sunflower root systems produced less kininlike activity than normal roots. Itai and Vaadia postulate that kinins may serve as chemical messengers to control shoot growth under varying soil moisture conditions. Kinetin and other kinins are known to sustain RNA and protein metabolism and to retard aging in leaves. In tobacco leaf discs Itai and Vaadia (1967) found water stress to reduce the capacity to incorporate L-leucine ^{14}C into protein; reduction was about 50% and could not be explained in terms of reduced uptake or isotopic dilution. Incorporation decreased progressively with leaf age in both control and stressed discs; at all ages tested, incorporation in stressed discs was lower than in controls; 72 hours were required after stress removal for the incorporation capacity to return to normal. Kinetin pretreatment partially restored incorporation capacity of stressed discs. Itai and Vaadia suggest that stressed plants may have a lower endogenous level of cytokinins than normal ones and that a normal supply of root cytokinins is important to shoot metabolism. Cytokinins have been found in xylem exudate (Kende, 1965).

Todd and Yoo (1964), studying the effects of detachment and desiccation on wheat leaves, found that both these treatments decreased saccharase activity rapidly—most rapidly in turgid leaves. Phosphatase activity decreased in both treatments—most rapidly in the dried leaves. Peptidase activity decreased slightly with drying; peroxidase activity increased in detached turgid leaves and decreased in dried leaves. The particulate enzyme dehydrogenase decreased slightly in turgid leaves but showed a slight rise followed by a sharp decline when leaves had lost 60% of their water. Cell viability decreased with increasing desiccation time; protein content decreased rapidly in both treatments: their content was 50% when leaves had lost 87% of their original water. In studies on the fate of protoplasmic constituents in wheat plants subject to drought, Todd and Basler (1965) found the nucleic acid content of leaf supernatant fraction to fall drastically with increasing water stress. The mitochondrial fraction of the leaves also declined, but less drastically; the chloroplast fraction declined even less. The Hill reaction activity of chloroplasts was unchanged by drought in old leaves; in young ones it increased in slightly wilted plants. Breakdown was much less marked in crown tissues than in leaves, possibly because of transfer from leaves. Since breakdown was found to be similar for intact or detached leaves, it seems that breakdown in crowns is low. Todd and Basler concluded from these studies that drought injury and death involve the breakdown of "synthetic machinery" rather than coagulation of the protoplasm. Tissues that resist drought probably have less hydrolytic enzyme activity on proteins and nucleic acids and possibly other structural materials. That desiccation is not always deleterious is evidenced by Alvim (1960), who found that moisture stress breaks dormancy in the flower buds of coffee and thus leads to fruiting.

Gaff and Levitt (1965) report a progressive decrease in the SH content of the soluble protein fraction and nonprotein fraction of cabbage leaves as they become water deficient, until the soluble protein had decreased by about half at the death point. Structural protein extracts experienced no change in SH until the number of water molecules per amino acid residue averaged only 2.5, where an appreciable conversion of SH to SS occurred. This was followed by death of the tissue.

Bozhenko (1965) reports work on the role of microelements in resistance to drought and high temperatures. Microelements influence the synthesis and transport of carbohydrates; they also increase viscosity and decrease permeability of the plasma. They increase the content and hydration of hydrophilic colloids (proteins and nucleoproteins), and they increase the content of bound (structured) water and thus bring about a decrease in transpiration during the hottest hours of the day. Microelements also increase ascorbic acid, and they influence metabolism of organic acids through decrease in ammonium content. Bozhenko used aluminum nitrate, cobalt nitrate, zinc sulfate, copper sulfate, and boric acid at 0.2 gm per liter to treat seeds of sunflower, soaking the seeds for 20 hours, drying, and planting. Drought conditions were created by withholding irrigation. Exposure to heat consisted of placing cut growing points in a wet chamber for 40 minutes at 50°C; ATP was determined by the method of Severin and Meshkova (1950). Heat treatment of root growing points proved that all microelements used increased the ATP content under temperature stress; combined temperature and moisture stress produced increased ATP in the case of aluminum, boron, and cobalt. The latter was particularly effective, leading to an ATP increase of 93% in growing points and 98% in whole root systems.

Henckel (1964) emphasizes the physical and chemical characteristics of drought resistance. These include high protoplasmic viscosity and elasticity coupled with low respiratory rates. He explains that heat and drought resistance often go together and that heat-resistant organisms may have high metabolic intensity providing they are rich in nucleoproteins. High DNA and

RNA stimulate protein synthesis and inhibit decomposition; where synthetic processes dominate hydrolytic ones, injury is retarded. Adenine may increase the level of nucleic acids, and seeds treated with this compound may have increased resistance to dehydration and high temperatures. Presowing hardening, a treatment involving soaking seeds for 2 days followed by air drying, stimulates resistance, enabling plants to survive heat and drought. Such plants carry on intensive protein synthesis, their protoplasm is high in viscosity and elasticity, and they carry high levels of bound water; salt and water extracts of their embryos and leaves have higher temperature coagulation thresholds than those of unhardened plants.

DISCUSSION

Mention has been made of the use of a liverwort by Slavik (1965) to study the relation between photosynthesis and water stress in a situation uncomplicated by stomatal activity. Mahmoud (1965) used mosses to study the ability of protoplasm to survive varying degrees of dehydration. Two species, *Antitrichia californica*, an epiphytic, drought resistant plant, and *Mnium hymenophylloides,* a terrestrial drought-sensitive one, were given cell vitality tests before and after 48-hour desiccation tests. Plasmolysis–deplasmolysis proved to be the most reliable method for determining vitality.

In the tests, *A. californica* survived a relative humidity approaching zero (96 vol % H_2SO_4), whereas *M. hymenophylloides* died when the relative humidity fell below 96%. Light microscopic observation proved that contrasting cellular characteristics are not the key to understanding the differences found. Under the electron microscope, the lamellar system of chloroplasts of the sensitive moss proved to be disorganized with many vacuoles whereas that of the resistant species was regularly arranged with no vacuoles. On rehydration the chloroplasts of the resistant desiccated moss regained their normal structure.

From these studies employing ultrasound, vital dyes, reduction of tetrazolium, fluorescent dyes, and autofluorescense and studies on plasmodesmata and callose as well as plasmolysis–deplasmolysis and the electron microscope, Mahmoud concluded that desiccation resistance is not a matter of mechanical–structural properties of plants but rather is a property of the protoplasm. A cell is resistant to drying when its protoplasm is readily stimulated by the loss of water to rearrange its structure into a stable, resistant molecular pattern—one from which a revival to normal growth is possible. A susceptible cell under like conditions breaks down and dies. While these two mosses represent extremes of drought resistance and susceptibility, the ability of higher plants to become adapted to limited water supply probably involves similar changes in the protoplasm.

There is probably no better way of drawing this review to a close than by calling attention to the excellent paper on the effect of water stress on plant growth by Gates (1964), which surveys current research on plant water relations. Gates contrasts the views of many workers in this country that plant growth is an expression of cell enlargement brought about by the uptake of water with that of many Russian plant scientists who emphasize the embryonic stage as the chief growth manifestation of an organism and that elongation resulting from water uptake is of secondary importance. Gates regards both aspects as important and regrets that our thinking so often has been influenced by the first of these views with the result that frequently water has been assigned merely a passive role in growth processes. He does not agree with Wood (1939), who proposes that " water does not exert an effect as such, that is, as a molecular species ... decrease in water content will only produce effects on metabolism by increasing the concentration of other substances."

Gates points out the relations of water to protoplasmic structure and cites work by a number of investigators to the effect that water stress brings about changes in structure that serve to modify enzymatic activity in an adaptive fashion. Thus, if protoplasmic stability depends not only on the degree of hydration but on the maintenance of specific centers of water binding at specified points on the macromolecular structure, then water may no longer be considered as playing merely a passive role; it must be looked upon as entering intimately into the functions as well as the structure of the living protoplasm. With this in mind Gates considers the ability of young embryonic tissues to resist water stress with little measurable injury. While such tissues are apparently highly resistant to water loss, there is evidence that their synthetic capacity may be affected; their continued development seems to require high levels of hydration.

Gates is tempted to consider the possibility of a complementary function relation between structure and behavior of a protein and its water envelope that might permit the ready suspension and resumption of growth processes with a minimum of damage, particularly by the adapted protoplast or the embryonic drought-tolerant form. He visualizes a possible role of water in the protoplast as that of linking macromolecules into a coordinated living whole, possibly by hydrogen bonding that may readily undergo changes of state so that bonding positions are partially satisfied within the macromolecules themselves, resulting in suspension of metabolic activity.

Attention has already been called to the possible fit of the DNA helix into the water lattice (Jacobsen, 1953). Thus, water structure might provide for the function of separation of the two polynucleotide chains in the reproductive process because of the ease with which hydrogen bonding positions in the lattice could accept the hydrogen bonds of the basic groups in the helical chains. It seems obvious from Chapter 2, Volume I that hydrophobic bonds might well be involved in these structure–function relations of drought-resistant protoplasm. It is these bonds that lend stability to the α-helix form of proteins. These bonds bring about a squeezing out of water from between aliphatic side chains (removal of partial clusters), thus adapting the protein structure to water stress; hydrophobic bonds induce curving and spiraling of otherwise straight backbone structures as water is extracted from the fully hydrated form; such bond forms are reversible and, hence, compatible with the shifts of water potential through repeated drying cycles. Petrie and Wood (1938a,b) noted a decrease in net protein formation from amino acids in water stressed plants; they found that water content affects protein synthesis and hydrolysis and, in turn, that protein affects water content.

Because the reversible hydration and dehydration of proteins alters their architecture, it must affect the specific function of enzymes; these, in turn, determine the rates of most metabolic reactions. Hence, it seems that water must have an effect *per se* upon the functions of biological molecules and their component systems. Water structure, then, may be the key to many vital processes of the living cell.

REFERENCES

Aldrich, W. W., Lewis, M. R., Work, R. A., Ryall, A. L., and Reimer, F. C., (1940). Anjou pear responses to irrigation in a clay adobe soil. *Oregon State Coll. Agr. Expt. Sta. Bull.* **374.**

Alvim, P. T. (1960). Moisture stress as a requirement for flowering of coffee. *Science* **132,** 354.

Ashton, F. (1956). Effects of a series of cycles of alternating low and high soil water contents on the rate of apparent photosynthesis in sugar cane. *Plant Physiol.* **31,** 266.

Badiei, A. A., Basler, E., and Santelmann, P. W. (1967). Aspects of movement of 2, 4, 5-T in blackjack oak. *Weeds* **14,** 302.

Barnett, N. M., and Naylor, A. W. (1966). Amino acid and protein metabolism in Bermuda grass during water stress. *Plant Physiol.* **41,** 1222.

Bernstein, L., and Gardner, W. R. (1961). Perspective on function of free space in ion uptake by roots. *Science* **132,** 1482.

Bernstein, L., Gardner, W. R., and Richards, L. A. (1959). Is there a vapor gap around plant roots? *Science* **129,** 1750.

Biddulph, O., and Cory, R. (1957). An analysis of translocation in the phloem of the bean plant using THO, P^{32} and $C^{14}O_2$. *Plant Physiol.* **32,** 608.

Biddulph, O., and Cory, R. (1960). Demonstration of 2 translocation mechanisms in studies of bidirectional movement. *Plant Physiol.* **35,** 689.

Biddulph, O., and Cory, R. (1965). Translocation of C^{14} metabolites in the phloem of the bean plant. *Plant Physiol.* **40,** 119.

Biddulph, O., Nakayama, F. S., and Cory, R. (1961). Transpiration stream and ascension of calcium. *Plant Phsysiol.* **36**, 429.

Bierhuizen, J. F., and Slatyer, R. O. (1965). Effect of atmospheric concentration of water vapor and CO_2 in determining transpiration-photosynthesis relationships of cotton leaves. *Agr. Meteorol.* **2**, 259.

Bonner, J. (1959). Water transport. *Science* **129**, 447.

Boyer, J. S. (1965). Effects of osmotic water stress on metabolic rates of cotton plants with open stomata. *Plant Physiol.* **40**, 229.

Bozhenko. V. P. (1965). The influence of microelements on ATP content in plants in the presence of water deficit and under the influence of high temperatures. *In* "Water Stress in Plants" (B. Slavik, ed.), p. 238. Proc. Symp. Prague, 1963. Czech. Acad. Sci., Prague.

Briggs, G. E., and Robertson, R. N. (1957). Apparent free space. *Ann. Rev. Plant Physiol.* **8**, 11.

Briggs, L. J., and Shantz, H. L. (1913). Relative water requirement of plants. *U.S. Dept. Agr. Bur. Plant Ind. Bull.* **284–285**.

Brix, H. (1962). The effect of water stress on the rates of photosynthesis and respiration in tomato plants and loblolly pine seedlings. *Physiol. Plantarum* **15**, 10.

Broyer, T. C. (1956). Current views on solute movement into plant roots. *Proc. Am. Soc. Hort. Sci.* **67**, 570.

Bukovac, M. J., and Norris, R. F. (1967). Foliar penetration of plant growth substances with special reference to binding by cuticular surfaces of pear leaves. *Agrochimica* **12**, 296.

Carolus, R. L., Erickson, A. E., Kidder, E. H., and Wheaton, R. Z. (1965). The interaction of climate and soil moisture on water use, growth, and development of tomatoes. *Mich. State Univ. Agr. Expt. Sta. Quart. Bull.* **47** (4), 542.

Čatsky, J. (1965). Water saturation and photosynthetic rate as related to leaf age in the wilting plant. *In* "Water Stress in Plants" (B. Slavik, ed.), p. 203. Proc. Symp. Prague, 1963. Czech. Acad. Sci., Prague.

Chen, D., Kessler, B., and Monselise, S. P. (1964). Studies on water regime and nitrogen metabolism of citrus seedlings grown under water stress. *Plant Physiol.* **39**, 379.

Collis-George, N., and Sands, J. E. (1962). Comparison of the effects of the physical and chemical components of soil water energy on seed germination. *Australian J. Agr. Res.* **13**, 575.

Crafts, A. S. (1939). Solute transport in plants, *Science* **90**, 337.

Crafts, A. S. (1961a). Absorption and migration of synthetic auxins and homologous compounds. *In* "Handbuch der Pflanzen physiologie" (W. Ruhland ed.), Vol. 14, p. 1044. Springer, Berlin.

Crafts, A. S. (1961b). "Translocation in Plants." Holt, New York.

Crafts, A. S. (1966). Bidirectional movement of labeled tracers in soybean seedlings. *Hilgardia* **37**, 625.

Crafts, A. S., and Broyer, T. C. (1938). The migration of solutes and water into the xylem of the roots of higher plants. *Am. J. Botany* **25**, 529.

Crafts, A. S., and Foy, C. L. (1962). The chemical and physical nature of plant surfaces in relation to the use of pesticides and their residues. *Residue Rev.* **1**.

Crafts, A. S., and Yamaguchi, S. (1964). The autoradiography of plant materials. *Calif. Univ. Agr. Expt. Sta. Ext. Serv. Manual.* **35**.

Crafts, A. S., Currier, H. B., and Stocking, C. R. (1949). "Water in the Physiology of Plants" Ronald Press, New York.

Curtis, O. F. (1935). "The Translocation of Solutes in Plants." McGraw-Hill, New York.

Darlington, W. A., and Cirulis, N. (1963). Permeability of apricot leaf cuticle. *Plant Physiol.* **38**, 462.

Davis, F. S., Merkle, M. G., and Bovey, R. W. (1967). Effect of moisture stress on the absorption and transport of herbicides in woody plants. *Weed Soc. Am. Meeting, 1967, Abst.* Washington, D.C., p. 44.

De Stigter, H. C. M. (1961). Translocation of C^{14} photosynthates in the graft muskmelon *Cucurbita ficifolia. Acta Botan. Neerl.* **10**, 466.

de Wit, C. T. (1959). Potential photosynthesis of crop surfaces. *Neth. J. Agr. Sci.* **7**, 141.

Dixon, H. H. (1914). "Transpiration and the Ascent of Sap on Plants." Macmillan, New York.

Doss, B. P., Bennett, O. L., Ashley, D. A., and Weaver, H. A. (1962). Soil moisture regime effect on yield and evapotranspiration from warm season perennial forage crops. *Agron. J.* **54**, 239.

Duloy, M., Mercer, F. V., and Rathgaber, N. (1961). Studies in translocation. II. Submicroscopic anatomy of the phloem. *Australian J. Biol. Sci.* **14**, 506.

Ehrler, W. L., van Bavel, C. H. M., and Nakayama, F. S. (1966). Transpiration, water absorption, and internal water balance of cotton plants as affected by light and changes in saturation deficit. *Plant Physiol.* **41**, 71.

Eidmann, F. (1962). Der Wasserverbrauch der Holzarten im Durchschnitt der Vegetationsperiode. *13th Congr. of the Intern. Union of Forest Res. Organ., Vienna, 1961.*

El-Sharkawy, M. S., and Hesketh, J. D. (1964). Effects of temperature and water deficit on leaf photosynthetic rates of different species. *Crop Sci.* **4**, 514.

Engleman, E. M. (1965). Sieve elements of *Impatiens sultanii.* 2. Developmental aspects. *Ann. Botany (London)* [N.S.] **29**, 103.

Esau, K. (1953). "Plant Anatomy." Wiley, New York.

Esau, K. (1965). Explorations of the food conducting system in plants. *Am. Sci.* **54**, 141.

Esau, K., and Cheadle, V. I. (1959). Size of pores and their contents in sieve elements of dicotyledons. *Proc. Natl. Acad. Sci. U.S.* **45**, 156.

Esau, K., and Cheadle, V. I. (1961). An evaluation of studies on ultrastructure of sieve plates. *Proc. Natl. Acad. Sci. U.S.* **47**, 1716.

Evert, R. F., and Murmanis, L., (1965). Ultrastructure of the secondary phloem of *Tilia americana. Am. J. Botany* **52**, 95.

Ferri, M. G., and Lex, A. (1948). Stomatal behavior as influenced by treatment with β-naphthoxyacetic acid. *Contrib. Boyce Thompson Inst.* **15**, 283.

Forde, B. J. (1963), Some aspects of phloem translocation in three grass species. Ph.D. Thesis, Univ. of Calif., Davis, California.

Fraser, D. A. (1962). Tree growth in relation to soil moisture. *In* "Tree Growth" (T. T. Kozlowski, ed.), Ronald Press, New York.

Freney, J. R. (1965). Increased growth and uptake of nutrients by corn plants treated with low levels of simazine. *Australian J. Agr. Res.* **16**, 257.

Frey-Wyssling, A. (1959). "Die pflanzliche Zellwand." Springer, Berlin.

Fry, K. E., and Walker, R. B. (1964). Relation of needle water stress and relative stomatal aperture to transpiration and net photosynthesis in Douglas fir (*Pseudotsuga menziesii*). *Plant Physiol.* **39**, (Suppl) 39, xlii.

Gaastra, P. (1959). Photosynthesis of crop plants as influenced by light, carbon dioxide, temperature and stomatal diffusion resistance. *Mededel. Landbouwhogeschool Wageningen* **59**, (13), 1.

Gaff, D. F., and Levitt, J. (1965). The sulfhydryl hypothesis in relation to drought tolerance. *Plant Physiol.* **40**, (Suppl.) xxxv.

Gale, J., and Hagan, R. M. (1966). Plant antitranspirants, *Ann. Rev. Plant Physiol.* **17**, 269.

Gardner, W. R. (1965) Dynamic aspects of soil-water availability to plants. *Ann. Rev. Plant Physiol.* **16**, 323.

Gardner, W. R., and Ehlig, C. F. (1962). Impedance to water movement in soil and plant. *Science* **138**, 522.

Gardner, W. R., and Ehlig, C. F. (1965). Physical aspects of the internal water relations of plant leaves. *Plant Physiol.* **40**, 705.

Gates, C. T. (1964). The effect of water stress on plant growth. *J. Australian Inst. Agr. Sci.* **30**, 3.

Gates, C. T., and Bonner, J. (1959). The response of the young tomato plant to a brief period of water shortage. IV. Effects of water stress on the ribonucleic acid metabolism of tomato leaves. *Plant Physiol.* **34**, 49.

Gingrich, J. R., and Russell, M. B. (1956). Effect of soil moisture and oxygen concentration on the growth of corn roots. *Agron. J.* **48**, 517.

Glinka, Z., and Reinhold, L. (1962). Rapid changes in permeability of cell membranes to water brought about by carbon dioxide and oxygen. *Plant Physiol.* **37**, 481.

Greenidge, K. N. H. (1954). Studies in the physiology of forest trees. I. Physical factors affecting the movement of moisture. *Am. J. Botany.* **41**, 807.

Greenidge, K. N. H. (1955a). Studies in the physiology of forest trees. II. Experimental studies of fracture of stretched water columns in transpiring trees. *Am. J. Botany* **42**, 28.

Greenidge, K. N. H. (1955b). Studies in the physiology of forest trees. III. The effect of drastic interruption of conducting tissues on moisture movement. *Am. J. Botany* **42**, 582.

Greenidge, K. N. H. (1957). Ascent of sap. *Ann Rev. Plant. Physiol.* **8**, 237

Hartt, C. E. (1965a). The effect of temperature upon translocation of ^{14}C in sugarcane. *Plant Physiol.* **40**, 74.

Hartt, C. E. (1965b). Light and translocation of ^{14}C in detached blades of sugarcane. *Plant Physiol.* **40**, 718.

Hartt, C. E. (1967). Effect of moisture supply upon translocation and storage of ^{14}C in sugarcane. *Plant Physiol.* **42**, 338.

Hauser, E. W. (1955). Absorption of 2,4-dichlorophenoxyacetic acid by soybean and corn plants. *Agron. J.* **47**, 42.

Hauser, E. W., and Young, D. W. (1952). Penetration and translocation of 2,4-D compounds *Proc. N. Central Weed Control Conf.* **9**, 27.

Heinicke, A. J., and Childers, N. F. (1935). The influence of water deficiency on photosynthesis and transpiration of apple leaves. *Proc. Am. Soc. Hort. Sci.* **33**, 155.

Heinicke, A. J., and Hoffman, M. B. (1933). The rate of photosynthesis of apple leaves under natural conditions. I. *Cornell Univ. Agr. Expt. Sta. Bull.* **577**.

Henckel. P. A. (1964). Physiology of plants under drought. *Ann. Rev. Plant. Physiol.* **15**, 363.

Hoagland, D. R. (1944). "Lectures on the Inorganic Nutrition of Plants." Chronica Botanica, Waltham, Massachusetts.

Hygen, G. (1951). Studies in plant transpiration. I. *Physiol. Plantarum* **4**, 57.

Hygen, G. (1953). On the transpiration decline in excised plant sample. *Skrifter Norske Videnskaps-Akad. Oslo, I: Mat.-Naturv. Kl.* **1**, 1.

Hygen, G. (1965). Water stress in conifers during winter. *In* "Water Stress in Plants" (B. Slavik, ed.), Proc. Symp. Prague, 1963. p. 89. Czech. Acad. Sci., Prague.

Hylmo, B. (1953). Transpiration and ion absorption. *Physiol. Plantarum* **6**, 333.

Iljin, W. S. (1957). Drought resistance in plants and physiological processes. *Ann Rev. Plant Physiol.* **8**, 257.

Itai, C., and Vaadia, Y. (1965). Kinetin-like activity in root exudate of water-stressed sunflower plants. *Physiol. Plantarum* **18**, 941.

Itai, C., and Vaadia, Y. (1967). Water and salt stresses, kinetin and protein synthesis in tobacco leaves. *Plant Physiol.* **42**, 361.

Ivanov, V. P. (1959). Effect of foliar nutrition and soil moisture on the growth and development of maize. *Fiziol. Rast.* **6**, 368.

Jacobsen, B. (1953). Hydration structure of deoxyribonulceic acid and its physicochemical properties. *Nature* **172**, 666.

Jarvis, P. G., and Jarvis, M. S. (1963). The water relations of tree seedlings. III. Transpiration in relation to osmotic potential of the root medium. *Physiol. Plantarum* **16**, 269.

Jarvis, P. G., and Jarvis, M. S. (1965). The water relations of tree seedlings. V. Growth and root respiration in relation to the osmotic potential of the root medium. *In* " Water Stress in Plants " (B. Slavik, ed.), Proc. Symp. Prague, 1963. p. 167. Czech. Acad. Sci., Prague.

Jenny, H., and Grossenbacher, K. (1963). Root-soil boundary zones as seen in the electron microscope. *Proc. Soil. Sci. Soc. Am.* **27**, 273.

Kende, H. (1965). Kinetin-like factors in the root exudate of sunflowers. *Proc. Natl. Acad. Sci. U.S.* **53**, 1302.

Kessler, B. (1959). Nucleic acids as factors in drought resistance of plants. *Proc. 9th Intern. Botan. Congr., Montreal,* **1959 2**, 190.

Kessler, B. (1961). Nucleic acids as factors in drought resistance in higher plants. *Recent Advan. Botany* **2**, 1153.

Kessler, B., and Frank-Tishel, J. (1962). Dehydration-induced synthesis of nucleic acids and changing of composition of RNA: A possible protective reaction in drought-resistant plants. *Nature* **196**, 542.

Kessler, B., Engelberg, N., Chen, D., and Greenspan, H. (1964). Studies on physiological and biochemical problems of stress in higher plants. *Volcani Inst. Agr. Res. Unit Plant Physiol. Biochem. Rehovot, Israel. Special Bull.* **64** *Project ALO-CR-7.*

Koch, W. (1957). Der Tagesgang der " Productivitat der Transpiration." *Planta* **48**, 418.

Kozlowski, T. T. (1958). Water relations and growth of trees. *J. Forestry* **56**, 498.

Kozlowski, T. T. (ed.) (1962). Photosynthesis, climate and tree growth. *In* " Tree Growth." p. 149. Ronald Press, New York.

Kozlowski, T. T. (1964). " Water Metabolism in Plants." Harper, New York.

Kozlowski, T. T., and Gentile, A. C. (1958). Respiration of white pine buds in relation to oxygen availability and moisture content. *Forest Sci.* **4**, 147.

Kozlowski, T. T., and Keller, T. (1966). Food relations of woody plants. *Botan. Rev.* **32**, 293.

Kramer, P. J. (1933). The intake of water through dead root systems and its relation to the problem of absorption by transpiring plants. *Am. J. Botany* **20**, 491.

Kramer, P. J. (1937). The relation between rate of transpiration and rate of absorption of water in plants. *Am. J. Botany* **24**, 10.

Kramer, P. J. (1938). Root resistance as a cause of the absorption lag. *Am. J. Botany* **25**, 110.

Kramer, P. J. (1949). " Plant and Soil Water Relations." McGraw-Hill, New York.

Kramer, P. J. (1957). Outerspace in plants. *Science* **125**, 633.

Kramer, P. J. (1962). The role of water in tree growth. *In* " Tree Growth " (T. T. Kozlowski, ed.), pp. 171–182. Ronald Press, New York.

Kramer, P. J., and Kozlowski, T. T. (1960). " Physiology of Trees." McGraw-Hill, New York.

Kursanov, A. L. (1963). Metabolism and the transport of organic substances in the phloem. *Advan. Botan. Res.* **1**, 209.

Larcher, W. (1965). The influence of water stress on the relationship between CO_2 uptake and transpiration. *In* "Water Stress in Plants" (B. Slavik, ed.), Proc. Symp. Prague, 1963. p. 184. Czech. Acad. Sci., Prague.

Laties, G. G., and Budd, K. (1964). The development of differential permeability in isolated steles of corn roots. *Proc. Natl. Acad. Sci. U.S.* **52**, 462.

Lebedev, G. V., and Askochenskaya, N. A. (1965). State of water in plant cell, mobility of water and factors which determine it. *In* "Water Stress in Plants" (B. Slavik, ed.), Proc. Symp. Prague, 1963. p. 81. Czech. Acad. Sci., Prague.

Leonard, O. A. (1958). Studies on the absorption and translocation of 2,4-D in bean plants *Hilgardia* **28**, 115.

Leonard, O. A. (1967). Private communication.

Leonard, O. A., and Crafts, A. S. (1956). Translocation of herbicides. III. Uptake and distribution of radioactive 2,4-D by brush species. *Hilgardia* **26**, 366.

Leonard, O. A., and Hull, R. J. (1965). Translocation of ^{14}C-labelled substances and $^{32}PO_4$ in mistletoe-infected and uninfected conifers and dicotyledonous trees. *In* "Isotopes in Weed Research," IAEA, FAO Symp. Proc., Vienna, 1966, p. 31.

Levitt, J. (1957). The significance of apparent free space (AFS) in ion absorption. *Physiol Plantarum* **10**, 882.

Loomis, W. E., Santa Maria, R., and Gage, R. S. (1960). Cohesion of water in plants. *Plant Physiol.* **35**, 300.

Loustalot, A. J. (1945). Influence of soil moisture conditions on apparent photosynthesis and transpiration of pecan leaves. *J. Agr. Res.* **71**, 519.

Magalhaes, A. C., and Foy, C. L. (1967). Aspects of the physiology of herbicidal action of dicamba in purple nutsedge (*Cyperus rotundus* L.). *Weed Soc. Am. Meeting, 1967, Abstr.* p.44

Mahmoud, M. I. (1965). Protoplasmatics and drought resistance in mosses. Ph.D. Thesis, Univ. of Calif. Davis, California.

Mason, G. (1960). The absorption, translocation and metabolism of 2,3,6-trichlorobenzoic acid in plants. Ph.D. Thesis, Univ. of Calif., Davis, California.

Mason, T. G., and Maskell, E. J. (1928). Studies on the transport of carbohydrates in the cotton plant. I. A study of diurnal variation in the carbohydrates of leaf, bark, and wood and of the effects of ringing. *Ann. Botany.* **42**, 1.

Maximov, N. A. (1923). Physiologisch-ökologische Untersuchungen über die Dürreresistenz der Xerophyten. *J. Wiss. Botany.* **62**, 128.

Merkle, M. G., and Davis, F. S. (1967). Effect of moisture stress on absorption and movement of picloram and 2,4,5-T in beans. *Weeds* **15**, 10.

Milburn, J. A., and Johnson, R. P. C. (1966). The conduction of sap. II. Detection of vibrations produced by sap cavitation in *Ricinus* xylem. *Planta* **69**, 43.

Moss, D., Musgrave, R. B., and Lemon, E. R. (1961). Photosynthesis under field conditions. III. Some effects of light, carbon dioxide, temperature, and soil moisture on photosynthesis, respiration, and transpiration of corn. *Crop Sci.* **1**, 83.

Mothes, K. (1956). Der Einfluss des Wasserzustandes auf Fermentprozesse und Stoffumsatz. *In* "Handbuch der Pflanzenphysiologie" (W. Ruhland, ed.), Vol. III, p. 656. Springer, Berlin.

Münch, E. (1930). "Die Stoffbewegungen in der Pflanze." Fischer, Jena.

Neuwirth, G., and Polster, H. (1960). Wasserverbrauch und Stoffproduktion der Schwarzpappel und Aspe unter Dürrebelastung. *Arch. Forstwesen* **9**, 789.

Ordin, L. (1960). Effect of water stress on cell wall metabolism of *Avena* coleoptile tissue. *Plant Physiol.* **35**, 443.

Ordin, L., and Gairon, S. (1961). Diffusion of tritiated water into roots as influenced by water status of tissue. *Plant Physiol.* **36**, 331.

Ordin, L., and Kramer, P. J. (1956). Permeability of *Vicia faba* root segments to water as measured by diffusion of deuterium hydroxide. *Plant Physiol.* 31, 468.

Ordin, L., Applewhite, T. H., and Bonner, J. (1956). Auxin-induced water uptake by *Avena* coleoptile sections. *Plant Physiol.* 31, 44.

Orgell, W. H. (1955). The isolation of plant cuticle with pectic enzymes. *Plant Physiol.* 30, 78.

Owen, P. C. J. (1952). The relation of germination of wheat to water potential. *J. Exptl. Botany* 3, 188.

Pallas, J. E., Jr. (1960). Effects of temperature and humidity on foliar absorption and translocation of 2,4-dichlorophenoxyacetic acid and benzoic acid. *Plant Physiol.* 35, 575.

Pallas, J. E., Jr., and Bertrand, A. R. (1966). Research in plant transpiration. *U.S. Dept. Agr. ARS Production Res. Rept.* 89.

Pallas, J. E., Jr., and Williams, G. G. (1962). Foliar absorption and translocation of P^{32} and 2,4-dichlorophenoxyacetic acid as affected by soil-moisture tension. *Botan. Gaz.* 123, 175

Petersen, H. I. (1958). Translocation of C^{14} labelled 2,4-dichlorophenoxyacetic acid in barley and oats. *Nature* 182, 1685.

Petrie, A. H. K., and Wood, J. G. (1938a). Studies on the nitrogen metabolism of plants. I. The relation between the content of proteins, amino acids, and water in the leaves. *Ann. Botany (London)* [N.S.] 2, 33.

Petrie, A. H. K., and Wood, J. G. (1938b). Studies on the nitrogen metabolism of plants. III. On the effect of water content on the relationship between proteins and amino acids. *Ann. Botany (London)* [N.S.] 2, 887.

Philip, J. R. (1957). The physical principles or soil moisture movement during the irrigation cycle. *Proc. 3rd Congr. Intern. Comm. Irrig. Drainage* 3, Question 8, 125. San Francisco.

Philip, J. R. (1958). The osmotic cell, solute diffusibility and the plant water economy. *Plant Physiol.* 33, 264.

Plaut, Z., and Ordin, L. (1964). The effect of moisture tension and nitrogen supply on cell wall metabolism of sunflower leaves. *Physiol. Plantarum* 17, 279.

Plaut, Z., and Reinhold, L. (1965). The effect of water stress on ^{14}C sucrose transport in bean plants. *Australian J. Biol. Sci.* 18, 1143.

Polster, H., Weise, G., and Neuwirth, G. (1960). Okologische Untersuchen über den CO_2-Stoffwechsel und Wasserhaushalt einiger Holzarten auf ungarischen Sand- und Alkali-(" Szik ") Boden. *Arch. Forstwesen* 9, 949.

Quinlan, J. D. (1966). The effect of partial defoliation on the pattern of assimilate movement in an apple root stock. *Ann. Rept. East Malling Res. Sta., Kent* 1965.

Quinlan, J. D., and Sagar, G. R. (1962). An autoradiographic study of the movement of C^{14}-labelled assimilates in the developing wheat plant. *Weed Res.* 2, 264.

Ragai, H., and Loomis, W. E. (1954). Respiration of maize grain. *Plant Physiol.* 29, 49.

Roberts, B. R. (1964). Effect of water stress on the translocation of photosynthetically assimilated carbon-14 in yellow poplar. *In* " The Formation of Wood in Forest Trees " (M. H. Zimmerman, ed.), p. 273. Academic Press, New York.

Roelofsen, P. A. (1965). Ultrastructure of the wall in growing cells and its relation to the direction of growth. *Advan. Botan. Res.* 2, 69.

Russell, R. S., and Shorrocks, V. M. (1957). The effect of transpiration on the absorption of inorganic ions by intact plants. *UNESCO Intern. Conf. Radioisotopes Sci. Res.* Vol. IV, p. 286. Pergamon Press, Oxford.

Schneider, G. W., and Childers, N, F. (1941). Influence of soil moisture on photosynthesis, respiration, and transpiration of apple leaves. *Plant Physiol.* 16, 565.

Scholander, P. F., Flagg, W., Walters, V., and Irving, L. (1952). Respiration of some arctic and tropical lichens in relation to temperature. *Am. J. Botany* **39**, 707.

Scholander, P. F., Love, W. E., and Kanwisher, J. K. (1955). The rise of sap in tall grapevines. *Plant Physiol.* **30**, 93.

Scholander, P. F., Ruud, B., and Leivestad, H. (1957). The rise of sap in tropical liana. *Plant Physiol.* **32**, 1.

Scholander, P. F., Hemmingsen, E., and Garey, W. (1961). Cohesive lift of sap in the rattan vine, *Science* **134**, 1835.

Scholander, P. F., Hammel, H. T., Bradstreet, E. D., and Hemmingsen, E. A. (1965). Sap pressure in vascular plants. *Science* **148**, 339.

Scholander, P. F., Bradstreet, E. D., Hammel, H. T., and Hemmingsen, E. A. (1966). Sap concentrations in halophytes and some other plants. *Plant Physiol.* **41**, 529.

Severin, S. E., and Meshkova, N. P. (1950). Effect of carnosine and anserine on carbohydrate-phosphate metabolism of red breast muscle of pigeon. *Dokl. Akad. Nauk. SSSR* **74**, 549.

Shah, C. B., and Loomis, R. S. (1965). Ribonucleic acid and protein metabolism in sugar beet during drought. *Physiol. Plantarum* **18**, 240.

Shimshi, D. (1963a). Effect of chemical closure of stomata on transpiration in varied soil and atmospheric environments. *Plant Physiol.* **38**, 709.

Shimshi, D. (1963b). Effect of soil moisture and phenylmercuricacetate upon stomatal aperture, transpiration, and photosynthesis. *Plant Physiol.* **38**, 713.

Slatyer, R. O. (1957). The significance of the permanent wilting percentage in studies of plant and soil water relations. *Botan. Rev.* **23**, 585.

Slatyer, R. O. (1960). Absorption of water by plants. *Botan. Rev.* **26**, 331.

Slatyer, R. O., and Bierhuizen, J. F. (1964). The influence of several transpiration suppressants on transpiration, photosynthesis, and water-use efficiency of cotton leaves. *Australian J. Biol. Sci.* **17**, 131.

Slavik, B. (1958). The influence of water deficit on transpiration. *Physiol. Plantarum* **11**, 524.

Slavik, B. (1965). The influence of decreasing hydration level on photosynthetic rate in the thalli of the hepatic *Conocephallum conicum. In* "Water Stress in Plants" (B. Slavik, ed.), Proc. Symp. Prague, 1963. p. 195. Czech. Acad. Sci., Prague.

Smith, A. E., Zukel, J. W., Stone, G. M., and Riddell, J. A. (1959). Factors affecting the performance of maleic hydrazide. *Agr. Food Chem.* **7**, 341.

Smith, D., and Buchholtz, K. P. (1962). Transpiration rate reduction in plants with Atrazine. *Science* **136**, 263.

Smith, D., and Buchholtz, K. P. (1964). Modification of plant transpiration rate with chemicals. *Plant Physiol.* **39**, 572.

Stanley, R. G. (1958). Gross respiratory and water uptake patterns in germinating sugar pine seed. *Physiol. Plantarum* **11**, 503.

Steward, F. C. (1964). "Plants at Work." Addison-Wesley, Reading, Massachusetts.

Stocker, O. (1960). Physiological and morphological changes in plants due to water deficiency. *In* "Plant-Water Relations in Arid and Semi-Arid Conditions. Reviews of Research," p. 63. UNESCO, Paris.

Stoddard, E. M., and Miller, P. M. (1962). Chemical control of water loss in growing plants. *Science* **137**, 224.

Stout, P. R., and Hoagland, D. R. (1939). Upward and lateral movement of salt in certain plants as indicated by radioactive isotopes of potassium, sodium and phosphorus absorbed by roots. *Am. J. Botany* **26**, 320.

Strugger, S. (1938). Fluoreszenmikroskopische Untersuchungen über die Speicherung und Wanderung des Fluoreszein Kaliums in pflanzlichen Geweben. *Flora (Jena)* **132**, 253.

Szabo, S. S., and Buchholtz, K. P. (1961). Penetration of living and non-living surfaces by 2,4-D as influenced by ionic additives. *Weeds* **9**, 177.

Takaoki, T. (1957). Relationships between plant hydrature and respiration. II. Respiration in relation to the concentration and the nature of external solutions. *J. Sci. Hiroshima Univ., Ser. B, Div. 2.* **8**, 73.

Ting, I. P., and Loomis, W. E. (1963). Diffusion through stomates. *Am. J. Botany* **50**, 866.

Todd, G. W., and Basler, E. (1965). Fate of various protoplasmic constituents in droughted wheat plants. *Phyton (Buenos Aires)* **22**, 79.

Todd, G. W., and Yoo, B. Y. (1964). Enzymatic changes in detached wheat leaves as affected by water stress. *Phyton (Buenos Aires)* **21**, 61.

Todd, G. W., and Webster, D. L. (1965). Effect of repeated drought periods on photosynthesis and survival of cereal seedlings. *Agron. J.* **37**, 399.

Tranquillini, W. (1963). Die Abhängigkeit der Kohlensäureassimilation junger Lärchen, Fichten und Zirben von der Luft und Bodenfeuchte. *Planta* **60**, 70.

Tumanow, J. J. (1927). Ungenügende Wassersorgung und das Welken der Pflanzen als Mittel zur Erhöhung ihrer Dürreresistenz. *Planta* **3**, 391.

Ulehla, J. (1965). The relationship between the sap exudation rate and the duration of exudation lag. *In* "Water Stress in Plants" (B. Slavik, ed.), Proc. Symp. Prague, 1963. p. 30. Czech. Acad. Sci., Prague.

Vaadia, Y., Raney, F. C., and Hagan, R. M. (1961). Plant water deficits and physiological processes. *Ann. Rev. Plant Physiol.* **12**, 265.

Van den Honert, T. H. (1948). Water transport in plants as a catenary process. *Discussions Faraday Soc.* **3**, 146.

van der Zweep, W. (1961). The movement of labelled 2,4-D in young barley plants. *Weed Res.* **1**, 258.

Vartapetyan, B. B. (1965). Water relations of plants in experiments with heavy isotope O^{18}. *In* "Water Stress in Plants" (B. Slavik, ed.), Proc. Symp. Prague, 1963. p. 72. Czech. Acad. Sci., Prague.

Wardlaw, I. F. (1967). The effect of water stress on translocation in relation to photosynthesis and growth. I. Effect during grain development in wheat. *Australian J. Biol. Sci.* **20**, 25–39.

Weatherley, P. E. (1965). Some investigations on water deficits and transpiration under controlled conditions, *In* "Water Stress in Plants" (B. Slavik, ed.), Proc. Symp. Prague, 1963. p. 63. Czech. Acad. Sci., Prague.

West, S. H. (1962). Protein, nucleotide and ribonucleic acid metabolism in corn during germination under water stress. *Plant Physiol.* **37**, 565.

Wiebe, H. H., and Wihrheim, S. E. (1962). The influence of internal moisture deficit on translocation. *Plant Physiol.* **37**, l–li.

Wills, G. D., and Davis, D. E. (1962). The influence of atrazine on the water uptake of plants. *Proc. Southern Weed Conf.* **15**, 210.

Wood, J. G. (1939). The plant in relation to water. *Australian, New Zealand Assoc. Advan. Sci., Rep.* **24**, 281.

Woodhams, D. H., and Kozlowski, T. T. (1954). Effects of soil moisture stress on carbohydrate development and growth in plants. *Am. J. Botany* **41**, 316.

Yamada, Y., Wittwer, S. H., and Bukovac, M. J. (1964). Penetration of ions through isolated cuticles. *Plant Physiol.* **39**, 28.

Yamaguchi, S. (1961). Absorption and distribution of EPTC-S^{35}. *Weeds* **9**, 374.

Yamaguchi, S. (1965). Analysis of 2,4-D transport. *Hilgardia* **36**, 349.

Yamaguchi, and Crafts, A. S. (1959). Comparative studies with labeled herbicides on woody plants. *Hilgardia* **29**, 171.

Yarosh, N. P. (1959). Effect of water supply on biochemical changes in cotton leaves and seeds. *Fiziol. Rast.* **6**, 211.

Yurina, E. V. (1957). Photosynthesis of woody plants under conditions of sufficient and insufficient moistures. *Fiziol. Rast.* **4**, 60.

Zelitch, I. (1961). Biochemical control of stomatal opening in leaves. *Proc. Natl. Acad. Sci. U.S.* **47**, 1423.

Zelitch, I., and Waggoner, P. E. (1962). Effect of chemical control of stomata on transpiration and photosynthesis. *Proc. Natl. Acad. Sci. U.S.* **48**, 1101.

Zholkevich, V. N. (1954). Use of labeled carbon under field conditions for observations on transport of assimilates. *Dokb. Akad. Nauk SSSR* **96**, 653. (English transl.)

Zholkevich, V. N., and Koretskaya, T. F. (1959). Metabolism of pumpkin roots during soil drought. *Fiziol. Rast.* **6**, 690.

CHAPTER 4

WATER DEFICITS AND
GROWTH OF HERBACEOUS PLANTS

C. T. Gates

COMMONWEALTH SCIENTIFIC AND INDUSTRIAL RESEARCH ORGANIZATION,
DIVISION OF TROPICAL PASTURES, BRISBANE, AUSTRALIA

I. THE COURSE OF DEVELOPMENT

The growth of plants is a dynamic, integrated process that is expressed in the pattern of development of individual organs. For a reasonable basic understanding of the role of water in the growth of herbaceous plants, therefore, it is important to study the responses of various plant parts and the pattern of their development through time under various levels of water stress.

This has not always been appreciated, nor was it necessary that it should be. Early studies were primarily colored by practical agricultural requirements, which might be met by simple levels of experimentation and rewarded by increases in crop yield. They were also influenced by a desire to describe growth attributes that enabled plant adaptation to environment and especially to dry environments. Accordingly, the complexities of patterns of plant growth in response to water stress were appreciated only gradually and became most apparent when metabolic responses of different plant organs at various stages of development were analyzed.

Because of the complexities of such detailed studies, very little real progress has been made to date. In contrast, certain aspects of plant water relations, such as factors affecting the entry of water into the plant and mechanisms whereby it is lost, have been studied in great detail. For example, in the *Encyclopedia of Plant Physiology* (Ruhland, 1956), which devotes approximately 1000 pages to water relations of plants, water uptake, storage, and loss are considered in about half the space and ecological aspects in another quarter. In contrast, water content and water balance of plants are covered in 150 pages, with only two out of 5 subsections dealing with effects of water shortage on plant growth. In one of these subsections which is only 18 pages long, 5 contributors consider such diverse topics as the significance of hydration to the state of protoplasm, water content and respiration, water balance and photosynthesis, the influence of water balance on respiration and synthesis, and the significance of water balance for growth. This carefully compiled appraisal of research in the field demonstrates that much research is still devoted to the study of water entry and loss and very little to its role within the plant.

It is important that the structuring of our knowledge of plant water relations is appreciated, because the aim of this chapter is to describe our knowledge of plant response to moisture stress and to deal with deficiencies. This chapter first considers the general characteristics of growth responses of plants, especially as they relate to agronomic practice and second seeks an understanding of the physiology of whole-plant response.

II. GROWTH RESPONSE OF VARIOUS HERBACEOUS PLANTS

A detailed description of plant response to water stress may have short-comings if it is that alone. It would be relatively easy to prepare such a statement for individual plant types over a range of environmental conditions. However, unless description increased our understanding of how and why a given response occurred, it would lose much of its value. Plant response to a set of environmental conditions is determined by the moisture status of the tissues, or more exactly by the levels of water activity in them. Hence, water balance of plant tissues may be affected by the atmosphere, soil moisture status, osmotic components of the substrate, or the control that the plant itself exerts over water loss. Unfortunately, many field studies fail to assess either these factors or prevailing tissue moisture levels, and so are of limited value in other situations, even for the same crop. Hence, a large body of field data exists that does not help to assess growth responses generally. In this regard, Richards and Wadleigh (1952) comment: "In view of the many factors that enter into field experiments, some controllable and some not, it is difficult to obtain reliable information from field tests concerning the relation of soil moisture stress to the growth rate of plants."

This section considers the useful principles that have emerged from research generally rather than reiterating the mass of accumulated data that is relevant only in its own locale.

Sachs (1859) appears to have been the first to clearly recognize the wide range in moisture content of different soils at the time of wilting of the plant cover. Subsequent workers concluded that different groups of plants differ widely in their ability to reduce the moisture content of a given soil. Briggs and Shantz (1912) made extensive determinations with plants from arid and semiarid regions to determine variations in their capacity to reduce soil moisture before permanent wilting took place. They concluded that the variation exhibited by different plants was much less than previous workers had supposed and was insignificant when compared with the range in moisture retentiveness exhibited by different soils. They found that plants native to dry regions were unable to reduce the water content of a soil to a lower point than was reached with other plants at the time of wilting. They also found only slight differences among crops in their ability to reduce soil moisture before wilting occurred.

The classic studies of Briggs and Shantz have been reconsidered by Furr and Reeve (1945) with sunflowers and by others, notably Hendrickson and Veihmeyer (1945) and Veihmeyer and Hendrickson (1949). Furr and Reeve partitioned permanent wilting percentage into the first PWP and ultimate PWP, whilst Veihmeyer and Hendrickson developed the sunflower test for determination of PWP. These studies are mentioned briefly because they are a point of reference for studies with tomato and lupin plants that will be referred to later.

A. Some Effects of Water Stress on Plant Metabolism

The literature on effects of water stress on plant metabolism is difficult to appraise. Many data are contradictory and much of the work relates to field-grown material. Data often are too variable for precise evaluation. Appraisal might now be made easier by new experimentation and the application of modern techniques for control of variability and the study of trends of plant growth with time. Much work has been reiterated in many reviews. A salient feature is that very little of it is new. Its only relevance here is to help appreciate the underlying characteristics of metabolism that result from water stress.

As soil moisture declines and internal water stress develops, the stomata begin to close and transpiration decreases. This affects gaseous exchange and photosynthesis. Several workers found water levels in leaves of various species to be important for net assimilation. Dastur (1925) found a linear relationship between leaf water content and photosynthesis; Dastur and Desai (1933) noted that photosynthesis was related more to water content than to chlorophyll content. Melville (1937) observed that gain in dry weight in seedling tomatoes increased as water content increased, and Goodall (1946) found that carbon assimilation was much lower in wilted tomato leaves than in turgid ones. These data suggest a dominant role of tissue water levels in carbon assimilation and possibly other synthetic processes.

There is general agreement that as water content of tissues decreases the amount of starch diminishes while the amount of sucrose increases. Maximov (1941) discussed this relationship and cited Mothes and Smirnov as claiming that plant proteins underwent similar changes. Possibly the increase in the net rate of starch hydrolysis with increasing moisture stress results from an increase in the amount of asparagine, because Petrie and Wood (1938a,b) showed that asparagine increased in amount as water content decreased, and it is known that asparagine activates amylases (Hartt, 1934). Wood (1939) reported that respiration rate was correlated with sucrose content. At first respiration increased as water content decreased, but later it decreased, presumably from lowered photosynthesis. Wood and Petrie (1938) concluded that the rate of respiration also appeared to increase with increasing amino acid content.

Petrie and Wood (1938a,b) noted decreased formation of protein from amino acids with a reduction in leaf water content of two grasses, *Phalaris tuberosa* L. and *Lolium multiflorum* Lam. They also noted that the water content of tissues not only affected protein synthesis and hydrolysis, but also that the amount of protein probably was a factor determining the water content. Petrie (1939) remarked that the protein nitrogen content of leaves seemed to be a function of both total free amino nitrogen content and water content, when these three variables were expressed on a dry weight basis. He regarded these interrelationships as a means of understanding the synthetic mechanisms involved, for example: "The change in water content may produce its effect merely by changing the concentrations of the proteins and the substances from which these are synthesized or the change in water content may specifically affect the rate of some reaction or reactions in the system."

The query was whether a relationship existed between the amounts of proteins and amino acids in plant tissues. Petrie and Wood (1938a,b) found such a relationship when working under as nearly as possible steady state conditions with reduced water content. They found that the curve for protein N on amino N was concave to the amino acid axis, thus suggesting that the relation was not an equilibrium one. Indeed, it might have resulted because certain amino acids increased less rapidly than others when the total concentration was increased. Cystine behaved in this way. Possibly the rate of oxidation of cystine increased as water content decreased.

It is of some interest that this early work, which focused attention on the sulfur-containing amino acids and protein in water shortage effects, has not been pursued further. Levitt *et al.* (1962) and Waisel *et al.* (1962) recently found changes in sulfydryls associated with increases in winter hardiness. Yet there may be no connection here, for Lugg and Weller (1941) showed that protein synthesis may be limited by the amount of methionine in the seed and this might merely indicate disturbed equilibrium in the protein–amino acid relationship for an entirely different reason.

The work discussed was undertaken in the light of an analysis of previous determinations of effects of water shortage on carbohydrate and protein metabolism and on the balance between respiration and photosynthesis, as outlined in the review statement of Wood (1939). It had as its aim an understanding of factors contributing to protein synthesis, as pointed out by Petrie (1943). However, the work undertaken from this viewpoint included the study of moisture shortage, respiration, nonreducing sugar levels, and amino acids. It was extended by Amos and Wood (1939), who showed a decrease in the amount of glucose–hemicellulose but an increase in fructosan in grasses with decrease in water content. Wood and Barrien (1939) also showed that the amount of protein sulfur in *Lolium* species declined with decreasing water content, while the amount of inorganic sulfate increased as the cystine content failed to increase.

Thus, there emerges from these interrelated studies of protein synthesis under conditions of moisture reduction the picture that almost all major aspects of metabolism are affected in a complementary, almost coordinated manner. The reasons for this are not clear. It must be remembered that these studies were designed (Petrie, 1943) to provide nearly steady state conditions so, although based on earlier research, they did not share the piecemeal character of much of that research. Accordingly, they demonstrated what earlier research had indicated, that all major aspects of metabolism are affected by water shortage. At the same time, there is very little indication of a drop in the level of any of the aspects of metabolism measured, for instance, as a result of active hydrolysis. For example, the rather cumbersome statement that "the cessation of increase in the amounts of cystine present" is used (Petrie and Wood, 1938b) to describe the case for that amino acid. Petrie commented (1943): "Although it is certainly probable, it is not yet certain that water content effects the velocity constants of synthesis or hydrolysis."

On the other hand, Oparin (1937) considered that a decrease in water content caused the release of adsorbed enzymes into the continuous phase of the protoplast where it caused hydrolysis; Sisakyan and Kobyakova (1947) claimed that protoplasmic "structures'' of wheat, peas, and sugar beets, on dehydration, lost their ability to bind invertase and other enzymes. Sisakyan and Kobyakova claimed that enzymes went into solution and stimulated intensive hydrolytic decomposition, but they presented data only for invertase and regarded observed changes as apparently resulting from alterations in the state of colloids and structural elements of cells. Certainly, in the experiments of Petrie and Wood there was little indication of true hydrolysis.

These are two important aspects, viz, many aspects of metabolism are affected, and this results in a block to synthesis rather than a change to hydrolysis. These aspects are discussed later.

B. Some Specific Plant Responses

In this section specific responses of various types of plants are described.

1. Native Species

It is difficult to consider the effects of moisture shortage on the growth of native plants without some mention of xerophytic qualities, because almost all measurements have been made with these in mind.

Schimper (1898) suggested transpiration criteria as a basis for drought resistance of plants. Maximov (1941) concluded that low transpiring xerophytes of the type Schimper described existed but there were also xerophytes that exhibited high transpiration. Maximov suggested that the capacity of plants to undergo tissue dehydration with the least possible harm was a suitable criterion of drought resistance. Earlier Maximov and Yapp (1929) had claimed that besides anatomical modifications that resulted from constricted development under moisture stress, xerophytes were distinguished physiologically by "an increase in the intensity of transpiration and assimilation, in osmotic pressure, and in the capacity to endure wilting."

Wood (1932) examined the carbohydrate metabolism of *Atriplex vesicaria* growing in the 8- to 5-inch annual isohyets of arid Australia and showed that, contrary to Maximov's views, succulent plants, such as *A. vesicaria*, (saltbush), did not show higher photosynthetic activity than mesophytes. The assimilation–temperature curve for saltbush differed from that of mesophytic plants, with the optimum rate of carbon assimilation of saltbush occurring between 40° and 45°C. Wood's measurements were made under clear, cloudless, dry conditions. Wood (1934) also found that measurements of stomatal frequency, transpiration, and osmotic pressures gave no basis for determining the cause of sclerophylly or of semisucculence. Maximov's contention that xerophytes could resist permanent wilting was borne out, but the structure of sclerophyll-leaved plants was attributed to the formation of uronic acid complexes and aromatic lignin compounds. By comparison, in plants of the succulent leaf type, the carbohydrate flux was directed toward pentosan formation. These metabolic changes gave their particular character to the plants concerned.

It is noteworthy that *A. vesicaria* and other succulents are important fodder plants in Australia. In the main sheep-producing state, New South Wales, 4–5 million sheep run on Atriplex-dominated pastures, and Williams (1960) has demonstrated the high level of stability of the pastoral industry based on *A. vesicaria*. The value of *Atriplex* species lies in the high protein content they can maintain under severe moisture stress. Trumble (1932) claimed that *Atriplex* species produced more protein-rich forage than lucerne did. Other work showed that leaf nitrogen contents of 3.2% (protein levels of approximately 20%) were maintained consistently in *A. vesicaria* throughout the year (Gates and Muirhead, 1967). Apparently some plants of high-protein content can maintain their vital properties, if not active growth, when exposed to severe water stress. Such plants are of great economic value for grazing, largely because of their unique properties of tolerance to water shortage, about which very little is known. Many agricultural and pastoral plants do not have such properties.

Acacia aneura (mulga) also is a protein-rich, valuable grazing plant in arid areas of Australia. Slatyer (1961) found that *A. aneura* could tolerate water potentials of the order of -130×10^6 ergs gm^{-1}. He also found that *Triodia basedowii* (spinifex) tolerated stresses of similar magnitude and might be in equilibrium with severe levels of water stress with only a slight fall in its relative turgidity. Slatyer (1960) compared tissue water relations of *A. aneura* phyllodes with tomato and privet leaves and found the former much more resistant to low tissue water potentials. In *A. aneura* phyllodes a DPD of 15 atm reduced relative turgidity to only 92%, whereas the same degree of water stress reduced relative turgidity of tomato leaves to 73% and of privet leaves to 87%. Furthermore, *A. aneura* tissue recovered from a DPD of 130 atm without injury, whereas DPD's of 45 and 90 atm proved lethal to tomato and privet, respectively. Slatyer claimed that tolerance to desiccation was directly related to the extreme drought resistance of *A. aneura*.

Oppenheimer (1947) found that transpiration of woody plants in Palestine often was reduced to near zero during the hot hours of bright summer days, following prolonged drought. The capacity of plants to survive under such conditions often poses a problem about the balance between respiration and photosynthesis. Wood (1939) commented that such plants could maintain net photosynthesis above the compensation point. He quotes Harder *et al.* (1932) measurements of CO_2 assimilation and respiration during 24-hour periods by several plants of the Sahara desert. After 14 months without rain the Chenopod *Anabasis* was growing and flowering whereas other species were dying. In *Anabasis*, Harder *et al.* (1932) found that during a 24-hour period carbohydrate synthesis exceeded respiratory losses, but in *Zollikoferia* and other plants the reverse was true.

Henrici (1946) found that most veld plants in the semiarid portions of South Africa usually were " perpetually in a state of incipient drying or even in a state of considerable water· deficiency throughout their vegetation period." They did not show wilting because of their peculiar anatomical structure. Henrici further commented: " In the semi-arid parts of South Africa the veld plants may be drought resistant, perhaps made so in the manner Tumanov and Maximov describe, but the low soil moisture after their first water loss scarcely allows them to display their acquired 'increased assimilation and transpiration'." The same may be said of *Atriplex* growing under the harsh conditions of the test site described by Gates and Muirhead (1967), where growth was very slow. In the greenhouse, however, growth rates of *A. nummularia* were comparable to those of tomato and were scarcely affected by raising the salt level of the soil solution to 400 ppm.

It would seem that the most significant propensities of native plants tolerant to moisture stress are their capacities to synthesize adequate protein and foods by maintaining photosynthesis above the compensation point. This implies that such plants can maintain the organization and functioning of protoplasts despite severe internal water deficits, although perforce at a low level of activity.

2. *Cultivated Species*

a. *Sugar Beet.* Watson and Baptiste (1938) compared growth and sugar accumulation of sugar beet and mangolds. They found that in all parts of the plant, water content of mangold was much greater than that of sugar beet. During the day the leaf lamina showed a marked decrease in water content and the petiole showed a similar but smaller decline. The difference in water content between sampling times in all parts of the plant could be attributed to variations in soil moisture availability.

Penman's (1952) experiments on the irrigation of sugar beet showed that maximum sugar yield was obtained when the soil-moisture deficit did not exceed about 2 inches in mid-July or about 4 inches in mid-September. Yield tended to a maximum when the soil moisture deficit at the end of 20 weeks was near zero.

Morton and Watson (1948) studied leaf growth in sugar beet under water stress. Growth in terms of dry matter increment was reduced, as were net assimilation rate, amount of sugars, leaf area and, in one variety, cell number. Sugar as percent dry weight was consistently higher in plants of the wet than the dry series. Uptake of nitrogen per plant, however, was as high in plants of the dry as the wet series and, because plant size in the latter group was smaller, the concentration of nitrogen was much higher, especially for plants receiving high nitrogen. However, such high nitrogen contents were accompanied by low net assimilation rates and low carbohydrate levels, so that plants of the dry series had an internal store of readily available nitrogen. When plants of the dry series were irrigated normally, carbohydrate synthesis increased, resulting in renewed protein synthesis, renewal of leaf production at the apex, and increase in yield.

Morton and Watson observed a sharp fall in the area of functional leaf tissue when plants of the high water series were switched over to low water supply. They attributed this to accelerated senescence of a number of the older, large leaves together with general decline because of nitrogen shortage. Conversely, the decline in leaf area was delayed when plants were switched from low to high water supply. They attributed this change to a delay in senescence of old leaves and an increase in the rate of formation of new leaves, at least under high nitrogen fertilization.

Morton and Watson also observed that nitrogen supply had a profound effect on meristematic activity and formation of new leaves, although they did not actually study the apex but inferred apical response from long-term trends in leaf production. Nitrogen supply had a direct effect on protein synthesis and cell division at the apex. However, even with high nitrogen supply, apical activity seemed remarkably insensitive to water supply. Over a period of more than 3 months, the rate of leaf production was similar in plants exposed to severe continuous drought and in plants that were regularly watered, although drought caused a marked reduction in the net assimilation rate.

Owen (1958) noted at Rothamstead that copious watering resulted in very large yields of sugar beet. Orchard (1963) reported an attempt to increase growth rates by controlled watering. Growth was greatly stimulated by watering after drought, due to an increase in net assimilation rate. However this was nullified to some extent by a decline in leaf area index.

b. Sugar Cane. Clements and Kubota (1942) developed a moisture index for sugar cane based on sampling elongating cane sheaths. This procedure was satisfactory for small plant populations as well as for mixtures of populations grown in different seasons or in different climatic areas. They took as their moisture index the moisture content of the elongating cane sheaths expressed as a percentage of the fresh weight. They found that the moisture index should not drop below 85 during the first 12 months of growth and the irrigation regime should be adjusted to suit this requirement. Subsequently,

Clements and Kubota (1943) extended the index to include nutrient needs of plants and called it the "primary index." They found that when the moisture level was high, the maximum level of nutrients was obtained from the soil. There was very high correlation between water and nitrogen, and potassium and phosphorus. Again, by varying nitrogen level, the moisture content of the plant also was altered, so that by this means a crop could be dried out or made to senesce during much of the year. Clements (1948) found that in order to maintain favorable moisture index values the soil had to be irrigated when the soil moisture tension was 0.25–0.3 atm and did not approach the PWP.

The interaction of favorable moisture relations and plant nutrition sometimes is overlooked, with attention usually concentrated on the effects of water stress on nutrient uptake. The importance of high tissue moisture levels to synthetic processes in sugar cane seems to be indicated by these interactions. Conversely, water stress effects on synthesis within the plant may occur even at very low internal moisture deficits, such as might result from changes in soil moisture that normally escape detection.

c. *Cereal Crops.* Peters (1960) examined the effects of soil moisture tension, moisture content and relative humidity on the growth of small corn seedlings. He confirmed that growth was limited by moisture stress in plant tissues.

Baker and Musgrave (1964) studied the effects of low level moisture stress on photosynthesis in corn. Their aim was to increase understanding of the term *stress* and to derive a meaningful description of the environmental conditions under which stress occurred so that they might determine the point at which moisture stress began to affect apparent photosynthesis. Signs of wilting were barely visible under the moisture stresses applied, but apparent photosynthesis was reduced by 40–50% when soil moisture tension was reduced to only 1 atm. Recoveries in photosynthetic efficiency always occurred immediately on reduction of the moisture stress. Effects of soil moisture stress on both photosynthesis and transpiration were evident at tensions well below 1 atm, with photosynthesis affected first by soil drying. It seemed that either stomatal control or increase in mesophyll resistance proportional to the increase in stomatal resistance caused the observed effects on transpiration and photosynthesis.

Begg et al. (1964) followed diurnal energy and water exchanges in bulrush millet (*Pennisetum typhoides*) in an area of high solar radiation in northern Australia. The soil moisture stress in 150 cm depth of soil was approaching -15 bars, and stomatal closure occurred during the middle of the day, with a corresponding marked reduction in transpiration, particularly in upper leaves, and an increased loss of sensible heat to the atmosphere. Transpiration from various crop layers was not a constant ratio to net radiation but was even in excess of 100% in the top layers and only 20–25% of net radiation in lower layers. These values were for a crop under a water stress of −15 bars but, despite this, at dawn each day relative water contents of the leaves rose to 97%, probably because of the absorption of water from deep soil layers at night.

Studies on effects of water stress on the growth of wheat have mostly been of local application. Some of these will be briefly summarized because wheat is well adapted to escaping and tolerating moisture stress.

Lehane and Staple (1962) examined the effects of early and late moisture stress on the growth of wheat in the greenhouse. Plant yield and grain production were highest under optimum moisture. When water stress occurred early in the life cycle of plants, more than twice as much grain was produced than when stress was applied late. The yields were respectively two thirds and one third of optimum. When late water stress occurred in plants on heavy clay the high moisture holding capacity tended to lessen the adverse effect.

Dubetz (1961) also grew wheat in the greenhouse on two soil types that were fertilized with nitrogen at rates of from 0 to 90 lb per acre. He imposed three levels of watering and maintained these regimes to the time of grain harvest. On loam the results differed from those on loamy sand, as previously observed by Lehane and Staple. Adding nitrogen increased grain yield in both soils, but protein content of the grain was increased only on loam soil.

Responses of spring wheat to applied phosphorus under different levels of moisture supply were examined by Power *et al.* (1961). Phosphorus addition resulted in increased grain production, but this was affected by soil moisture and available phosphorus already in the soil.

d. Potato and Truck Crops. Cykler (1946) assessed the effects of available water on yield and quality of potatoes grown in pots in the greenhouse. Plants were irrigated when 52, 64, or 88% of water in the available range was depleted. Potato yields were significantly affected by treatment but size of tubers was not.

Portyanko (1948) observed in the Sea of Azov region, USSR, that the greatest water loss from growing tissues of the potato plant occurred in the early afternoon but tissues rehydrated overnight. These trends were influenced greatly by available soil moisture, and the depletion of water in the afternoon was partly replaced by water loss from the tubers. In fact, under drought conditions large scale wilting of young tubers was observed. Water content decreased in successively higher leaves of the plant.

The growth and water balance of leaves of gladiolus in response to moisture stress was assessed by Halevy in Palestine (1960). The aim was to determine physiological indexes for irrigation. Corms grown in the field were tested at the 5-leaf stage for transpiration, stomatal opening, water content, osmotic value, and water saturation deficit. Transpiration dropped sharply at a soil moisture tension of about 4 atm, but stomatal closure commenced at 0.7 atm. Leaf water content also dropped to 80 % of fresh weight at 0.8 atm tension. Under severe moisture stress, some water translocation occurred from lower, old leaves to high ones, leading to a sudden rise in the water saturation deficit of the lower leaves. Osmotic values rose slowly as moisture stress increased. There appeared to be no influence of moisture stress on the pattern of leaf elongation.

Data of both Halevy (1960) and Portyanko (1948) showed a plant to be an integrated system, which responded as a whole to moisture stress. Halevy's studies suggest that some phases of this response may be detected when soil moisture is in the upper portion of the available range. The best physiological indicator of incipient moisture stress appeared to be an infiltration test on specified leaves to determine whether stomatal closure was affected.

Many examples suggest that limitations to yield resulting from internal moisture stress may occur within the range of available soil moisture. Halevy (1960) and Clements and Kubota (1943) developed tests in which the water stress of plant tissues interacted with other plant attributes. Hobbs *et al.* (1963) assessed the effects on crop growth of depleting soil moisture to 25, 50, and 75 % of the available range before irrigating. Potatoes benefited by irrigation at the highest level of moisture when water was depleted by only 25 % of the available range. Sugar beets and sweet clover benefited from irrigation only when water loss had reached 75 % of the available range. Alfalfa, wheat, barley, corn, and peas benefited by irrigation when 50 % of the available water had been depleted.

Schwalen and Wharton (1930) reported that production of the maximum number of heavy heads of lettuce was favored by high, uniform soil moisture contents. On the other hand, burst heads resulted from soil moisture deficiency that was followed by irrigation, especially just before harvest.

e. Conclusions. Formation of rapidly proliferating or developing tissues appears to be adversely affected by the soil moisture shortage in the available range. This was observed for heads in lettuce (Schwalen and Wharton, 1930); for flowers in gladioli (Halevy, 1960) and roses (Post and Seeley, 1947); for the synthesis and storage of sugar in sugarcane (Clements, 1948) and sugar beet (Morton and Watson, 1948); and for fruiting in tomato (Salter, 1958). Attributes dependent on active synthesis appear to be adversely affected by moisture shortage. The tissue moisture levels that are the genesis of these effects vary with atmospheric conditions, as was shown for bulrush millet and corn and not merely with soil moisture levels. Therefore, it has been difficult to establish the importance of irrigating at a certain point in the available range of soil moisture. There has been some success in determining time of irrigation in relation to the energy balance, as Penman (1952) demonstrated. However, the really central point that emerges is that the internal water balance of plant tissues must not be impaired, and water stresses must not develop if growth processes are to proceed normally.

3. *Pasture Species*

Investigations with pasture species often are conducted in containers in a greenhouse because of the difficulty of controlling soil moisture in the field. McKell *et al.* (1960) compared two species of orchard grass, *Dactylis glomerata*, subspecies *lusitanica* from Portugal and subspecies *judaica* from Palestine, to determine whether their ecological differences were reflected in physiological adaptations to moisture stress. Leaf elongation was reduced when 1 bar of soil moisture suction was exceeded; it was small at 3 bars, and it ceased at 5 bars.

When soil moisture suction exceeded 5 bars, subspecies *lusitanica* produced a limited amount of leaf tissue, but *judaica* leaves ceased to grow. Clipping to 1 inch at 3-day intervals while losing water affected *lusitanica* adversely but not *judaica*. The response of both subspecies to water stress was similar even though their ecological origin was different.

Bourget and Carson (1962) grew lucerne and oats at 4 moisture levels within the available range. In both crops, yields of plant tops decreased as moisture stress increased. When no fertilizer was added, phosphorus content (percent dry weight) of alfalfa was not affected by moisture stress, but nitrogen content was increased. In oats, phosphorus content decreased without fertilizer but increased with it when moisture stress was increasing. Variations in phosphorus content were attributed to species differences in the ability to use soil phosphorus. Values for water use efficiency increased for oats as available water decreased, and they were variable for alfalfa, although both species had a higher efficiency with added fertilizer.

Doss *et al.* (1962) irrigated individual plots of *Paspalum dilatatum*, *Lespedeza cuneata*, *Paspalum notatum*, and coastal and common Bermuda grass (*Cynodon dactylon*) to field capacity when 85%, 65%, and 30% of available soil moisture had been removed to a depth of 24 inches. The average yield increased in all species except *Lespedeza cuneata* as available soil moisture increased, and yield was highest for *L. cuneata* at the 65% level of depletion. The average annual yield of all species over a 3-year period was 8630 lb per acre dry matter per annum at the 85% level, 9440 lb at the 65% level, and 9950 lb at the 30% level of soil moisture depletion.

Henzell and Stirk (1963a) studied the effects of nitrogen deficiency and soil moisture stress on *Paspalum dilatum*, *P. commersonii*, and *Chloris gayana* (Rhodes grass) in southeast Queensland, Australia. The experiments were conducted with supplemental irrigation and 2 levels of nitrogen. Soil moisture stress was determined with gypsum blocks.

As all 3 grasses showed similar responses, Henzell and Stirk presented data for Rhodes Grass only. Nitrogen fertilizer caused large increases in dry weight, with maximum annual yields exceeding 10,000-lb dry matter per acre. Control plots yielded only 3000 lb of dry matter. Nitrogen-fertilized grass used water more rapidly and from greater soil depths than grass without added nitrogen. The response to treatment may be attributed to more complete interception of incoming radiation and to deeper root penetration. However, nitrogen deficiency was more important than soil moisture stress in limiting growth under natural rainfall, although the yield of nitrogen-fertilized grass was reduced by as much as a third in a dry season.

Henzell and Stirk (1963b) calculated a water budget for the expected frequency of dry periods that would reduce yields of nitrogen-fertilized grass. By this means, the value of supplemental irrigation of nitrogen-fertilized grass could be assessed with known probability. They concluded that yield reduction to less than 50% of that with adequate water might be expected as a result of drought with an average frequency of slightly less than 1 year in 10. Following Penman's (1956) procedures, meteorological records for a 100-year period were used for the calculations.

The useful approach by Henzell and Stirk has obvious advantages and would seem a more profitable use of data than the customary procedure of calculating water use efficiency ratings for small plots or containers. The latter are open to the criticism that the well-known " oasis effect " may occur, whereby advective energy may greatly influence evaporation. McIlroy and Angus (1964) established that on the coastal fringe of southern Australia the oasis effect may be large and lead to long-term evaporation some 20% per annum (up to 50% for individual months) greater for well-watered grass than for a small free-water surface. There is a corresponding energy requirement greater than net radiation, which is met from external sources. The roughness of the surface and large-scale subsidence or advection were largely responsible for this effect at the comparatively large lysimeter installation.

III. GENERAL CHARACTERISTICS OF THE GROWTH RESPONSE AT VARIOUS STAGES OF ONTOGENY

Ballard and Petrie (1936) pointed out that the growth and development of an organism are the result of a large number of directed and integrated metabolic reactions. These are subject to temporal drifts in metabolism that can be recognized as characteristic features of ontogeny. They find expression in alterations in cell number and shape, protein content, patterns of enzyme activity, etc. Morgan and Reith (1954) stated: "Normally the cell changes from a small nonvacuolated isodiametric system to one which has a twenty-fold greater volume, is provided with a large central vacuole, and a permanent cell-wall. In the course of this development the capacity to divide is suppressed." Robinson and Brown (1952) discussed cell growth in roots and pointed out that "the growth of the cell requires to be interpreted in terms of processes that tend to promote it and of others that tend to arrest it ... the promotion mechanism involves at least partly the synthesis and activity of enzymes involved in growth, and the arresting mechanism the processes tending to reduce the activities of these and other systems." Robertson and Turner (1951) and Robertson et al. (1962) characterized the pattern of change in developing apple fruits and pea seeds and demonstrated the interacting pattern of enzyme synthesis and degradation. Robertson et al. (1962) even suggested "that the interrelated processes of protein synthesis, enzyme synthesis, sucrose decomposition, and starch synthesis, all affect the stage at which the seed ceases to increase in water."

It is clear that the ontogeny of plants and the stages of development of their organs afford a set of systems of changing metabolic states, the patterns of which respond variously to controlling influences and so present a matrix of change from which possible modes of action of the influence may be deduced. Enough has been said in previous sections to indicate that tissue moisture levels of almost all plant organs have a controlling influence of the kind referred to here. Accordingly, a study of the role of tissue moisture levels in ontogeny of plants and their separate organs may be most rewarding, even though difficult.

This section considers available information on water stress and development of the embryo and seedling and on the actively growing and senescing plant. Interrelations among the developing organs of two test plants are considered in response to moisture stress, and deductions concerning plant tolerance to internal moisture stress are made.

A. The Embryo and Seedling

Negbi and Evenari (1961) discussed the means of survival of some desert summer annuals in Palestine. *Salsola volkensii* exhibits a remarkable capacity for survival on germination after the first rains. It possesses dispersal units consisting of a heterocarpic fruit enclosed in a winged perianth. Two forms exist: those having green and those having albino embryos. In the "albino" units, the hypocotyl and rootlet of the germinated seed develop and penetrate the seed coats of the dispersal units, but the rootlet may not penetrate the soil due to lack of moisture and the cotyledons remain within the dispersal units. The first rains may be followed by a dry period during which the seedlings dry out, but they revive if placed in water. By drying germinated seeds over sulfuric acid, Negbi and Evenari reduced the water content to their preimbibition level. They found that the germinated seedlings resumed growth normally when rewetted. This could be done up to 26 hours from first imbibition. In the desert, of course, desiccation would not be severe for some time after the first rains, so considerable ecological significance may be attached to this ability to withstand desiccation. Negbi and Evenari compared germinating lettuce seeds with *Salsola* seedlings and found that lettuce could not stand desiccation even when exposed only 8 hours after imbibition.

Milthorpe (1950) assessed desiccation injury of a drought-resistant variety of wheat, Bencubbin, and a drought susceptible one, Charter, by drying seedlings over sulfuric acid. Treatments consisted of drying seedlings in the embryo stages for 3, 9, 27, and 81 hours and at intervals up to 17 days from the beginning of soaking. Three distinctly different degrees of susceptibility, related to seedling age at time of drying, were found. Seedlings were completely resistant from the dormant embryo stage until coleoptiles were 3 to 4 mm long. Until the first leaf emerged, 98 % of tissue water could be lost and only the elongating roots were killed. At later stages, however, growth was permanently impaired by small water losses (Fig. 1). No varietal differences were shown, as both Bencubbin and Charter were highly resistant to drying.

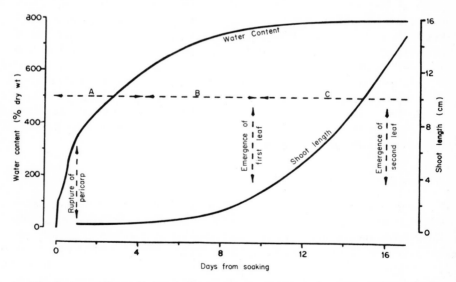

Fig. 1. Approximate duration of three phases of drought resistance (A, B, and C) in relation to shoot length and water content of developing wheat seedlings. Drought stress was imposed by drying over sulfuric acid solutions for varying times. Data of Milthorpe (1950).

Milthorpe commented that although susceptibility to drought increased markedly after rapid extension of the first leaf began; "the capacity to recover from severe drying is not altogether lost. So long as some root primordia remain in a meristematic condition, eventual recovery of the plant from drought is still possible, for the leaf primordia can resume growth if functional roots are present to effect the necessary water uptake." He considered that the phases of differential drought resistance appear to be related to the proportion of nonvacuolate cells to elongated, vacuolate cells, with the former apparently completely resistant.

The wheat embryos tested by Milthorpe were excised from fully developed seeds, but Hubac (1961) reported similar characteristics in embryos of very immature seeds prior to ripening, because they were already adapted to anhydrobiosis or reanimation. A seed or embryo could be subjected to several cycles of desiccation and rewetting without harm. Hubac claimed that this prolonged the period of natural resistance, as has been claimed by Soviet scientists, notably by Henkel (1946, 1961), and critically evaluated by May *et al.* (1962). Hubac worked with embryos of wheat and swede turnip and, like Milthorpe, found that the adaptability to anhydrobiosis was lost with hydration of vacuoles and the beginning of extension growth. As the radicle commenced auxesis earlier than the shoot, the radicle died sooner under dehydration, as also noted by Milthorpe.

These two studies indicate that the meristematic and vacuolate stages of embryo development differ in their response to water stress. Whereas the meristematic stage is tolerant to water stress, the vacuolate stage is not. Tolerance of the meristematic stage, however, involves only capacity for recovery as growth does not occur during desiccation, although the tissue can completely recover with some delay after loss of as much as 85% of its moisture. The susceptible tissue dies.

Data of Owen (1952a) suggest that early stages of embryo development can proceed at very low water potentials. Owen supplied water to wheat seeds by equilibrium with sodium chloride solutions of predetermined vapour pressures. The water potential ranged from -113 to -322 meters of water. A high level of temperature control ensured that only vapor transfer occurred, and it eliminated condensation on the seeds. The critical level for the complete inhibition of germination was not reached. A potential of -320 meters of water (pF 4.5) allowed 20% germination after 20 days. This is drier than permanent wilting, where pF is 4.2 and water potential is -160 meters, but many seeds germinated normally when transferred to moist filter paper after 20 days at a water potential of -320 meters. The relation of time and solution concentration to germination is shown in Fig. 2 over the range -205 to -322 meters water potential. Owen (1952b) examined the nature of his water absorption curves, comparing the behavior of living seeds and seeds whose germination was inhibited by propylene oxide treatment (termed dead seed by Owen). He identified two components of the water uptake curves, the first apparently representing the physical process of imbibition and the second, the initiation and progress of starch hydrolysis.

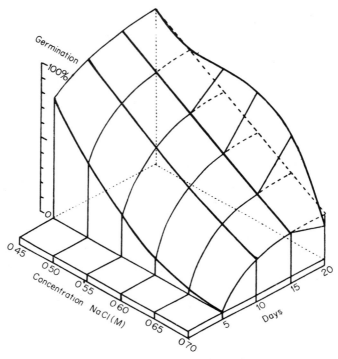

Fig. 2. Relation of time and solution concentration to germination of wheat seed at controlled low water potential. Data of Owen (1952b).

Owen's criterion for germination was emergence of the primary root from the embryo. Seeds placed on wet filter paper germinated in 48 hours at 20°C, so the high water potentials of the trial retarded growth processes of the seed considerably but did not prevent their occurrence. Obviously, more than two components were involved in germination, but it appeared that their operation and coordination were delayed and not deranged. The only hint that this was not so for some seeds was the observation that the lower the water potential the greater the variability in the capacity of seeds to germinate. There was, of course, no indication that growth could not continue just as slowly at low water potentials following emergence and then continue slowly to the 3–4-mm-long coleoptile stage of Milthorpe (stage A in Fig. 1); but this was not determined.

B. During the Growth Cycle

Tiver (1942) found that low moisture availability, at the time of the appearance of the inflorescence, depressed net assimilation rate and adversely affected the thickening of inner fiber cells of flax. Periodic changes during seven cycles of water shortage were followed. Under water stress, the normal course of development of fibers was disturbed, so that thickening of fiber cells, which occurs progressively toward the inner boundary of the fiber bundle, was not achieved. Lignification of fiber cells also was delayed, with fiber yield depressed by 40%. Development of the inflorescence was inhibited more than that of the stem, largely because the inflorescence developed after drought was imposed. In fact, many of the observed effects were expressions of delayed development under low water supply. The uppermost leaves of droughted plants, for example, remained green right up to final harvest, whereas, with high moisture treatment, these leaves had senesced.

Tiver and Williams (1943) exposed the linseed variety—*Linum usitatissimum*, Punjab—to ten cycles of water shortage at the time of the appearance of the inflorescence. The final four harvests of a series of eight were studied. Low water treatment depressed seed yield to 65% of controls, but the 1000 seed weight was scarcely affected. As inflorescence weight, after removal of seed bolls, was reduced to 49% of controls, it was deduced that the seed competed successfully with the inflorescence during development. With low water treatment, the demand of the developing seeds for the limited supply of carbohydrates etc. presumably tended to depress the activity of the secondary cambium of the vascular tissues of the inflorescence branches. This competitive demand occurred even though seed yield was reduced greatly by water stress. The oil content of the seed was also slightly reduced, and oil yield at maturity was reduced by nearly 40%.

Petrie and Arthur (1943) described the effects of a single period of early and late drought on dry weight and leaf area during ontogeny of tobacco plants. Besides the adverse effects of recurring cycles of water stress, significant effects of a single period of water shortage were shown on growth and total nitrogen, and total protein nitrogen, whether water stress was imposed at early or late stages of development. In plants subjected to temporary drought, the percentage of protein content was much higher than in control plants at harvests following the drought period, although this effect declined with time. Petrie and Arthur concluded that low water content resulted in a greater uptake of nitrogen from the soil. Following temporary drought the extra nitrogen was used to form proteins, so that the total protein content of leaves became greater than that of control plants. However, Gates (1957) pointed out that Petrie and Arthur's data may be somewhat misleading with respect to the time of nitrogen uptake, because in each case their harvests were made several days after returning the plants to normal soil moisture levels. Petrie and Arthur also noted that, for all treatments and harvests except the last harvest, leaf area was highly correlated with the absolute water content of leaves, and up to the time of maximum leaf area, water content showed a high correlation with protein content. This correlation, therefore, applied to all drought treatments as well as controls. This is one of many instances of consistency in the relationship between water content and protein content in tissues of varying age and type.

Petrie and Arthur also noted that net assimilation rates were depressed by temporary drought but were quickly restored upon rewatering. In plants subjected to a succession of water shortages to the wilting point, net assimilation rates were not restored nor was there an increase in the protein content of leaves.

Williams and Shapter (1955) made a comparative study of growth and nutrition in barley and rye as affected by low water treatment. Harvests were made after each of five cycles of water shortage, all to levels below the permanent wilting percentage. They found highly significant yield reductions at each harvest as well as depression of net assimilation rate on a leaf nitrogen basis. The most actively growing plant parts suffered the greatest growth inhibition during the period of low water treatment. In fact, many of the effects of drought on ratios of plant parts were only expressions of this fact. The necessity of submitting plants to intermittent periods of water shortage complicated the interpretation of results, because growth characteristics of the species compared undoubtedly modified the severity of one or more of the periods of water shortage.

C. Reproductive Growth

Because many studies of plant water relations seek information on more than one phase of plant development, references have already been made to the effects of water stress on the reproductive growth of herbaceous plants. These need only be recalled here when they help to arrive at logical conclusions. In one sense, the section of this chapter that deals with the embryo and seedling has a distinct place in any consideration of reproduction. The present section largely deals with seed set and fruiting.

Tiver (1942) and Tiver and Williams (1943) analyzed the relation of seed yield to inflorescence development during moisture stress. Newman (1965) described factors influencing the seed production of *Teesdalia nudicaulis* in the East Anglian Breckland and reported that droughting hastened rosette senescence and so reduced fruit and seed set. Senescence was estimated as the percentage of plants still having some green leaves, an attribute significantly affected by soil moisture availability. Prolonged drought hastened the onset of senescence, or it was produced by a short drought at the peak of flower production. Prolonged drought caused cessation of flowering earlier than normally, but a short drought delayed flower production without altering the ultimate number of flowers produced. The development of fruits and seeds was closely linked to the state of the rosette and, hence, a drought that increased the rate of senescence reduced the number of fruits and seeds that formed. There was no evidence that soil moisture directly influenced the development of flowers, fruits, or seeds apart from the delay of flowering. All other direct effects of moisture shortage seemed to be exerted on vegetative tissues.

Newman's results seem to differ from those of Tiver (1942) and of Tiver and Williams (1943), but Newman's findings might easily have applied in both cases, had not Tiver and Williams determined the weight ratios of stem, inflorescence, and seeds. Tiver (1942) also described the effects of water stress as a delay in the development of tissues growing during the stress.

Salter's (1957) data indicated that maximum growth and yield of tomato fruits were obtained from plants grown in soil maintained near field capacity and that a change of water regime during picking resulted in reduced growth and yield. Salter commented that wet regimes led to shallow rooting, in contrast to the deeper development of roots under dry regimes. The adverse effect on growth would be greatest when a change was made from a wet to a dry regime, since this would subject many of the roots formed under the wet regime to subsequent severe moisture stresses. Woodhams and Kozlowski (1954) observed that before the first onset of wilting the content of sugars and starch decreased in bean and tomato roots. Leaves and stems were similarly affected, but the root response indicated that roots did not become sinks for carbohydrates from tops but were themselves sensitive to moisture stress.

Whiteman and Wilson (1965) observed that severe water stress caused cessation of morphological changes in both vegetative and reproductive tissues of *Sorghum vulgare*. Moderate droughting early in the life cycle caused flowering to occur at a slightly reduced leaf number on return to normal watering, but droughting late in the life cycle did not prevent vegetatively advanced plants from making the floral transition. Rewatering then led to immediate floral development without further leaf initiation. It was apparent that the floral apex might develop at a water stress sufficient to prevent leaf expansion.

These studies may suggest that once the inflorescence is formed its development is not as sensitive to moisture stress as is that of vegetative tissues. This may be so, but a full assessment of the effects of water stress requires that studies also relate directly to the apex. This was not so in the trials described here. Studies by Tiver and Williams (1943) of the linseed inflorescence indicate that while overall development of the inflorescence may be restricted, once the seed is formed it can make a significant competitive demand on the rest of the inflorescence. In the studies discussed above, it would appear that the critical point may well be whether or not droughting that occurs at a particular growth stage can impair differentiation and so adversely limit subsequent development. Such an influence of drought could impair particular stages of differentiation, depending on the stage at which moisture stress occurred; similarly, its relief might allow subsequent stages to be resumed readily. This possibility has been examined by Russian workers. Maximov (1941) drew attention to the importance of the effect of drought on embryonic stages of growth. He compared these stages with those of cell enlargement in relation to water supply and described them as stages primarily of synthesis, which were characterized by the increase in the quantity of protoplasm. He stated: "The embryonic stage is the chief growth process of the organism, and although elongation due to incoming water is a very important phenomenon, it is to a certain extent of secondary importance."

Henkel (1961) suggested that during the critical periods of grain formation an "extensive inner reconstruction of the plant" occurs. "At the same time, there is a change in colloido-chemical properties, the elasticity and viscosity of the protoplasm and the metabolism (respiration) all decreasing." Henkel placed considerable emphasis on the importance of colloidal and chemical properties of protoplasm in resistance to the dehydration and overheating that accompany drought, and he considered that viscosity and elasticity of the protoplast along with the amount of its bound water play a major part in this regard during the different developmental phases of plant growth. Henkel regarded drought resistance as a dynamic quality that does not continue to increase with age but declines sharply during reproductive growth stages. He noted two types of protoplasmic viscosity: (1) that due to cations, particularly calcium, termed "hydrophil viscosity," and (2) that due to organic acid anions, termed "structural viscosity." Henkel's relatively recent statement of the views of workers in the USSR is not without support among workers in Eastern Europe. Moreover, it is typical of the views of Russian workers generally.

Aspinall et al. (1964) studied the tillering, stem elongation, and grain yield of barley in response to soil moisture stresses at varying stages of plant development. All growth stages were affected by either long or short periods of stress, but there was evidence of a sensitive stage between the completion of spikelet formation and anthesis. The growth processes depressed most by stress were in organs growing fastest. The effects were more severe at the beginning of a particular growth process, suggesting impairment of cell division by tissue moisture stress.

Asana and Saini (1958, 1962) studied the effects of soil moisture supply on grain development of Indian wheats. During the first 3–4 weeks after anther dehiscence, grain weight per ear increased at a more or less similar rate under normal water supply and intermittent drought. In the subsequent stage, grain weight increased at a faster rate under normal water supply than under drought, and the ear yellowed at a slower rate. The dry weight of the stem decreased considerably during the grain-filling stage, with the decline more or less similar under normal water supply and intermittent drought. These observations suggested that the ultimate difference between grain yields under normal watering and drought might result from the relatively larger functional surface of the ear when adequately watered at the grain-filling stage. This hypothesis was tested with two wheat varieties. Both behaved similarly with respect to an increase in grain weight until 3–4 weeks after anthesis, but subsequently the photosynthetic surface gradually diminished due to yellowing of the stem and ear. At this stage, the grain was filling either with currently produced carbohydrates or by transfer of reserves from the stem. Stem sugars were lost during drought in early stages of growth, but the hastening of yellowing of ears by drought in later stages decreased the photosynthetic capacity of the ear, with the result that grain weight decreased. The varieties differed in this regard. Such a difference suggested that varieties might be screened on the basis of the yellowing rate of the ear and of grain development for tolerance to high temperature and drought.

Asana and Saini studied the development of grain that had already differentiated, so their data do not directly help in evaluating the conclusions of Henkel (1961). However, Henkel cites work by Jolkevitch that in irrigated plants the translocation of labeled carbohydrates from leaves was $2\frac{1}{2}$ times as rapid as in nonirrigated plants. The incorporation of assimilated substances into the ear began earlier and was more intense in irrigated plants than in nonirrigated ones.

Both Kursanov (1956) and Henkel (1961) stated that by a presowing treatment of grain, protoplasmic qualities favoring drought resistance could be conferred on subsequent crop plants. Such plants are said to be distinguished by a high viscosity and elasticity of the protoplasm and by intensive photosynthesis and respiration. A similar presowing treatment with a solution of NaCl is recommended for crops on saline land. It is claimed that the special properties of hardening conferred in this way may be enhanced by heredity. The appraisal by May *et al.* (1962) suggested that the greater drought resistance may result primarily from a higher rate of root growth relative to tops.

Jarvis and Jarvis (1964) found no obvious advantages of presowing treatment on growth, relative turgidity, and leaf water potential of *Sorghum cernuum*, grown under osmotically induced water stress.

IV. DETAILED STUDIES WITH SUITABLE TEST PLANTS

Several points emerged from studies of the ontogenetic development of plants when under water stress:

1. Actively developing tissues were markedly retarded but readily resumed growth on rewatering.
2. Metabolism was affected generally, as shown by changes in assimilation and respiration and by decline in protein synthesis accompanied by hydrolysis.
3. Resumption of growth on rewatering was at rates comparable to those of control plants, provided that only one cycle of water stress occurred.
4. Single cycles of water stress might be easier to interpret than recurrent cycles, provided the effects of a single cycle can be measured. The nature of plant response should be compared both during and after plants are subjected to water stress.

Further study of these points requires suitable test plants. Tomato is particularly useful because its high growth rate makes detailed measurements of growth response possible both during and following a single cycle of moisture stress. Lupin is useful also because its apical parts may be easily manipulated so that comparative studies of tissues of differing age that include meristematic parts may then be made.

A. TOMATO

Gates (1955a,b, 1957) assessed the growth response to water stress of the tomato plant as a whole, of its principal parts, and of individual laminae and petioles of the first eight leaves. Observations were made in the green-house of dry weight and water content, total protein and soluble nitrogen, phosphorus content and leaf area under both moderate (WM) and severe (WS) water shortage. Soil moisture content did not drop to the permanent wilting percentage in either treatment. Plants were harvested and their roots washed out at 38, 42, 48, 49, 55, and 61 days after sowing. Water usage and leaf area changes were recorded.

1. *The Whole Plant.* Figure 3 shows changes in whole-plant dry weight in response to soil moisture availability. Plant responses were very similar in two experiments, with both treatments markedly retarding plant growth during water shortage, even though only one cycle of wilting was imposed for the more severe treatment. When water deficit was more severe (WS),

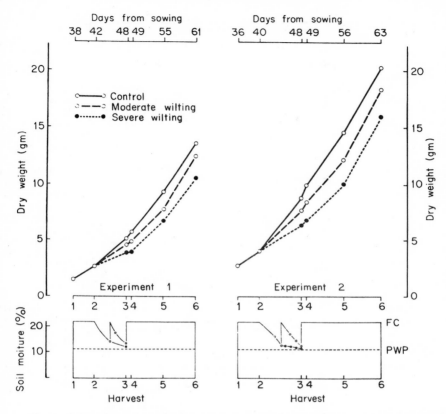

Fig. 3. Dry weight of whole plant and diagram of soil moisture regime. FC, field capacity; PWP, permanent wilting percentage, as determined by standard sunflower test.

growth was retarded about twice as much as it was under moderate water deficit (WM). The effect on the whole plant, of course, was reflected in its principal parts, but to varying degrees (Fig. 4).

Subdivision of the whole-plant response shows that the principal effects of drought were on proportions of leaf and stem. At the beginning of the experiments, more than half the dry matter of the plant was in the leaf laminae. Thereafter, the ratio of lamina to whole plant weight (lamina weight ratio) in control plants fell steeply with time whereas that for the stem rose. In water-stressed plants there were significant departures from this control pattern. During wilting, the lamina weight ratios decreased below those of controls and the stem weight ratio increased to a value above it. After rewatering soil to field capacity this trend was again reversed, because the lamina weight ratio increased to values above those of controls, whereas

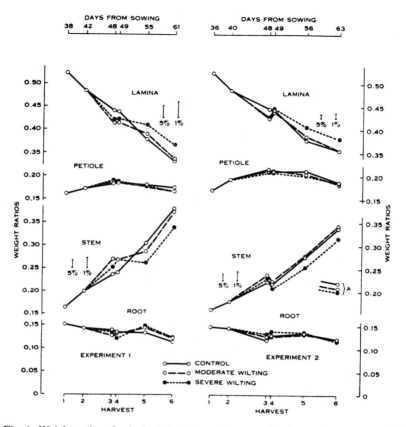

Fig. 4. Weight ratios of principal plant parts. The stem fraction includes inflorescences.
A; The ratio for stem minus inflorescences at harvest 6. Minimum significant differences
for lamina and stem weight ratios are indicated in the usual way on the graph.

that for the stem decreased to a value below it. Translocation of dry matter
from the laminae to the stem was enhanced during wilting by comparison
with controls, whereas on rewatering a higher proportion of the dry matter
assimilated was retained by the laminae.

There was a similar contrast for growth rates during and following water
stress. During wilting, relative growth rate and net assimilation rate were
considerably and significantly reduced (Fig. 5). Following rewatering, however, growth rates rose to levels significantly greater than those of controls.
The net assimilation rates showed this pattern whether they were based on
dry weight, lamina area, or protein content of the laminae.

The pattern of dry weight accumulation in whole plants subjected to
water stress was interpreted as a tendency toward senescence during wilting

and return toward a more juvenile condition upon rewatering. They were, of course, the result of changes in the levels of tissue moisture, in both WM and WS, rather than the direct result of changes in soil moisture itself, for the levels of tissue moisture were assessed in these experiments at harvest times. The determinations at harvest were not critical assessments of tissue moisture levels necessary to maintain optimal growth rates. Further research is necessary to make such critical assessments.

The effects of water stress on efficiency and rate of transpiration have often been discussed in the literature. High values for efficiency of transpiration were observed during these wilting treatments, but growth characteristics under water shortage were impaired. There was only slight growth increment in the early stages of the cycle when tissue water levels were high. On rewatering, transpiration on a unit leaf basis returned to normal or slightly above it. The period of wilting did not effect any economy in subsequent water usage

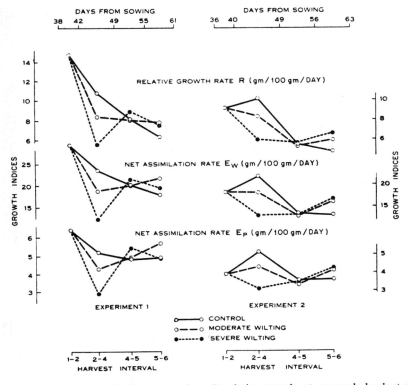

Fig. 5. Growth indices for the whole plant. R, relative growth rate on a whole plant dry weight basis; E_w, net assimilation rate on a lamina dry weight basis; E_p, net assimilation rate on a lamina protein nitrogen basis.

per unit of leaf weight or leaf area. The findings were contrary to the commonly held view that wilting "trains" plants to economize on water use.

2. The Individual Leaves

A better understanding of the internal factors involved in whole-plant responses was obtained by studying the growth of individual laminae and petioles, which were in various stages of development when treatments were imposed.

As shown in Fig. 6, treatment effects, which were not apparent in the cotyledons, occurred in lamina 1 and increased in magnitude for each lamina in passing up the plant. The development of all laminae was retarded by

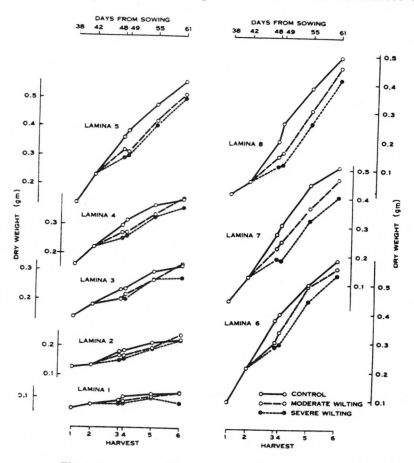

Fig. 6. Dry weights of the laminae of the first eight leaves.

water stress, especially the upper four, which were growing most actively at the time of treatment. The response of petioles differed from that of laminae, because in older leaves the petioles showed higher dry weights and the laminae showed lower ones than controls. On recovery from wilting, the four lower laminae either regained dry weight levels of controls in WM or closed the gap somewhat in WS treatment, whereas the four upper laminae did not, probably because their development had been more retarded.

Relative growth rates (RGR) of the first eight leaves are given in Fig. 7. Response to wilting was a modification of the control pattern, in which RGR increases with leaf position in passing up the plant. During water stress, RGR values were depressed for all laminae with those of WM intermediate between controls and WS. On rewatering, however, the relationship of RGR in treated plants to controls was the inverse of that during wilting, especially in the interval immediately following rewatering. The depressing effect of

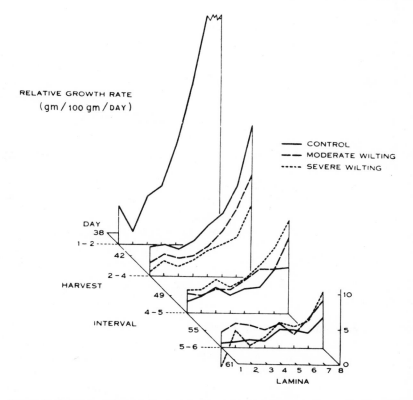

Fig. 7. Relative growth rates *R* of the laminae of the first eight leaves. Oblique parallel projection.

wilting on RGR was greatest for upper laminae, but these showed more
significant response and longer recovery than lower ones, even though the
gap with controls was not fully closed.

These changes in the growth of individual leaves in response to water
stress merely resulted from changes in tissue moisture levels of the laminae
and petioles. Both relative and absolute water content fell during wilting
(Fig. 8). Recovery on rewatering was rapid. One day after rewatering there

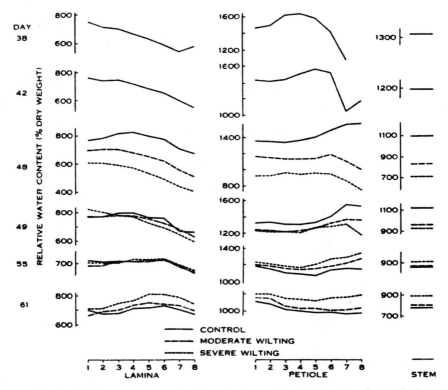

Fig. 8. Relative water content of the laminae and petioles of the first eight leaves
and of the stem.

were no significant differences in relative water content among any of the
laminae. Thereafter, the curves for laminae and petioles of WM and WS
rose above controls, with high significance in the severely wilted plants
evident at final harvest. However, the greatest differences in relative water
content were in the groups of younger laminae and petioles, although in
opposite senses with respect to controls during and following wilting. Young
leaves were affected more than old ones by water shortage, but the response

of the former to rewatering was more intense. The old leaves, however, behaved in a similar manner to the young ones even if not to the same degree. Hence, it seems likely that it was the younger portions of old leaves that contributed most to that response.

In the second experiment, only the second and sixth leaves were studied as representing old and young leaves, respectively. The data confirmed the results described for the first experiment.

In both experiments, leaves differently situated and so in various developmental stages when treatment was initiated, responded to differing degrees both during and after wilting. This was apparent from their dry weights and relative growth rates and from relative water data. It was established that translocation of dry matter from laminae to stem increased during wilting. This was because the normal pattern of translocation of dry matter from lower to upper leaves was impaired during wilting, but continued to the stem and may even have been enhanced by increased hydrolytic activity in lower leaves. On rewatering, translocation of dry matter from the lower laminae to the upper ones apparently was resumed.

In both experiments, rates of leaf growth in the two groups of laminae also contrasted in the degree of response to treatment. The greater changes in RGR values in upper laminae over lower ones indicated that development during and after wilting was affected most in laminae that were growing most actively at the time water shortage occurred. In tomato, the growth of mature leaves occurs in localized regions. It may be that it was the young portions of all leaves that contributed very largely to decline in the active development of the whole plant during water shortage. On recovery from water shortage, these young portions were important centers for the response characteristic of the physiologically young condition of the whole plant.

The cause of this response could not be determined in these experiments. The young, actively growing reproductive tissues were also important. Data for flowering and fruit set showed that at final harvest, the number of parts was not visibly affected by wilting whereas the stage of development was. In young, actively developing tissues, water shortage apparently did not impair the capacity for development but delayed it temporarily.

3. Nitrogen and Phosphorus Content

a. In the Whole Plant. Further analysis of growth response required that some knowledge of the internal factors contributing to it be obtained. An indication of these factors may be derived from changes in phosphorus and nitrogen content because these elements play a dominant role in the synthesis of protein and essential macromolecular compounds.

The effects of water shortage on nitrogen and phosphorus were shown to be the result of two differing trends: that during and that subsequent to wilting. However, these were expressed most strongly in the nutrient uptake of very actively growing organs. This was so for both experiments. The findings of the more detailed experiment only will be discussed here.

Total amounts of nitrogen and phosphorus in plant tops and their principal parts are shown in Fig. 9. During WM and WS treatments the levels of

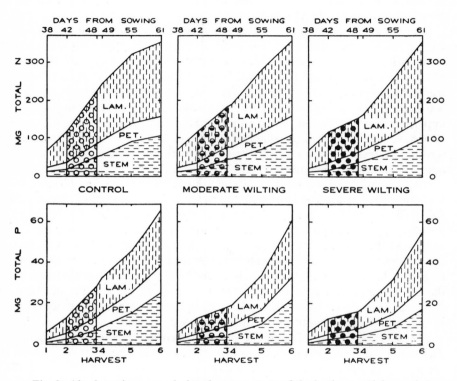

Fig. 9. Absolute nitrogen and phosphorus contents of the laminae, petioles, and stem in response to treatment. The data are plotted additively. The period of wilting is denoted by the blocks of circles and its severity by the intensity of their shading.

both elements, but especially phosphorus, were decreased relative to controls. On rewatering, however, the pattern was reversed so that 13 days later, at final harvest, the amount of nitrogen in the tops was approximately equal for all treatments, whereas phosphorus was increasing more rapidly in wilted plants than in controls. This tendency to regain control levels by the final harvest was most pronounced in the laminae and least in the stem, because the former regained control levels in both nitrogen and phosphorus, whereas the stem failed to do so, except for the amount of nitrogen in WM treatment.

The decline in accumulation of nitrogen and phosphorus in plant tops during wilting suggested further examination for WM and WS treatments. Either the uptake of these nutrients was impaired in early stages of water shortage or nutrients taken up in the early stages were lost in the late stages of wilting. Both may have occurred, but the effects on nutrient accumulation in the tops were marked. It would be possible for nutrients absorbed during wilting to be stored in roots, but wilting restricted root development, so it seemed unlikely that roots then became storage sinks. Rates of intake of nitrogen and phosphorus per unit weight of root during wilting are shown in Table I. Neither the amount nor rate of absorption of nitrogen and phosphorus was increased by wilting treatments.

It was previously observed that the accumulation of phosphorus was inhibited more than that of nitrogen. If the distribution of these elements is considered (Fig. 10), there were marked differences between the stem and

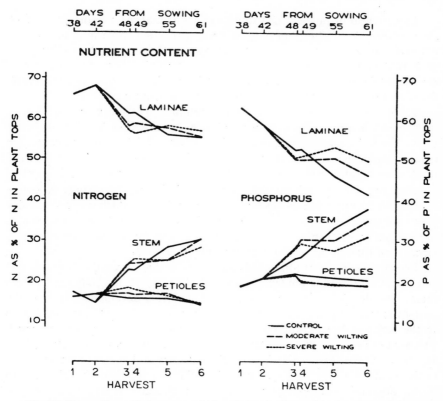

Fig. 10. Proportions of nitrogen and phosphorus in the various fractions of the plant tops, expressed as a percentage of the total.

laminae. During wilting, the proportion of each element in the laminae fell in comparison with controls, whereas that in the stem rose. On rewatering, the reverse was the case, with the laminae having a greater proportion of these elements than the stem. The trend in each case was more marked for phosphorus than for nitrogen.

TABLE I

RATE OF INTAKE OF NITROGEN AND PHOSPHORUS
PER UNIT ROOT WEIGHT

Nutrient	Control	WM	WS
Nitrogen (mg gm^{-1} day^{-1})	33.9	23.4	14.1
Phosphorus (mg gm^{-1} day^{-1})	4.9	1.9	1.2

This pattern of nutrient distribution suggests that the proportions of each element, especially phosphorus, that were distributed to various plant parts differed markedly in the two treatment phases, before and after wilting, for both WM and WS. However, it was observed previously that the pattern of dry weight distribution was similar to that for nutrient distribution. Therefore, the extent to which one paralleled the other was examined.

Table II expresses increments in nitrogen and phosphorus per unit increment in dry weight in the laminae and stem in successive harvests. It is apparent from the relative magnitude of the effects of wilting on the values for laminae and for stem, that translocation of nitrogen and phosphorus occurred to an even greater extent than translocation of dry matter both during and after wilting, but in opposite directions. Furthermore, the change following rewatering occurred as early as harvest interval 3 to 4, which was a single day interval immediately following rewatering, because at that stage the increments in nitrogen and phosphorus per unit increment in dry weight were relatively large, even though the absolute increments were small.

For all these observations, WM caused effects that were either intermediate to control and WS or tended toward WS. This suggested that the changes must have developed progressively as water stress developed. At first, the pattern in treated plants followed that for controls and most nutrients moved to the laminae, but as water shortage developed, the pattern was reversed. The greater response of phosphorus over nitrogen changes to moisture stress suggests that the former was clearly affected when soil moisture was still high in the available range. The synthesis of soluble nitrogen and phosphorus into highly organized compounds may have been checked. It might be expected that the extent to which this inhibition occurred would differ among organs such as individual leaves at different stages of maturity. Therefore, the effects

TABLE II

INCREMENTS IN NITROGEN AND PHOSPHORUS IN THE LAMINAE AND STEM FOR
SUCCESSIVE HARVEST INTERVALS[a]

Nutrient	Plant part	Treatment	Harvest interval				
			1–2	2–3	3–4	4–5	5–6
Nitrogen	Laminae	Control	64.9	60.4	52.0	29.0	17.1
		WM		47.5	36.4	49.6	31.2
		WS		26.5	101.4	53.3	48.5
Nitrogen	Stem	Control	17.4	50.2	25.5	24.4	7.2
		WM		40.0	16.7	26.8	15.6
		WS		47.0	42.3	33.0	19.3
Phosphorus	Laminae	Control	6.8	7.7	8.9	3.9	6.6
		WM		2.7	3.4	7.3	9.9
		WS		1.8	21.5	7.2	9.9
Phosphorus	Stem	Control	5.0	6.8	6.8	4.6	4.2
		WM		3.5[b]	9.9[b]	4.8	4.7
		WS		4.3	6.9	5.1	4.6

[a] Increment per unit increment of dry weight (mg gm^{-1}).
[b] Values obtained by interpolation for stem phosphorus content at WM 3.

of water stress on synthetic growth processes at varying stages of plant development were studied.

b. In Individual Leaf Parts. Extensive data are available elsewhere on the amount and concentration of nitrogen and phosphorus at various times in eight individual laminae and petioles, other laminae and petioles, and the stem during plant development (Gates, 1957). To relate such data to whole-plant response, concentrations of phosphorus and nitrogen in the laminae alone are shown in Fig. 11. Nutrient contents of all laminae were influenced by water stress, and the effect was greatest in the upper leaves. Conversely, on recovery from water stress the concentration of nutrients in upper laminae rose above that in controls to a greater extent than in lower laminae. Thus, both during and after wilting the individual parts responded in a similar fashion but to a differing extent, with the degree of response of a plant part related to the stage of development.

Changes in nutrient concentrations paralleled those for dry weight changes of the laminae. This indicated that the total amounts of both elements behaved somewhat similarly, but effects of water stress were greater on amounts of phosphorus than of nitrogen. After wilting, control levels were gradually regained and were even exceeded in laminae 5 and 6, with the

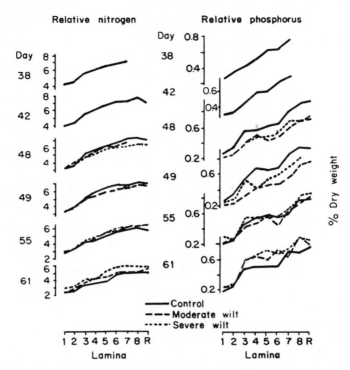

Fig. 11. Relative nitrogen and phosphorus contents (percent dry weight) of the laminae of the individual leaves and the rest of the leaves (R).

response occurring first in the lower-leaf fractions and later in the upper ones. The whole-plant response reflected a pattern of differing intensities of response of various organs, with the level of intensity related to the age of each organ when water stress was imposed.

The nature of the whole-plant response is also shown by the pattern of nutrient translocation from the laminae to the stem that was referred to above. If the period of wilting is examined in detail, it is seen that translocation to the stem was primarily from the lower laminae.

In Fig. 12, for harvests 2, 3, and 4, which precede and follow wilting, the logarithms of the milligrams of nitrogen and phosphorus are given for laminae and petioles of all leaves as they decrease in age up the plant. During wilting, depression in nitrogen uptake was so marked in the laminae as to give very few individual increases relative to controls. In fact, there was a slight loss of phosphorus and possibly of nitrogen in lower laminae, and only the uppermost laminae showed real increases. All petioles showed increases in nitrogen, and all but the lowest petioles increased in phosphorus. As the proportions of both nutrients were higher in stems and lower in laminae during wilting, their translocation from lower laminae through petioles and into the stem is indicated.

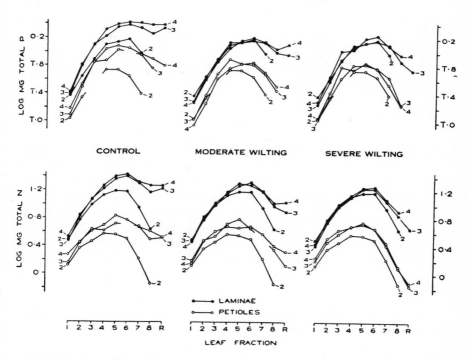

Fig. 12. Logarithm of nitrogen and phosphorus contents (mg) of the laminae and petioles, grouped into treatments. Harvests are denoted by numerals, the comparison being for the second, third, and fourth harvests only.

The data suggested that although the major decrease in nutrient uptake was in the more actively developing, young leaf parts, their ability to function was not completely lost. Rather, the capacity to resist the depressing effects of water shortage on nutrient uptake by maintaining some degree of synthesis seemed to be greatest in the young leaf tissues and least in older ones.

Changes during water shortage suggest that the normal tendency toward synthesis of soluble nitrogen and phosphorus into highly organized compounds may have been checked, especially in the older laminae, thus giving rise to an increasing proportion of hydrolytic breakdown products. This suggestion gains support from observed reductions in the uptake of nitrogen and especially of phosphorus during water shortage. These reductions commenced early in a drying cycle before effects on dry weight occurred, apparently as a result of enhanced hydrolysis. The plant appeared to reduce uptake of these elements in response to lowered internal demand, rather than because of changes in the availability of elements.

Conversely, the ready resumption of nutrient uptake after rewatering suggests there was increased demand within the plant for supplies of available substrates, which resulted from the upsurge of growth in juvenile tissues. The young tissues resisted the drastic effects of water shortage better than old ones, and the former were able to mobilize reserves and maintain high metabolic rates upon rewatering.

4. Conclusions

Detailed studies with tomato indicated that the underlying causes of the effects of water stress on plant growth were still far from clear. In particular, the high resistance of juvenile tissues to damage by water stress suggested further study of the effects of water stress on apical tissues and old leaves. Also, as both net assimilation rate and nutrition were simultaneously affected, the nature of water stress on the factor first impaired, namely, phosphorus nutrition, needed further examination. This was especially so as phosphorus intermediaries to metabolism are closely associated with synthetic processes such as protein synthesis, dry weight increase, and nucleic acid synthesis.

B. *Lupinus Albus* AND APICAL DEVELOPMENT

Morton and Watson (1948) concluded that " synthetic activity at the apex would seem to be less easily disturbed by changes in water tension than in mature organs such as leaves." Their conclusion was based on estimates of the rate of formation of new leaves in sugar beet over more than 3 months, where the rate was the same in plants exposed to severe and continuous drought as in fully watered plants. They commented that " the relative insensitivity of apical activity to water supply is rather remarkable." Morton and Watson's estimates were therefore of a long-term character, and they did not report direct assessments on the apex itself of short-term changes.

Gates (1955a) applied single periods of moisture stress to tomato plants because short-term changes might be readily interpreted. He also studied the effects of single cycles of water stress on apical development of *Lupinus albus*.

The apex in *Lupinus albus* is an easily manipulated, simple structure. As it may become floral at an early stage of development, young seedlings were used. Treatment usually began when the number of leaf structures, including the youngest foliar primordium (collectively called "foliar units") numbered 13, counting from the basal, oldest leaf. The methods were those of Sunderland *et al.* (1957) and Sunderland and Brown (1956). Harvests were made daily or at two daily intervals.

1. As Stress Develops

a. Primordial Initiation. Initiation of primordia was very sensitive to moisture stress, even at early stages in the drying cycle. The results of six trials are summarized in Table III. The comparisons of numbers of primordia initiated

TABLE III

PRIMORDIAL INITIATION DURING MOISTURE STRESS—NUMBER OF FOLIAR UNITS

Trial	Treatment	Days from watering						Comparison Cv.W significance[a]
		0	1	2	3	4	5	
1	C	14.1			15.4		16.3	
	W				14.4		14.3	***
2	C	13.7				15.5		
	W					14.2		**
3	C	12.5				14.8		
	W					13.8		***
4	C	13.6	14.1	14.7	15.1			
	W		13.8	14.1	13.9			*
5	C	13.2	14.5					
	W		13.6	14.5	14.1	14.6		**
6	C	12.8	13.8	14.0	14.7			
	W		13.1	13.8	13.5			***

[a] $*P < 0.05$; $**P < 0.01$; $***P < 0.001$.

between controls (C) and plants from which water was withheld (W) were significant for all trials and, where several harvests were made, regression slopes for water-stressed plants differed significantly from controls. In all trials, harvests commenced early in a drying cycle and plants usually were rewatered when soil moisture approached permanent wilting percentage. The effects of withholding water were initiated early in the drying cycle, suggesting that the initiation of primordia was very sensitive to moisture stress.

b. Volume and Protein Nitrogen. For plants of trial 1 in Table III, the volume of apical fragments bearing 5 foliar primordia was compared for treatment effects. However, the volume of such a fragment decreases in a developing plant as the apex ages, so that it becomes difficult to make meaningful comparisons with a plant whose apex has suspended its development by reason of moisture shortage. This may be seen in Table IV, where other relevant data may also be compared. To avoid this complication, in all further trials the apical fragment that was located above a certain foliar unit (usually the eighth from the base) was dissected. In a watered plant, this meristematic region was developing rapidly, whereas in a droughted plant it was not.

TABLE IV

RESPONSE TO WATER SHORTAGE OF AN APICAL FRAGMENT BEARING 5 FOLIAR UNITS

		Days from sowing		
	Treatment	15	18	20
Volume of apical fragment	C	54.2	56.3	35.0
($mm^3 \times 10^{-3}$)	W		41.7	32.6
Volume of lamina 2	C	0.238	0.286	0.318
(fresh wt gm)	W		0.178	0.143
Relative water lamina 2	C		700	769
(% dry wt)	W		526	373

Volume and protein nitrogen content for 3 successive days of withholding water to almost the permanent wilting percentage are shown in Table V for apical fragments above the eighth foliar unit. Protein content did not increase during water shortage and so differed significantly from controls. The volume of the apex differed with high significance from controls in water-stressed plants. Protein nitrogen per unit volume was quite stable in droughted plants, but fell markedly in controls. These trends indicated a block to development in the apex in response to moisture stress.

TABLE V

VOLUME AND PROTEIN NITROGEN CONTENT OF APICAL FRAGMENTS ABOVE FOLIAR UNIT 8

		Days from sowing			Comparison Cv.W significance[a]
	Treatment	13	14	15	
Volume of fragment	C	165	313	560	
($mm^3 \times 10^{-3}$)	W	149	193	162	***
Protein nitrogen γ per	C	1.9	2.5	3.2	
apical fragment	W	1.4	1.8	1.6	*
Protein nitrogen per	C	11.5	6.6	5.3	
unit volume	W	9.6	9.5	9.9	
($mg \times 10^{-3}$ per mm^3)					

[a] $*P < 0.05$; $***P < 0.001$.

Protein nitrogen content of the apex is compared with that of the second oldest leaf in Table VI, where data are presented for slightly younger plants than those just considered. Regression slopes for the apex were 0.39 γ protein

TABLE VI

PROTEIN NITROGEN CONTENT OF AN APICAL FRAGMENT ABOVE FOLIAR UNIT 8 AND OF LEAF 2 FROM THE BASE OF THE PLANT

		Days from sowing				Comparison Cv.W significance[a]
	Treatment	13	14	15	16	
γ protein nitrogen per apical fragment	C	0.91	1.4	1.7	2.1	
	W		1.0	1.3	1.0	**
mg protein nitrogen in leaf 2	C	1.45	1.50	1.75	1.83	
	W		1.59	1.65	1.50	***

[a] $**P < 0.01$; $***P < 0.001$.

nitrogen per day for controls and 0.018 γ per day as stress developed, the difference in slope being significant ($P < 0.01$). For leaf 2, the regression slopes were 0.14 mg protein nitrogen per day for controls and -0.041 mg per day as stress developed, differences in slope again being significant ($P < 0.001$). These comparisons indicate that protein synthesis was severely restricted in the apex and second leaf during moisture stress, whereas in controls it continued in both organs. The magnitude of inhibition of protein synthesis was greater in the more juvenile structure.

This result prompted the query that drying soil beyond the permanant wilting percentage may lead to protein hydrolysis in the plant apex. For the plants of Fig. 13, wilting was carried through to extreme desiccation, so that the leaves were finally stuck together and could not be separated at final harvest. In this case, the magnitude of the difference in protein content of the apex by comparison with controls was larger, but this was because the time of drying was longer and protein nitrogen content hardly increased during moisture shortage. The regression slopes were 0.013 γ protein N per apex per day for water-stressed plants and 0.69 γ per day for controls (difference of comparison significant at $P < 0.001$). Protein synthesis in both trials (Table VI and Fig. 13) appeared to be blocked when soil moisture was still high in the available range, indicating the apex was very sensitive to even low internal moisture stress. However, when severe soil moisture stress was imposed, the protein content of the apex did not appear to be lowered by active hydrolysis. This was so even though the water content of leaf 2 was lowered by half and soil moisture was depleted far below the permanent wilting percentage. In fact, under these extreme conditions, the apex appeared visually to be the most viable part of the plant.

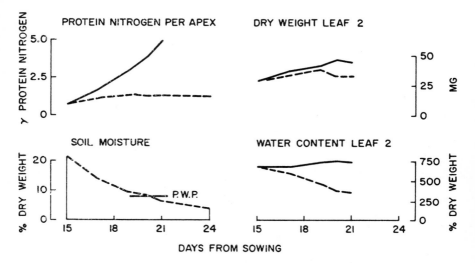

Fig. 13. Protein nitrogen during prolonged moisture stress in an apical fragment above foliar unit 8; dry weight and water content of the second leaf from the base of the plant; and diagram of the soil moisture regime.

2. Upon Rewatering

Relief of soil moisture stress was followed by rapid resumption of apical development. Primordial initiation was actively resumed, but as the plant initiated floral primordia at approximately the fifteenth foliar unit, counts had to include floral primordia, which are initiated more quickly than foliar primordia. Table VII shows counts of foliar and floral primordia during and after moderate moisture stress. The plants were rewatered at noon when their outer leaves were becoming flaccid.

Regression slopes for primordial initiation during moisture stress differed significantly ($P < 0.001$), even though moisture stress was moderate. Active resumption of development followed rewatering, and it appeared that the control condition was attained both for primordial number and dry weight of leaf 2.

Protein synthesis was also actively resumed following its cessation during moisture stress (Fig. 14). After rewatering, protein synthesis was resumed, although the regression slopes for the increase in protein nitrogen differed significantly ($P < 0.05$) between controls and wilted plants. The volume of

TABLE VII

PRIMORDIAL INITIATION DURING AND FOLLOWING A MODERATE MOISTURE STRESS

Treatment	Moisture stress (days)[a]				Days following rewatering				
	0	1	2	3	1	2	3	5	7
Total leaf and floral structures									
C Foliar Units	12.8	13.7	14.0	14.7	15.2	15.0	14.9	15.2	14.7
Floral Units						3.5	4.2	7.1	15.2
W Foliar Units		13.1	13.8	13.5	13.8	13.4	14.1	14.9	14.3
Floral Units						0.7	3.0	5.0	15.2
Leaf 2 from base									
Water (% dry wt)									
C	698	716	736	751	767	757	773	719	726
W		651	588	503	706	715	747	704	689
Dry wt (mg)									
C	29	36	44	49	53	60	62	68	66
W		36	38	38	43	50	55	61	66

[a] Cv.W differences during moisture stress were significant at $P < 0.001$.

the apical fragment rose markedly, and protein per unit volume resumed its normal fall with time. Thus, while development of the apex was suspended during moisture stress, it was resumed upon rewatering in a relatively unimpaired fashion.

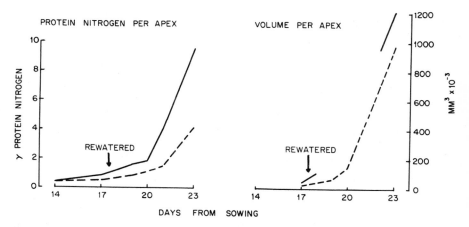

Fig. 14. Protein nitrogen during and following a moderate soil moisture stress in an apical fragment above foliar unit 8.

If this is a general characteristic of apical development, it might afford an explanation for the apparent insensitivity of the apex to moisture stress observed by Morton and Watson (1948). However, the conditions were so different in the two studies, that this cannot be suggested with certainty. A more important consideration is that development in the most juvenile tissue of the lupin plant apparently is extremely sensitive to moisture stress, as mentioned previously. Plants responded even when soil moisture was still high in the available range and no visual symptoms of moisture stress could be detected.

The dominant role of the young organs of the tomato plant in maintaining its organization when exposed to moisture stress suggested the need for a similar study of lupin. Once again the most juvenile tissues, which may be regarded as being akin to embryonic tissues in young seedlings, appeared to be most sensitive to even mild moisture stress, and yet most resistant, in their ability to resume active development upon relief from water stress. Severe moisture stress did not seem to disorganize juvenile, actively developing tissues, because with a long period of stress there was no indication of protein hydrolysis in the apical fragment. Suspension but not impairment of function was apparent in both tomato and lupin in response to moisture stress.

Nicholls and May (1963) examined the effects of water availability on primordium formation, apex length, and spikelet development in barley. Soil moisture stress consistently reduced the slope of the apex length–time curves, and the difference between stressed and unstressed plants became significant at pF values as low as 3.0. Nicholls and May did not report on tissue moisture, but as their plants were grown in artificially illuminated cabinets at 17°C and 85% relative humidity the atmospheric component contributing to internal moisture stress was minimal. Under various lighting regimes, soil moisture stress down to a limit of 5 atm always reduced the number of primordia and increased the proportion of primordia forming spikelets instead of leaves. Depending on the stage of apical development, water-stressed plants sometimes failed to develop a spikelet and so did not subsequently produce a grain at the position of its omission on the axis. Furthermore, on rewatering, the plane of insertion was sometimes rotated about the longitudinal axis, thus causing discontinuity at that point in the place of symmetry upon reaching maturity. Nicholls and May did not measure protein nitrogen or volume, but their data indicated that moisture stress may affect the overall pattern of development in addition to delaying it. Richards (1948) described an aberrant Mangold plant with several united leaf pairs. These occurred (personal communication) as a flush of growth in response to rain after drought, when four or five leaves became united along the 3-parastichies, with no unions along the 5-parastichies. Richards (1948) commented that considerable displacement of the united leaves occurred, but phyllotaxy was nonetheless normal.

C. Ribonucleic Acid Synthesis in the Tomato Plant

Gates and Bonner (1959) determined the effect of water stress on RNA metabolism of tomato leaves. Water stress suppressed net RNA increase in leaves. This is shown in Fig. 15, where RNA as percent dry weight of water

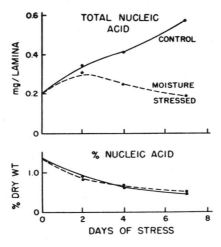

Fig. 15. Concentration and amount of nucleic acid in the laminae of the third pair of leaves of tomato plants during a single cycle of moisture stress under greenhouse conditions.

stressed plants and controls shows little difference, but absolute amounts differ markedly between them. However, leaves of moisture-stressed plants had the ability to incorporate labeled phosphate in RNA even though they did not exhibit net synthesis of the material. The lamina of leaf 5, which was at the stage of rapid expansion at the time treatment began, gave rise to the results of Fig. 16, where total and specific activity of the laminae are presented. Total activity remained practically constant in controls, whereas specific activity decreased with time. Thus, in controls, phosphorus was incorporated into RNA immediately after the application of the isotopic material and RNA synthesis continued at the expense of unlabeled phosphorus taken up by the roots. The labeled RNA was degraded only slightly during the experiment. In wilted plants, RNA remained constant as plants depleted soil moisture. Specific activity of these plants, however, decreased with time and total activity also decreased, until after 5 days only about two thirds of the original activity was measured.

The data indicated that RNA synthesis in moisture-stressed plants had continued at the expense of the unlabeled pool, and the labeled RNA made immediately after application of labeled phosphate was being gradually destroyed. Figure 16 also includes data for osmotically induced moisture stress. Incorporation of ^{32}P into RNA of leaflets of leaf 4 took place initially at the

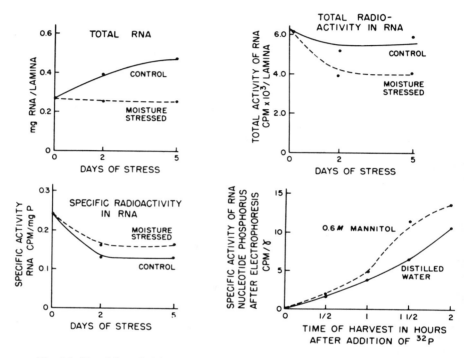

Fig. 16. (Top left and right and bottom left). RNA content and ^{32}P activity in RNA of tomato leaves during a single cycle of moisture stress. ^{32}P labeled phosphate incorporation into leaf RNA for 3 hours before the initial harvest. (Bottom right). Specific activity of RNA of tomato leaves during incubation in ^{32}P labeled phosphate solution and in either water or 0.6 M mannitol.

same rate in controls and in those in 0.6 M mannitol, but the rate of incorporation gradually increased with time in leaves incubated at the higher osmotic concentration.

The fact that ^{32}P was incorporated into leaves of moisture-stressed plants indicated that inhibition of net accumulation was not a block in the ability to synthesize RNA. On the contrary, net RNA synthesis under moisture stress reflected accelerated destruction of RNA.

D. WATER STRESS AND SALINITY

The imposition of moisture stress inevitably must lead to some increase in osmotic pressure in plant tissues. The full extent of such an effect may be expected to reach a maximum in halophytes, such as *Atriplex*, which can tolerate up to 30% of their dry weight as salt (Wood, 1924). Atriplex apparently can maintain high protein levels in leaves despite this high salt content. Gates and Muirhead (1967) described the saline soils and harshness of the habitat of three *Atriplex* species that had high protein contents during a 12-month period. The growth rate of *Atriplex* in its natural habitat is low, as is the growth rate of many semiarid plants, and it is sometimes considered that this is natural to plants that can withstand severe environments. Figure 17 depicts growth curves for three species of *Atriplex* growing under greenhouse conditions. The relative growth rates were similar to those of tomato plants of comparable age. The differences in total dry weight, then, represented primarily initial differences and not differences in growth rate. The author has unpublished data showing that *A. nummularia* grew very fast even in a soil solution having 400 mEq per 1 NaCl. On relief from water stress, the capacity to recover was greater than that of most species. Under natural conditions the restriction to growth would appear to be water shortage and not high salt content. Yet the plants still maintained a high protein content. It is a remarkable feature of such types of arid zone plants that they can so

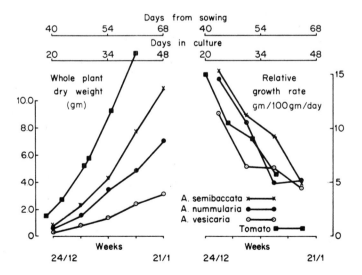

Fig. 17. Growth under glasshouse conditions of three species of *Atriplex* and of tomato (*Lycopersicon esculentum* Mill. v. Pearson) at comparable age.

successfully remain viable and protein-rich in their severe environment and yet can grow quickly in the absence of water stress. In this sense they appear to retain the embryonic capacity to resist stress until maturity.

V. CONCLUSIONS REGARDING GROWTH AND MOISTURE STRESS

A. LEAF FORM AND DEVELOPMENT

The response of herbaceous plants during and following wilting may be best described as a senescent decline in growth during wilting and development of a physiologically young condition on rewatering. The young portions of the leaves were important centers for the development of whole plant responses (Gates, 1955a,b, 1957; Morton and Watson, 1948). Williams and Shapter (1955) found that many effects of water stress on ratios of plant parts only reflected the fact that the most actively growing parts were those that suffered the greatest growth inhibition during a period of water stress. The same observation was made by Aspinall *et al.* (1964). Obviously, the net result of such effects is to alter the structure of water-stressed plants. Although obvious, the reasons for it are not always appreciated. The high degree of tolerance of young tissues to desiccation often is overlooked.

Wood (1932, 1933, 1934) showed that the structure of the sclerophyll and tomentose succulent-leaved plants of semiarid parts of Australia might be attributed to deep-seated metabolic changes, which conferred an ecological advantage upon them.

B. PHYSIOLOGICAL AGE

The old parts of a plant represent by far the greater bulk of its tissues. Accordingly, when the response of a plant as a whole to water stress is studied, the greater tolerance shown by young tissues is likely to be overlooked. Reports of extensive hydrolysis may be colored by the failure to appreciate this fact and to realize that very active synthesis may occur in control plants while wilting is in progress. When individual leaves are studied in relation to the development of the whole plant, the extent of proteolysis that occurs is rather small, unless the cycle of wilting is repeated several times or unless physiologically old leaves are considered. A central observation is that growth is suspended during moisture stress and resumed upon its elimination. The extent to which damage to plants occurs is influenced by their physiological age, the degree of water stress, and the species concerned.

C. The Role of Water in Plant Growth

Evidence is available of a relation between water content and protein level in tissues of varying age. This suggests an interesting connection between tissue moisture levels and protein synthesis.

Robertson *et al.* (1962) suggested for the pea seed " that the interrelated processes of protein synthesis, enzyme synthesis, sucrose decomposition, and starch synthesis, all affect the stage at which the seed ceases to increase in water." Robertson and Turner (1951) observed a linear relationship between protein nitrogen per cell and cell surface in apples, implying a correlation between protein nitrogen per cell and cell volume. Petrie and Wood (1938a,b) also noted that the water content of leaf tissue not only affected protein synthesis and hydrolysis but also that the amount of protein probably was a factor determining tissue water content. Finally, Sunderland *et al.* (1957) studied apical development in *Lupinus albus* and showed that on passing from the first to the seventh primordium, total protein of successive primordia increased by a factor of about 30. By comparison, protein concentration decreased regularly, but only by a factor of 2. These relationships implied a high correlation between water and protein content. They regarded primordial volume as indicative of water content (Sunderland and Brown, 1956) and thus protein was linearly correlated with water content for all seven primordia. There is a high degree of correlation between water and protein content for such juvenile structures as the foliar primordia of the apex in both normally watered lupin plants of Sunderland and Brown (1956) and those discussed above.

In view of the apparently close relation between protein and water content, the reasons for the greater degree of injury to vacuolated tissues by water stress, when compared with nonvacuolated embryonic structures, would seem worthy of comment.

It often is claimed that the extent of injury resulting from water stress is directly related to the degree of vacuolation of the plant cells that has taken place. However, injury may just as easily be associated with changes in structure and composition of protein complexes leading to alterations in relative activities of different enzymes as growth proceeds (Morgan and Reith, 1954). Robinson and Brown (1952) described changes in the enzyme complement of cells of growing roots of broad bean. They suggested that protein may have an indirect effect on growth through the determination of the level of enzyme activity as a cell matures. They found that cells increased in volume some 20 times during maturation. Changes in the enzyme complement as a cell increases in volume are likely to be disturbed by water shortage, to the point of permanent injury. A tolerant plant would be one whose metabolism suffered least from a given volume change. More research is needed to determine specific reasons for changes in tolerance of vacuolated tissues as physiological age advances.

It is not the function of this chapter to discuss the biochemistry of water in fulfilling its role in the physiology of the plant. There are, however, certain basic requirements for any biochemical or biophysical explanations of the role of water in growth and metabolism.

First, tissue moisture levels must be high if active synthesis is to proceed. Throughout the studies described here, the need for high water levels in tissues has become very apparent when experimental conditions were carefully controlled. At the most precise level of control, the rate of growth was directly dependent on water potential (Owen, 1952a). Many aspects of synthetic processes of plants are markedly influenced by a decline in water potential. All such aspects that the author has been able to measure have been arrested by a fall in tissue moisture levels.

Second, a further consistent attribute of the data reviewed here is that although development and synthesis cease when water stress ensues, they are resumed freely on the restoration of turgor.

Third, all aspects of metabolism do not appear to cease completely when a plant is exposed to water stress. Rather, net synthesis apparently ceases, at least in some cases. Gates and Bonner (1959) observed the turnover of RNA without net synthesis, and Owen (1952b) discussed the implications of slow development of wheat embryos at very low water potentials.

Fourth, the role of water in plant development does not seem to be greatly different for different plant tissues despite a wide disparity in their physiological age. The response of plants to moisture stress is an integrated one and different in extent rather than in nature for different plant parts.

Fifth, resistance to moisture stress has common features in embryonic tissues and in drought resistant, mature plants.

Therefore, it will be necessary to propose a role for water that is so general in character as to fit many types of metabolism (Section V,A) in tissues of differing physiological age for diverse plant types. Gates (1964) discussed both the dynamic role of tissue water in plant development and possible mechanisms that may help in meeting these basic requirements from the metabolic viewpoint.

This chapter began on the note that the aim would be to describe present knowledge and to grapple with deficiencies. To some extent this has been achieved. However, as most research has been biased toward a description of factors that influence plant growth rather than toward an analysis of the growth response, it is remarkable that the parameters of the role of water in plant development can be suggested to the present extent. The descriptive approach that avoids analysis of the problem has long gone in a number of kindred sciences and has little place in further study of plant water relations.

Instead, the value of study at an integrated level has been emphasized here. This applies not only to the whole plant but to the nature of the protoplast with which plant water is intimately connected (Gates, 1964). However, the knowledge of how the relationship between plant water and the protoplast may occur is extremely limited for obvious reasons. This chapter, then, ends on the note that in clarifying this relationship, the sets of systems of changing metabolic states that are associated with plant ontogeny should be manipulated to advantage, as already suggested. With modern techniques of chemical and statistical analysis the matrix of change that results from gradations of treatment may be assessed and interpreted, and so the parameters may be restated for the role of tissue moisture in effects of water stress on plant growth.

REFERENCES

Amos, G. L., and Wood, J. G. (1939). The effects of variation in nitrogen-supply and water content on the amounts of carbohydrates in the leaves of grass plants. *Australian J. Exptl. Biol. Med. Sci.* **17**, 285.

Asana, R. D., and Saini, A. D. (1958). Studies in physiological analysis of yield. IV. The influence of soil drought on grain development, photosynthetic surface and water content of wheat. *Physiol. Plantarum* **11**, 666.

Asana, R. D., and Saini, A. D. (1962). Studies in physiological analysis of yield. V. Grain development in wheat in relation to temperature, soil moisture and changes with age in the sugar content of the stem and in the photosynthetic surface. *Indian J. Plant Physiol.* **5**, 128.

Aspinall, D., Nicholls, P. B., and May, L. H. (1964). The effects of soil moisture stress on the growth of barley. I. Vegetative development and grain yield. *Australian J. Agri. Res.* **15**, 729.

Baker, D. N., and Musgrave, R. B. (1964). The effects of low level moisture stresses on the rate of apparent photosynthesis in corn. *Crop Sci.* **4**, 249.

Ballard, L. A. T., and Petrie, A. H. K. (1936). Physiological ontogeny in plants and its relation to nutrition. I. The effect of nitrogen supply on the growth of the plant and its parts. *Australian J. Exptl. Biol. Med. Sci.* **14**, 135–163.

Begg, J. E., Bierhuizen, J. F., Lemon, E. R., Misra, D. K., Slatyer, R. O., and Stern, W. R. (1964). Diurnal energy and water exchanges in bulrush millet in an area of high solar radiation. *Agr. Meteorol.* **1**, 294.

Bourget, S. J., and Carson, R. B. (1962). Effect of soil moisture stress on yield, water use efficiency and mineral composition of oats and alfalfa grown at two fertility levels. *Can. J. Soil Sci.* **42**, 7.

Briggs, L. J., and Shantz, H. L. (1912). The relative wilting coefficients for different plants. *Botan. Gaz.* **53**, 229.

Clements, H. F. (1948). Managing the production of sugar cane. *Repts. Hawaiian Sugar Techologists 6th Meeting, Honolulu, 1948.* p. 31. Honolulu Sugar Planters Assn., Honolulu.

Clements, H. F., and Kubota, T. (1942). Internal moisture relations of sugar cane—The selection of a moisture index. *Hawaiian Planter's Record* **46**, 17.

Clements, H. F., and Kubota, T. (1943). The primary index, its meaning and application to crop management with special reference to sugar cane. *Hawaiian Planter's Record* **47**, 257.

Cykler, J. F. (1946). Effect of variations in available soil water on yield and quality of potatoes. *Agr. Eng.* **27**, 363.

Dastur, R. H. (1925). The relation between water content and photosynthesis. *Ann. Botany (London)* **39**, 769.

Dastur, R. H., and Desai, B. L. (1933). The relation between water-content, chlorophyll-content, and the rate of photosynthesis in some tropical plants at different temperatures. *Ann. Botany (London)* **47**, 69.

Doss, B. D., Bennett, O. L., Ashley, D. A., and Weaver, H. A. (1962). Soil moisture regime effect on yield and evapotranspiration from warm season perennial forage species. *Agron. J.* **54**, 239.

Dubetz, S. (1961). Effect of soil type, soil moisture, and nitrogen fertiliser on the growth of spring wheat. *Can. J. Soil Sci.* **41**, 44.

Furr, J. R., and Reeve, J. O. (1945). Range of soil-moisture percentages through which plants undergo permanent wilting in some soils from semiarid irrigated areas. *J. Agric. Res.* **71**, 149.

Gates, C. T. (1955a). The response of the young tomato plant to a brief period of water shortage. I. The whole plant and its principal parts. *Australian J. Biol. Sci.* **8**, 196.

Gates, C. T. (1955b). The response of the young tomato plant to a brief period of water shortage. II. The individual leaves. *Australian J. Biol. Sci.* **8**, 215.

Gates, C. T. (1957). The response of the young tomato plant to a brief period of water shortage. III. Drifts in nitrogen and phosphorus. *Australian J. Biol. Sci.* **10**, 125.

Gates, C. T. (1964). The effect of water stress on plant growth. *J. Australian Inst. Agri. Sci.* **30**, 3.

Gates, C. T., and Bonner, J. (1959). The response of the young tomato plant to a brief period of water shortage. IV. Effects of water stress on the ribonucleic acid metabolism of tomato leaves. *Plant Physiol.* **34**, 49.

Gates, C. T., and Muirhead, W. (1967). Studies of the tolerance of *Atriplex* species. 1. Environmental characteristics and plant response of *A. vesicaria*, *A. nummularia* and *A. semibaccata*. *Australian J. Exptl. Agr. Animal Husbandry* **7**, 39.

Goodall, D. W. (1946). The distribution of weight change in the young tomato plant. II. Changes in dry weight of separated organs, and translocation rates. *Ann. Botany (London)* [N.S.] **10**, 305.

Halevy, A. H. (1960). The influence of progressive increase in soil moisture tension on growth and water balance of gladiolus leaves and the development of physiological indicators for irrigation. *Proc. Am. Soc. Hort. Sci.* **76**, 620.

Harder, R., Filzer, P., and Lorenz, A. (1932). Über Versuche zur Bestimmung der Kohlensäure assimilation immergrüner Wustenpflanzen während der Trockenzeit in Beni Unif (Algerische Sahara). *Jahrb. Wiss. Botan.* **75**, 45.

Hartt, C. E. (1934). Some effects of potassium upon the amounts of protein and amino nitrogen, sugar, and enzyme activity of sugar cane. *Plant Physiol.* **9**, 453.

Hendrickson, A. H., and Veihmeyer, F. J. (1945). Permanent wilting percentage of soils obtained from field and laboratory trials. *Plant Physiol.* **20**, 517.

Henkel, P. A. (1946). Ustoichivost rastenii K zasukhe i puti ee povysheniya. *Tr. Inst. Fiziol. Rast. Akad. Nauk SSSR* **5**, 237.

Henkel, P. A. (1961). Drought resistance in plants: methods of recognition and of intensification. *Proc. Madrid Symp. Plant-Water Relationships in Arid and Semi-Arid Conditions, 1959* pp. 167–174. UNESCO, Paris.

Henrici, M. (1946). Effect of excessive water loss and wilting on the life of plants (with special reference to karoo plants and Lucerne). *Union S. Africa, Dept. Agr. Forestry Bull.* **256.**

Henzell, E. F., and Stirk, G. B. (1963a). Effects of nitrogen deficiency and soil moisture stress on growth of pasture grasses at Samford, south-east Queensland. 1. Results of field experiments. *Australian J. Exptl. Agr. Animal Husbandry* **3,** 300.

Henzell, E. F., and Stirk, G. B. (1963b). Effects of nitrogen deficiency and soil moisture stress on growth of pasture grasses at Samford, south-east Queensland. 2. Calculation of the expected frequency of dry periods by a water-budget analysis. *Australian J. Exptl. Agr. Animal Husbandry* **3,** 307.

Hobbs, E. H., Krogman, K. K., and Sonmor, L. G. (1963). Effects of levels of minimum available soil moisture on crop yields. *Can. J. Plant Sci.* **43,** 441.

Hubac, C. (1961). La "reviviscence" ou "Aptitude a l'anhydrobiose" et ses variations naturelles et expérimentales chez les embryons et les plantules. *Proc. Madrid Symp. Plant-Water Relationships in Arid and Semi-Arid Conditions,* 1959 pp. 271–274. UNESCO, Paris.

Jarvis, P. G., and Jarvis, M. S. (1964). Pre-sowing hardening of plants to drought. *Phyton (Buenos Aires)* **21,** 113.

Kursanov, A. L. (1956). Recent advances in plant physiology in the U.S.S.R. *Ann. Rev. Plant Physiol.* **7,** 401.

Lehane, J. J., and Staple, W. J. (1962). Effect of soil moisture tensions on growth of wheat. *Can. J. Soil Sci.* **42,** 180.

Levitt, J., Sullivan, C. Y., and Johansson, N. O. (1962). Sulfhydryls—a new factor in frost resistance. III. Relation of SH increase during hardening to protein, glutathione, and glutathione oxidizing activity. *Plant Physiol.* **37,** 266.

Lugg, J. W. H., and Weller, R. A. (1941). Protein metabolism in seed germination. *Biochem. J.* **35,** 1099.

McIlroy, I. C., and Angus, D. E. (1964). Grass, water and soil evaporation at Aspendale. *Agr. Meteorol.* **1,** 201.

McKell, C. M., Perrier, E. R., and Stebbins, G. L. (1960). Responses of two subspecies of orchard grass (*Dactylis glomerata* subsp. *lusitanica* and *judaica*) to increasing soil moisture stress. *Ecology* **41,** 772.

Maximov, N. A. (1941). Influence of drought on physiological processes in plants. "Collection of Papers on Plant Physiology in Memory of K. A. Timiryazev," pp. 299–309. Inst. Fiziol. Rast. Timiryazeva. Acad. Sci. U.S.S.R.

Maximov, N. A., and Yapp, R. H. (1929). "The Plant in Relation to Water," p. 451. Allen and Unwin, London.

May, L. H., Milthorpe, E. J., and Milthorpe, F. L. (1962). Pre-sowing hardening of plants to drought. An appraisal of the contributions by P. A. Genkel. *Field Crop Abstr.* **15,** 93.

Melville, R. (1937). The influence of environment on the growth and metabolism of the tomato plant. II. The relationship between water content and assimilation. *Ann. Botany (London)* [*N.S.*] **1,** 153.

Milthorpe, F. L. (1950). Changes in the drought resistance of wheat seedlings during germination. *Ann. Botany (London)* [*N.S.*] **14,** 79.

Morgan, C., and Reith, W. S. (1954). The compositions and quantitative relations of protein and related fractions in developing root cells. *J. Exptl. Botany* **5,** 119.

Morton, A. G., and Watson, D. J. (1948). A physiological study of leaf growth. *Ann. Botany (London)* [*N.S.*] **12,** 281.

Negbi, M., and Evenari, M. (1961). The means of survival of some desert summer annuals. *Proc. Madrid Symp. Plant-Water Relationships in Arid and Semi-Arid Conditions,* 1959, pp. 249–259. UNESCO, Paris.

Newman, E. I. (1965). Factors affecting the seed production of *Teesdalia nudicaulis*. II. Soil moisture in spring. *J. Ecol.* **53**, 211.

Nicholls, P. B., and May, L. H. (1963). Studies on the growth of the barley apex. I. Interrelationships between primordium formation, apex length, and spikelet development. *Australian J. Biol. Sci.* **16**, 561.

Oparin, A. J. (1937). Enzyme systems as the basis of physiological characters in plants. *IZV Akad. Nauk SSSR Ser Biol.* **6**, 1733.

Oppenheimer, H. R. (1947). Studies on the water balance of unirrigated woody plants. *Palestine J. Botany Rehovot Ser.* **6**, 63.

Orchard, B. (1963). The growth response of sugar beet to similar irrigation cycles under different weather conditions. "The Water Relations of Plants." (A. J. Rutter and F. H. Whitehead, eds.), *Brit. Ecol. Soc. Symp.*, pp. 340–355. Blackwell, Oxford.

Owen, P. C. (1952a). The relation of germination of wheat to water potential. *J. Exptl. Botany* **3**, 188.

Owen, P. C. (1952b). The relation of water absorption by wheat seeds to water potential. *J. Exptl. Botany* **3**, 276.

Owen, P. C. (1958). The growth of sugar beet under different water regimes. *J. Agri. Sci.* **51**, 133.

Penman, H. L. (1952). Experiments on irrigation of sugar beet. *J. Agri. Sci.* **42**, 286.

Penman, H. L. (1956). Evaporation: An introductory survey. *Netherlands J. Agri. Sci.* **4**, 9.

Peters, D. B. (1960). Growth and water absorption by corn as influenced by soil moisture tension, moisture content, and relative humidity. *Soil Sci. Soc. Am. Proc.* **24**, 523.

Petrie, A. H. K. (1939). Protein synthesis in plants. *Australian Chem. Inst. J. Proc.* **6**, 43.

Petrie, A. H. K. (1943). Protein synthesis in plants. *Biol. Rev.* **18**, 105.

Petrie, A. H. K., and Arthur, J. I. (1943). Physiological ontogeny in the tobacco plant. The effect of varying water supply on the drifts in dry weight and leaf area and on various components of the leaves. *Australian J. Exptl. Biol. Med. Sci.* **21**, 191.

Petrie, A. H. K., and Wood, J. G. (1938a). Studies on the nitrogen metabolism of plants. 1. The relation between the content of proteins, amino acids and water in the leaves. *Ann. Botany (London)* [*N.S.*] **2**, 33.

Petrie, A. H. K. ,and Wood, J. G. (1938b). Studies on the nitrogen metabolism of plants. III. On the effect of water content on the relationship between proteins and amino-acids. *Ann. Botany (London)* [*N.S.*] **2**, 881.

Portyanko, V. F. (1948). Water regime of the growing parts of the potato. *Dokl. Akad. Nauk SSSR* **59**, 375.

Post, K., and Seeley, J. G. (1947). Automatic watering of roses, 1943–1946. *Proc. Am. Soc. Hort. Sci* **49**, 433.

Power, J. F., Brown, P. L., Army, T. J., and Klages, M. G. (1961). Phosphorus responses by dry land spring wheat as influenced by moisture supplies. *Agron. J.* **53**, 106.

Richards, F. J. (1948). The geometry of phyllotaxis and its origin. *Symp. Soc. Exptl. Biol.* **2**, 217.

Richards, L. A., and Wadleigh, C. H. (1952). Soil water and plant growth. In "Soil Physical Conditions and Plant Growth" (Byron T. Shaw, ed.), pp. 13–251. Academic Press, New York.

Robertson, R. N., and Turner, J. F. (1951). The physiology of growth in apple fruits. II. Respiratory and other metabolic activities as functions of cell number and cell size in fruit development. *Australian J. Sci. Res. Ser. B* **4**, 92.

Robertson, R. N., Highkin, H. R., Smydzuk, J., and Went, F. W. (1962). The effect of environmental conditions on the development of pea seeds. *Australian J. Biol. Sci.* **15**, 1.

Robinson, E., and Brown, R. (1952). The development of the enzyme complement in growing root cells. *J. Exptl. Botany* **3**, 356.

Ruhland, W. (ed.) (1956). "Handbuch der Pflanzenphysiologie," Vol. III: Pflanze und Wasser, pp. 1073. Springer, Berlin.

Sachs, J. (1859). Berichte über die physiologische Thätigkeit an der Versuchstation in Tharandt. *Landwirtsch Ver. Sta.* **1**, 235.

Salter, P. J. (1957). The effects of different water-regimes on the growth of plants under glass. III. Further experiments with tomatoes (*Lycopersicom esculentum* Mill). *J. Hort. Sci.* **32**, 214.

Salter, P. J. (1958). The effects of different water regimes on the growth of plants under glass. IV. Vegetative growth and fruit development in the tomato. *J. Hort. Sci.* **33**, 1.

Schimper, A. (1898). "Pflanzengeographie auf physiologischer Grundlage." Jena. (Translated edition published by Oxford at the Clarendon Press in 1903.)

Schwalen, H. C., and Wharton, M. F. (1930). Lettuce irrigation studies. *Ariz. Univ. Agr. Expt. Sta. Bull.* **133**.

Sisakyan, N. M., and Kobyakova, .A M. (1947). Directivity of enzymatic action as a sign of drought resistance in agricultural plants. VI. Adsorption of invertase by plant tissues during wilting. *Biokhimiya* **12**, 377.

Slatyer, R. O. (1960). Aspects of the tissue water relationships of an impotrant arid zone species (*Acacia aneura* F. Muell.) in comparison with two mesophytes. *Bull. Res. Council Israel, Sec. D* **8**, 159.

Slatyer, R. O. (1961). Internal water balance of *Acacia aneura* F. Muell. in relation to environmental conditions. *Proc. Madrid Symp. Plant-Water Relationships in Arid and Semi-Arid Conditions*, 1959 pp. 137–146. UNESCO, Paris.

Sunderland, N., and Brown, R. (1956). Distribution of growth in the apical region of the shoot of *Lupinus albus*. *J. Exptl. Botany* **7**, 127.

Sunderland, N., Heyes, J. K., and Brown, R. (1957). Protein and respiration in the apical region of the shoot of *Lupinus albus*. *J. Exptl. Botany* **8**, 55.

Tiver, N. S. (1942). Studies of the flax plant. I. Physiology of growth, stem anatomy and fibre development in fibre flax. *Australian J. Exptl. Biol. Med. Sci.* **20**, 149.

Tiver, N. S., and Williams, R. F. (1943). Studies of the flax plant. 2. The effect of artificial drought on growth and oil production in a linseed variety. *Australian J. Exptl. Biol. Med. Sci.* **21**, 201.

Trumble, H. C. (1932). Preliminary investigations on the cultivation of indigenous salt-bushes (*Atriplex* spp.) in an area of winter rainfall and summer drought. *J. Council Sci. Ind. Res.* **5**, 152.

Veihmeyer, F. J., and Hendrickson, A. H. (1949). Methods of measuring field capacity and permanent wilting percentage of soils. *Soil Sci.* **68**, 75.

Waisel, Y., Kohn, H., and Levitt, J. (1962). Sulfhydryls—a new factor in frost resistance. IV. Relation of GSH-oxidising acitivity to flower induction and hardiness. *Plant Physiol.* **37**, 272.

Watson, D. J., and Baptiste, E. C. D. (1938). A comparative physiological study of sugar-beet and mangold with respect to growth and sugar accumulation. 1. Growth analysis of the crop in the field. *Ann. Botany* [*N.S.*] **2**, 347.

Whiteman, P. C., and Wilson, G. L. (1965). Effects of water stress on the reproductive development of *Sorghum vulgare* Pers. *Univ. Queensland Papers, Dept. Botany* **4**, 233.

Williams, O. B. (1960). The selection and establishment of pasture species in a semi-arid environment—an ecological assessment of the problem. *J. Australian Inst. Agr. Sci.* **26**, 258.

Williams, R. F., and Shapter, R. E. (1955). A comparative study of growth and nutrition in barley and rye as affected by low water treatment. *Australian J. Biol. Sci.* **8**, 435.

Wood, J. G. (1924). The selective absorption of chlorine ions, and the absorption of water in the genus *Atriplex*. *Australian J. Exptl. Biol. Med. Sci.* **2**, 226.

Wood, J. G. (1932). The physiology of xerophytism in Australian plants. The carbohydrate metabolism of plants with tomentose succulent leaves. *Australian J. Exptl. Biol. Med. Sci.* **10**, 89.

Wood, J. G. (1933). The physiology of xerophytism in Australian plants. Carbohydrate changes in the leaves of sclerophyll plants. *Australian J. Exptl. Biol. Med. Sci.* **11**, 139.

Wood, J. G. (1934). The physiology of xerophytism in Australian plants. The stomatal frequencies, transpiration and osmotic pressures of sclerophyll and tomentose-succulent leaved plants. *J. Ecol.* **22**, 69.

Wood, J. G. (1939). The plant in relation to water. *Australian New Zealand Assoc. Advan. Sci. Rept.* **24**, 281.

Wood, J. G., and Barrien, B. S. (1939). Studies of the sulphur metabolism of plants. III. On changes in amounts of protein sulphur and sulphate sulphur during starvation. *New Phytol.* **38**, 265.

Wood, J. G., and Petrie, A. H. K. (1938). Studies on the nitrogen metabolism of plants. II. Interrelations among soluble nitrogen compounds, water and respiration rate. *Ann. Botany* [*N.S.*] **2**, 729.

Woodhams, D. H. and Kozlowski, T. T. (1954) Effects of soil moisture stress on carbohydrate development and growth in plants. *Am. J. Botany* **41**, 316.

CHAPTER 5

WATER DEFICITS AND GROWTH OF TREES

R. Zahner

DEPARTMENT OF FORESTRY, UNIVERSITY OF MICHIGAN, ANN ARBOR, MICHIGAN

I. INTRODUCTION

This chapter is concerned primarily with the development of tissues and organs in woody plants under internal water stress. Because of the perennial nature, long life, and potentially large size of trees, these plants require special consideration in the study of their growth and development under environmental stress. For example, water stresses of several separate growing seasons affect each year's growth increment and the effect of a given water stress is different in a tree seedling than a mature tree. Drought at a critical period one year may result in reduced food storage for utilization in growth the following year, and the effect on development of wood tissue or of flowers and fruits can be appreciable for several succeeding years. The root/shoot ratio may be seriously affected by water deficits in tree seedlings, while in large trees the more important effect of the same deficits may be in the distribution of growth along the annual sheath of wood.

There is no experimental evidence that annual or other periodic water stress is a requirement for the vegetative growth of trees, in the same sense that cold hardening and photoperiodic conditioning are necessary for the perennial growth pattern of many trees in temperate and cold regions. However, some lignified tissues with exceptional secondary wall thickening differentiate only during periods of water stress. Thus, it can be argued that some degree of water stress is essential for the development of normal vegetative structure in woody plants. It is probable that many trees would not attain large size or survive the rigors of high wind and heavy ice if they were not subject to periods of water stress during which differentiating woody tissues are strengthened. On the other hand, seasonal "drought hardening" is not known to play an essential role in the internal biochemical mechanisms that regulate vegetative growth in trees.

It is an established fact that apical, radial, and reproductive growth of trees is highly correlated with environmental water stress. Botanists, foresters, and horticulturists in every decade since the middle of the eighteenth century (reviewed by Studhalter et al., 1963) have reported increasingly convincing evidence that tree growth responds more to natural changes in sap flow, and therefore environmental water stress, than to any other normal, perennial factor in a forest or an orchard. In temperate climates, where the duration of the growing season is controlled primarily by photoperiod, water deficits during the middle of the season have almost continual influence on the growth of forest trees and determine how much and when new tissues can be formed. In cold climates, where temperature largely controls the span of the growing season, internal water stress during the dormant season often determines whether a tree survives when the ground is frozen and stems are exposed to desiccation. In warm equatorial regions, water deficits themselves delineate the growing season, because the dry season is the dormant period that marks annual rings and annual shoots of trees.

Most evidence of effects of water stress on tree growth has been indirect, such as the traditional correlations of rainfall and annual ring width. In addition, gross responses of shoots and radial enlargement under measured soil water regimes in the field have received considerable attention. Tissue development under controlled levels of soil water has received little attention, however, and studies are rare that measure directly the water stress within the tree. Controlled water experiments have usually utilized tree seedlings grown in containers, because the control of water in a forest or orchard is difficult and only partially attainable for trees beyond the seedling stage. Seedlings also provide tissue samples for analysis more readily than do large trees, because there are physical limitations to replicated destructive sampling of developing tissues in the main stems of large trees.

It is presumptuous, however, to assume that internal correlation mechanisms remain constant during development from a seedling to sapling, to pole size, and on to a large tree. Whereas the entire crown of a young sapling may react more or less uniformly to water deficits, the upper branches and foliage of large trees are in a different environment and compete internally under dissimilar conditions than the lower branches and foliage. The same is true of development in the main stem and roots. Thus, it is obvious that a study of carbohydrate supply to the terminal meristem of a seedling grown in the greenhouse under water stress, for example, will not necessarily reveal any important quantitative relationships about bud and shoot development in a pole-size tree under drought conditions in the field. The larger the tree, the more important it is to stratify, by position within the tree, the study of tissue response to water stress.

The great majority of growth and anatomical measurements that have been made of meristematic activity in large trees have been accomplished by one or two methods: (1) by instruments employed on the outside of the stem of an intact, growing tree, which periodically reflect total cambial activity and total shoot elongation or (2) by reconstruction of what probably happened through the interpretation of annual rings and terminal shoots on boles of cut trees or from increment borings and lengths of branch whorl internodes on standing trees. Either method is unsatisfactory from the morphogenic and physiological standpoints. The first, although measuring response at the time it occurs, cannot observe directly the morphogenesis of individual tissues. The second, although it measures directly the anatomical end result, can only surmise which cells were produced under past moisture conditions.

There is much literature concerned with the effects of internal water stress on physiological processes in woody plants, which in turn provides excellent indirect evidence for the effects of water deficits on growth and development. For example, inferences can be drawn from Brix's (1962) study of photosynthesis in seedling *Pinus taeda* or from Helms' (1965) study of photosynthesis in large *Pseudotsuga menziesii* trees concerning the reduction in substrate for tissue development during periods of water stress. Physiological processes *per se* will not be reviewed in this chapter, however, except to be drawn upon to explain growth responses peculiar to the perennial habit and size of trees. Other chapters in this book discuss thoroughly the subject of water stress and physiological processes in plants.

Meristems, growth, and development in woody plants have been reviewed many times, most notably by Büsgen and Münch (1939) and Kramer and Kozlowski (1960) and most completely by Romberger (1963). Other reviews have appeared from time to time relating tree growth specifically to water

(Kozlowski, 1958; Kramer and Kozlowski, 1960; Kramer, 1962; Gaertner, 1963). These have been either generalized or too brief to include in depth many aspects of the role of water in the growth of woody tissues and organs. The review that follows is organized into direct and indirect evidence that water stress affects the development of primary and secondary meristems. Further distinctions are made following Stanhill's (1957) classification of the relative degree of experimental control over both water and plants, by indicating whether studies were made under field conditions or with container-grown trees in the greenhouse. Cambial meristems and wood formation have special significance in trees, and therefore the section on growth in girth makes up about half of this chapter. The literature reviewed throughout this chapter is limited to that published in the English language.

II. WATER DEFICITS AND EXTENSION GROWTH

There is convincing indirect evidence that shoot growth in trees is sensitive to water stress. Hundreds of published forest soil studies agree that trees growing on dry sites do not reach the heights attained on moist sites. Forest mensurationists find tree height the most sensitive growth parameter for measuring site productivity, to which the generally accepted key is soil moisture (White, 1958; Spurr, 1964; Ralston, 1964). Correlations between rainfall and extension growth have been attempted with various degrees of success for many decades (for example, Pearson, 1918; Motley, 1949), and there is little doubt that wet years produce longer shoots than dry years for many tree species on upland sites.

Most species of trees exhibit an intrinsic seasonal pattern of shoot flushing (Kramer, 1943; Tepper, 1963; Kozlowski, 1964a; Kozlowski and Clausen, 1966). The influence of weather and site, and thus of water stress, is limited by genetic variation in the timing of active elongation and bud set. Some species are capable of continuing extension growth for the complete growing season, as are *Liriodendron tulipifera* and *Pinus taeda*, for example, and much new terminal tissue is produced from current photosynthesis and thus is affected by late season water deficits. Many other species, however, complete height growth and set buds by midsummer, as do *Fraxinus americana* and *Pinus resinosa*, for example, and late season droughts do not affect the current year's height increment. Kozlowski and Keller (1966) discuss these patterns in relation to utilization of food in woody plants.

Direct observation of expanding shoots has been confined to small trees, mostly seedlings, because of the physical difficulty of measuring precisely stem elongation on tall trees. However, the sigmoid shape of the normal height-age curve of cumulative tree growth reveals that morphogenesis at the shoot apex is not uniform as a tree increases in height. A condition of water stress inherent in trees is that a hydrostatic gradient exists from the top of the tree to the ground (Zimmermann, 1964b; Scholander *et al.*, 1965). Thus, development of buds, stems, and leaves at the main stem apex occurs at ever-increasing, negative hydrostatic pressure as the tree grows taller and older. The gradient is perhaps negligible in young trees, a magnitude of only 0.1 atm at the tip of a tree 1 m tall. However, in a tree 30 m tall the growing tip exists continually at a negative pressure of at least 3 atm greater than that at the roots. It is probable that this tension gradient, because of the weight of the column of water in the main stem, is a contributing factor in the slowing and eventual cessation of elongation at the top as a tree grows older.

The structural development of shoot apices in trees is fairly well understood, especially in conifers (e.g., Sacher, 1954; Duff and Nolan, 1958; Romberger, 1963). Hence, a basis for interpretation of the effect of water stress on extension growth is available, as reviewed in the following sections.

A. TERMINAL MERISTEMS, BUDS, AND LEAVES

Soil water stress reduces the number, rate of expansion, and final size of leaves. In species whose entire leaf crop for 1 year is present as preformed primordia in the overwintering bud, water stresses of the preceding year regulate the numbers of leaf primordia that form in the developing bud (Duff and Nolan, 1958; Kozlowski, 1964a). Unpublished data of the author indicate that needle primordia of terminal buds in 20-year-old *Pinus strobus* were reduced as much as 40% during mild droughts of midsummer and late summer when buds were forming. In such species, water stress during the period of shoot flushing has no significant effect on the numbers of leaves that mature (Lotan and Zahner, 1963) but results in smaller leaves spaced closer together along the shoot. In other species, which have the capacity to add new foliage throughout the growing season, not all leaf primordia are preformed in overwintering buds, and water deficits during the season of flushing probably reduce total leaf production as much as deficits of the previous season. Few quantitative data are available on these relationships, however, and no conclusions can be made at present regarding the magnitude of stress and the corresponding reduction in leaf numbers.

Direct measures of leaf water potential were correlated with the elongation of young needles on potted *Pinus taeda* seedlings in a study by Miller (1965). Four watering regimes resulted in average water potentials in needle tissue of -3.6, -5.1, -7.0, and -10.2 atm, maintained over a period of 30 days. Growth was directly proportional to water stress in leaves, with total 30-day elongation at -10.2 atm of water potential being less than half that at -3.6 atm (Fig. 1). This growth was largely meristem development at bases of needles. Evidently, no permanent damage resulted to needle growth in this species by 30 days of water stress, for rates of elongation recovered to those of control seedlings during 60 days of watering following the various stress treatments.

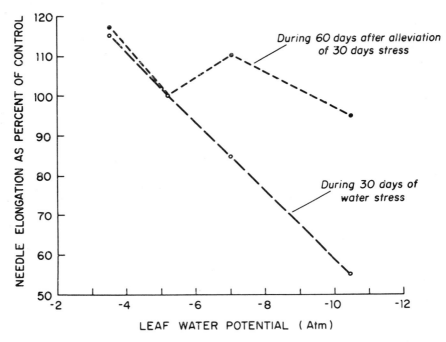

Fig. 1. The relation between internal water potential and elongation of needles in *Pinus taeda* seedlings maintained at four watering regimes for 30 days (lower curve). The upper curve shows total elongation for the same plants during 90 days, including the treatment period and 60 days following rewatering. Points are plotted at the average water potentials of foliage measured periodically at midday. Redrawn from Miller (1965).

Wadleigh and Gaugh (1948) found that elongation rates of leaves of woody plants were reduced exponentially in relation to induced soil water stress. These leaves ceased elongating at 15 atm total soil moisture stress, and expansion resumed original rates on alleviation of the stress by irrigation. Lotan and Zahner (1963) measured elongating foliage on 20-year-old *Pinus resinosa* trees under conditions of imposed drought and irrigation during the entire period of needle elongation. Elongation of internodes ceased several weeks earlier than that of the leaves themselves in this species. Needles on irrigated trees expanded for several weeks longer, at a 30% faster rate, and reached a 40% greater length than needles of trees undergoing drought. Needles near the tops of shoots and on uppermost branches were affected to a greater degree than needles lower on shoots or on lower branches. Fritts *et al.* (1965) studied the pattern of needle elongation in *P. edulis* in a semiarid climate. Needles growing on trees exposed to the normal dry weather of early summer did not elongate at all during late June and early July, but they grew rapidly in late July and August following a mid-July rain. That the renewal growth was not an intrinsic pattern is shown by alleviation of the early season drought by irrigation of one tree, on which needles elongated at a relatively rapid rate during the period when no elongation was occurring on trees under natural conditions.

Therefore, it can be concluded that previous year water deficits affect tree species whose leaf primordia are all preformed in the overwintering bud by reducing leaf numbers and that current year deficits reduce leaf size and their spacing along the shoots. These conclusions are supported by unpublished data of the author and of Clemments (1966), who independently found that numbers of needle fascicles on current shoots of 20- and 5-year-old *P. resinosa* were directly proportional to the frequency of irrigation during the previous growing season. They found further that current season watering had no effect on the numbers of needles produced, but that needle density along shoots was inversely proportional to frequency of irrigation during the current season.

Anatomical changes in needles of *Pinus* have been attributed to growth under severe water stress. Parker (1952) measured changes in needle anatomy of potted *P. nigra* and *P. strobus* seedlings as desiccation approached the lethal level and found marked shrinkage of chlorenchyma cells that did not recover turgor on rehydration. Although the outward morphology of needles was not permanently changed by temporary water stress, permanent damage resulted to various tissues within the leaf. Thames (1963) evaluated genetic adaptations that tend to reduce water loss from needles of *P. taeda* growing on a geographically isolated dry site. He found that mature needles on seedlings of the dry site seed source had thicker epidermis, fewer stomata, and more hypodermal scherenchyma cells between stomata than did needles of moist site seed sources.

B. Seedling Shoot Growth and Dry Weight

Studies with potted seedlings have shown consistently that shoot elongation and dry weight production are related directly to watering frequency. Pessin (1938) and Wenger (1952) both found that various species of the southern pines (*Pinus palustris, P. elliottii, P. taeda, P. echinata*) were sensitive in shoot growth characteristics to watering levels from soil maintained near " field capacity " to soils allowed to approach " wilting point " before rewatering. It is not possible to establish quantitative relations with soil moisture tension from such studies, but Wenger, for example, reported that shoot elongation was twice as great in soils allowed to dry to 60% available water than in soils allowed to dry to 20% available water before rewatering and that seedlings in soils maintained near field capacity elongated 3 times as much as those in soils that dried to 20% available water before they were rewatered.

Shoot growth in seedlings of woody plants is apparently affected by relatively low levels of soil moisture tension. Sands and Rutter (1959) found dry weight production by first-year seedlings of *Pinus sylvestris* sensitive to soil moisture conditions of only 0.5 atm tension, and production of shoots was reduced over 50% in soils allowed to dry to tensions of 1.5 atm. Similar results were obtained by Kenworthy (1949) with *Prunus* seedlings, in which shoot production was reduced by moisture tensions beginning at about 0.5 atm, as measured by tensiometers.

Effects of low tensions on dry weight production were studied by Jarvis and Jarvis (1963), who grew first-year seedlings of four tree species under varying soil moisture tensions for a period of 20 days. *Betula verrucosa* and *Populus tremula* were less sensitive than *Pinus sylvestris* and *Picea abies* to changes in average soil moisture tension up to about 3 atm. The maximum growth of all four species occurred in soils allowed to dry to about 0.5 atm before watering. Dry weight in the two conifers was reduced about one third in soils that dried to 1.7 atm, whereas dry weight was reduced only 10–20% in the two angiosperms by tensions up to 2 atm. The external resistance to vapor loss from the leaves of the conifers was about one third that from leaves of the angiosperms, largely because of mutual interference between leaves on the same plant. The result was less water deficit within the seedling and greater net assimilation in aspen and birch than in pine and spruce.

Further evidence is presented by Stransky and Wilson (1964) that relatively low soil moisture tension affects shoot growth in tree seedlings. Terminal elongation of potted *Pinus taeda* and *P. echinata* was inhibited by tensions not greater than 2 atm, and elongation stopped completely as tensions approached 3.5 atm (Fig. 2). Leaf moisture content was not well correlated with shoot elongation rates or with soil moisture tension, because needles lost only 15% of initial moisture (about 210% by weight) at the time terminal extension ceased. Evidently, such slight decreases in water content of foliage have large effects on growth, probably because leaf water potential decreases rapidly with the initial loss of turgor (Hammel, 1967). It was not until 5 atm soil moisture tension was reached that leaf water content began to drop appreciably in these pine seedlings, long after terminal elongation ceased.

In the experiments cited above, shoot development was studied in relation to soil water tension. The magnitude of water stress measured in the soil, however, is not to be equated with that of water stress in the tissues of the tree. As Miller (1965) noted, nightly watering of potted seedlings to "field capacity," equivalent to the low tension treatments in the above studies, resulted in average leaf water potentials of less than −5 atm, considerably greater water stress than implied by a soil water tension measured at 0.5 atm.

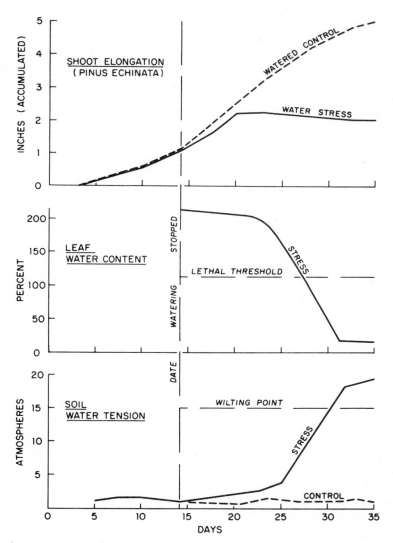

Fig. 2. Average trends of height development and leaf water content in *Pinus echinata* seedlings as related to soil moisture tension in a drying soil and in a well-watered soil. Redrawn from Stransky and Wilson (1964).

As discussed elsewhere in this volume, the average moisture stress in a volume of soil is no measure of the stress at the root-soil interface, which is considerably greater.

In a greenhouse study that controlled the depth to the water table precisely in large tanks, Mueller-Dombois (1964) tested shoot growth response of *Pinus banksiana* and *Picea mariana* seedlings to a soil moisture gradient in dry to saturated sandy soils. Both species reacted similarly, making twice the height growth at optimum depth to ground water—where roots were growing in the capillary fringe—than where roots were in dry soil above the effect of capillary water. At the other extreme, both species made least shoot growth when roots were in saturated soil.

The development of most tree seedlings is impeded in soils with excess water (McDermott, 1954; Hunt, 1951; Walker, 1962; Hosner and Leaf, 1962). Although the physiological mechanism of water stress probably is the same in trees subjected to drought or flooding of roots (reviewed by Kramer and Kozlowski, 1960), the subject of excess water in relation to growth will not be reviewed in detail here. Other factors, such as disease organisms in roots and toxicity under anaerobic conditions, usually interact to cloud the true effect of "physiological drought" resulting from too much water in the soil.

C. Trees Beyond Seedling Stage

The relative effect on shoot growth of previous-year vs. current-year water deficits has received attention in several controlled experiments with pine, in which irrigation vs. induced drought has been the method of water control. In sapling size *Pinus taeda*, grown outdoors in large containers, the 2-year effect on elongation of the terminal leader of well-watered trees was almost twice that of trees subjected to extreme drought conditions (Zahner, 1962). Water was withheld from the latter trees from May to October in each of two consecutive growing seasons, with a 6-month soil moisture recharge during the dormant seasons. *P. taeda* is capable of several recurring flushes of terminal tissue late in the season, which are not preformed in the bud. Drought during the first year of the experiment reduced terminal elongation only during the later part of the season, because the initial shoot flush was equal on trees in both treatments. Drought in the subsequent year reduced elongation in the initial flush by 20%, and in later flushes up to 100%, indicating the effect of previous plus current season's drought on bud formation and overall tree vigor. In both years the watered trees added four flushes on the terminal shoot while the trees subjected to drought ceased elongation after only two flushes. Thus, most of the difference in height growth due to current-year drought in this species was attributed to shoot development during the middle and later part of the current season.

By contrast, shoots of *Pinus resinosa* are largely preformed in the over-wintering bud (Duff and Nolan, 1958), and extension of bud tissues produces a single growth flush each year. Controlling water in the current year only, Lotan and Zahner (1963) found that severe artificial drought during the period of shoot expansion reduced main stem elongation by one third in sapling-size trees of this species. Periodicity of leader growth was not affected by current water deficits, because trees under both drought and irrigation ceased elongating and set new buds at the same time, in early July in northern Michigan. Thus, the rate of elongation of preformed tissue is affected by current deficits independently of the previous year's conditions. Shoots on lateral branches in the lower portion of the crown were not measurably affected by drought. Under natural conditions severe droughts are not common during the brief period of internode extension in this species, and water stress of the previous year probably has a far greater net effect on shoot development than does that of the current year (Kozlowski, 1964a).

Pinus strobus is another species characterized by single flushes of annual shoot growth, but it commonly produces short second flushes in midsummer. Owston (1968), studying the anatomy of developing shoot tips in 17-year-old trees of this species, reports that primordial tissues for these second flushes are preformed in overwintering buds and are subject to control by the previous year's water deficits. He found a significant reduction in both the frequency and size of second shoots on trees under controlled drought the previous year.

Summer shoots in *P. resinosa* have also been attributed to extreme changes in environmental water conditions. Carvell (1956) related proleptic shoots from lateral buds that had set weeks earlier to a sequence of severe early drought followed by extensive late season rains. In this case, terminal buds remained dormant, and there was no indication that the second shoots from lateral buds were preformed, as with the study of *P. strobus* cited above. Thus, lateral proleptic shoots may be more the result of water relations of the current season, whereas some terminal late shoots result when conditions of the previous year are conducive to primordia formation in terminal buds.

No study is known to have measured directly the internal water stress of trees beyond the seedling stage and to have related such measured stress to shoot development. Measurement on seedlings cannot be extrapolated to leaf and stem tissue high in the crown of a large tree, because water stress in the latter is regulated by a tall column of xylem resistance under the pull of gravity (Zimmermann, 1964b). Measures of soil water tension in the root zone of large trees are likewise meaningless for estimating accurately internal water stress in the crown, for reasons already discussed.

Diurnal differences in elongation rates of terminal shoots on *Pinus radiata* 20 feet tall were reported by Fielding (1955) from Australia. During spring and early summer, daily terminal elongation averaged three times greater for the 4 PM to 10 AM period than for the 10 AM to 4 PM period. No direct measures of water stress were made, but it was assumed that the slow rates of shoot elongation, often dropping to zero or becoming negative during the midday hours, were the result of transpirational stress in the shoots. Fielding correlated this diurnal elongation pattern almost perfectly with shrinking and swelling in girth of the lower bole, a phenomenon repeatedly associated with internal water stress during the noon hours (Kozlowski and Winget, 1964) and with measured xylem sap tension (Worrall, 1966). Haasis (1932), however, reported terminal shoot elongation in early spring, even while the stem was undergoing shrinkage. Haasis did not clarify this information exactly as to when the respective enlargement and shrinkage processes were occurring, and it is likely that he noted diurnal midday stem contraction and overnight shoot elongation, which agrees with Fielding's observations.

D. Shoot Growth Correlations with Rainfall and Other Indirect Measures

As a tree reaches the sapling stage of development, it enters the "grand period of growth" in which annual shoot elongation is at its fastest rate and thus can be fairly well correlated with annual variation in environmental stress. Much reduction in height growth because of dry climates or dry sites occurs during the sapling stage of development.

Many correlations between precipitation records and annual shoot increments of forest-grown trees have been reported during the past half century (see examples cited below). Although most of these have dealt with northern conifers that exhibit single, preformed shoot flushes, few have attempted to analyze rainfall records in sufficient detail to correlate the full 2-year effect of water deficits on shoot elongation. For example, Pearson (1918), with *Pinus ponderosa*, Bell (1957), with *Pseudotsuga menziesii*, and Mitchell (1965), with seven coniferous genera, report positive correlations between shoot increment and the amount of current (spring and early summer) rainfall but no correlation with the previous season's rain records. Only Motley (1949), with *Pinus strobus*, and Bell (1957), with *Picea sitchensis*, found strong effects of the previous year's rainfall on current shoot growth. Kirkwood (1914), with *P. ponderosa*, claimed that height increment was reduced by a lack of rainfall of the preceding year, but his conclusion has been criticized by those who have not found the same correlation (Tryon *et al.*, 1957).

Duff and Nolan (1958), Kozlowski (1958, 1964a), Kozlowski and Keller (1966), and others theorize that the 2-year effect of weather on shoot growth should be quite significant in species that set terminal buds by midsummer. A recent multiple regression approach to this problem proves that water stress in each of two successive years can be equally important (Zahner and Stage, 1966). Seventy-two percent of the variation in annual shoot growth in five stands of young *Pinus resinosa* was accounted for by water deficits calculated from rainfall records of the previous and current growing seasons together. The water deficit during the summer of the previous year (June 15 through October) accounted for as much reduction in annual height increment as the deficit from May 1 to July 15 of the current year.

Tree species that exhibit continued shoot flushing throughout the summer would be expected to show little correlation between total annual elongation and the previous year's rainfall. Tryon *et al.* (1957) found no effect of warm and dry weather of the previous year on height increment in *Liriodendron tulipifera*, but May–June rainfall of the current year was positively related to current growth. Shoot growth in recurrently flushing pines (i.e., *Pinus taeda*, *P. rigida*, and *P. radiata*) has been studied intensively in the field, but no significant correlations have been reported with precipitation records or water deficits of the previous year.

High temperatures have usually been interpreted as enhancing the growth of shoots rather than contributing to water stress and thus reducing growth. However, this conclusion is based on studies in cool, moist climates (Hustich, 1948). High temperature unquestionably contributes to serious water deficits on dry sites in warm climates, but reductions in shoot growth have not been related quantitatively to this effect directly from weather records.

Husch (1959) found that annual height growth in young *Pinus strobus* trees was affected more by available soil moisture than any other environmental factor measured. On sites that were rapidly depleted of available water, shoot elongation slowed early in the season. This process, repeated season after season, for many species in many parts of the world, results in the general conclusion that soil water is the key to forest site productivity. Many studies have shown consistently that soil characteristics associated with the quantity of soil water available to tree roots account for the most variation in height growth from site to site. Coile (1952) and Ralston (1964) reviewed many studies that agreed that height growth of at least 35 species of trees was reduced on soils with texture, structure, and depth properties associated with low availability of soil water. Equally important, as found by such studies, is the effect of topographic position on water regimes, where trees on steep, elevated sites with southerly aspects exhibit reduced height growth. Typical examples of the influence of soil and topography on the availability of water, and thus on height growth of trees, include studies with several species of *Quercus*. Most variation in height of *Q. rubra*, *Q. velutina*, *Q. coccinea*, and *Q. alba* from site to site is related to the fact that shallow, coarse-textured soils on steep ridges are conducive to annual water deficits (Trimble and Weitzman, 1956; Doolittle, 1957; Carmean, 1961; McClurkin, 1963).

E. Shoot Growth Under Perennial Drought

A final aspect of water stress and apical growth of forest trees should be mentioned in the closing paragraphs of this section. This is the survival and gross morphology of trees that grow on sites or in climates conducive to perpetually recurring water deficits. The adaptive mechanisms in trees that permit recovery following an occasional extended period of drought, or that tolerate frequent, brief periods without rain, are not necessarily the same as those that permit survival under annaully recurring, long-term droughts.

Species of trees that do not become established on perennially dry sites probably fail to survive past the early stages of seedling development. This reasoning has led to much ecological study of drought tolerance in seedlings. For example, Stransky (1963) and Pharis (1966) demonstrated that four species of *Pinus* had a remarkably similar capacity to recover from periods of drought up to 30 days, during which time water content of older foliage may be reduced to nearly half that at full turgor. Only one of the four species, *P. ponderosa*, is associated with perennially dry sites, but all of them are often subjected to brief periods of drought under normal conditions, and usually they recover. Stone (1957b) has suggested that dew is an important factor in the establishment and survival of *P. ponderosa* seedlings on dry sites, and it is likely that the internal distribution of dew absorbed through foliage is equally effective in survival of many other species (Stone, 1957a).

There are no tall trees on dry sites. Species that become established under conditions of perennial drought grow very slowly, often adding annual increments of growth so small as to be literally microscopic. A well-known extreme example of this condition, a stand of *Pinus aristata* in the White Mountains of California (Schulman, 1958), is composed of short trees with annual terminal growth of as few as two needle fascicles per shoot. The total amount of living tissue in these trees is very small, even though some individuals are thousands of years old. Water stress is virtually ever present on this site, following rapid depletion of snow melt water in the spring. Nevertheless, total photosynthesis is adequate to produce a small annual net dry weight increase.

Once established, therefore, trees on dry sites may be long lived, largely because of their proportionally small amount of living tissue, which is the result of very small annual increments. Water stress in such trees does not bring on early senescence and death, as hypothesized for rapidly growing, large trees on moist sites (Went, 1942; Kramer and Kozlowski, 1960). The so-called water requirements of trees (Roeser, 1940), transpiration ratios (Kramer and Kozlowski, 1960), or photosynthetic efficiencies of crop plants

(Lemon, 1965) have little meaning for tree growth on dry sites, because the efficiency of dry weight increment per unit of transpired water usually is evaluated with plants maintained in moist soil. Jarvis and Jarvis (1963) and Larcher (1965) emphasize the complexity of such relationships when tree seedlings are grown under varying conditions of water stress. Species with the most efficient transpiration ratios are not necessarily those which survive the rigors of perennial and severe environmental water stress.

J. Parker discusses related aspects of this type of drought resistance in volume I of this book.

III. WATER DEFICITS AND GROWTH IN GIRTH

Studhalter *et al.* (1963) reviewed early investigations of radial growth in trees and its response to the rise of sap in the spring and its cessation during periods of drought. Foresters have been aware of such gross responses to environmental water for several centuries and find their greatest challenge in wood production by controlling diameter growth of trees. Through regulation of tree spacing for root development, the forester controls the availability and depletion of soil water and thus regulates annual ring widths (Hartig, 1897, 1898, annotated by Larson, 1962; Craib, 1929; Savina, 1956; Zahner, 1959; Bassett, 1964a). The science of dendrochronology is based on the relation between dry weather and narrow growth rings in trees (Glock, 1955). Mac-Dougal (1924) recorded early in this century that radial dimensions of trees respond directly and quickly to changes in soil moisture, shrinking under drought conditions and immediately swelling following rain recharge.

It is in the perennial production of xylem tissue that woody plants differ most markedly from herbaceous plants, and therefore this aspect of tree growth is given detailed treatment here.

Initiation of cambial activity in the spring has received considerable detailed study (reviewed by Lodewick, 1928; Priestly, 1930; Wareing, 1951, 1958; Digby and Wareing, 1966; and by various authors in Kozlowski, 1962, and in Zimmermann, 1964a) but not with major emphasis on the effects of water stress. It is generally assumed that limiting water stress does not develop in cambial tissue early in the growing season in temperate regions (Kramer, 1964). Earlier workers attributed initiation of cambial activity to an increase in water content of the bole of the tree (e.g., Priestley and Scott, 1936), but it is probable that early spring rehydration of stem tissues occurs independently of growth. In normal temperate climates, photoperiod and temperature are apparently the controlling factors (stimulating hormone production, Digby and Wareing, 1966) in breaking the winter dormancy in cambial meristems after the initial swelling of its tissues (Fraser, 1952; Kozlowski and Peterson, 1962). Winter shrinkage occurs in stems of trees as response to low temperature (Winget and Kozlowski, 1964; Small and Monk, 1959) and to desiccation (Daubenmire and Deters, 1947). Both effects normally are past when growth is initiated in the spring, and even if stems remain partly dehydrated, Gibbs (1958) and many others have shown that late winter decreases in water content of stem wood are not as great as those that occur early in the summer when cambial activity is high.

It is still probable that the initiation of cambial growth is often delayed in forests of extremely cold regions by internal water deficits that develop on warm days when there is a lag in the absorption of water by roots in cold or frozen soil. Hygen (1965) and others present evidence that internal water stress in evergreen conifers during winter becomes severe, and recovery in early spring may be so slow that initiation of annual growth is affected. More-over, at the other extreme, in dry equatorial forests, it must be certain that water deficits regulate the annual initiation of radial growth, because periods of rainfall determine the growing season there. Little experimental evidence is available on the effect of water stress on the initiation of tree growth in any climate, however, and further comment here would be speculative.

It is after radial growth is well underway that water stress plays a highly effective role in the development of an annual ring. General reviews of the role of water in wood formation have stressed physiological controls of photo-synthesis, carbohydrate translocation, and hormone synthesis and transport (Zahner, 1963; Kramer, 1964). The present discussion will be limited largely to a review of the effect of water stress on the development of secondary xylem in trees and of the gross influence of drought and water deficits on radial expansion and tree ring dimensions.

In transverse section the prominent features across annual rings are the gradients in relative sizes of cells and the thickness of cell walls, regardless of whether the species observed is a vessel-containing angiosperm or a gymno-sperm with tracheids comprising most of the xylem (Priestly and Scott, 1936; Chalk, 1937). In ring-porous angiosperms, the production of large vessels of earlywood is completed during the first few weeks of the growing season. Some of the so-called "latewood" portion of annual rings of ring-porous species, therefore, may be formed relatively early, with a gradual change in appearance of this tissue throughout the rest of the season. In most diffuse porous species the transition across the ring is gradual, from large vessels and relatively large other elements, to the late-formed small vessels and other elements. In conifers, most species display well-defined zones of large-diameter and small-diameter tracheids, with thin and thick secondary walls, respectively, but the transition between these two zones varies from abrupt (as narrow as one cell wide) to extremely gradual across the entire ring. Harris (1955) relates the abrupt transition to summer drought. In turn, conifers with abrupt earlywood–latewood transitions vary from species such as *Pinus palustris*, which often contain wide bands of latewood, to species like *Sequoia sempervirens*, with latewood bands only one or two cells wide. It is well to keep in mind these diverse patterns of ring formation when considering the effect of water stress on cambial growth. Some generalities are ventured in the discussions below, but principles often must be modified to fit the ring architecture of individual species.

Development of phloem tissue in trees has not been studied specifically in relation to water stress, and thus this aspect is omitted here. Several important differences in the timing of phloem and xylem differentiation and in the rate of production of phloem and xylem mother cells make it difficult to draw analogies from one side of the cambium to the other. For example, in conifers and diffuse porous angiosperms annual initiation of sieve cell differentiation precedes that of xylem differentiation by a month or more (Cockerham, 1930; Abbe and Crafts, 1939; Evert, 1963; Tucker and Evert, 1964; Alfieri and Evert, 1965). Only in ring-porous angiosperms is there evidence that tissues on both sides of the cambium initiate differentiation together in the late winter or early spring (Lodewick, 1928; Artschwager, 1950). In some species phloem production continues in the fall long after xylem mother cells are dormant, and Tucker and Evert (1964) have found that sieve elements in *Acer* gradually differentiate over the winter domant period.

Except for the difference in the timing of initiation and cessation of differentiation in the two tissues, it generally can be considered that early maturing derivatives on either side of the cambium function largely in long-distance transport, and those derivatives maturing in the summer function largely in imparting strength to the respective tissues. Lodewick (1928) and more recently Evert (1963) and others report that formation of phloem sclereids and sclerified parenchyma occurs coincident with maximum production of xylem elements in early summer, following the formation of the earlywood portion of the annual xylem ring. Thus, summer drought should have little effect on the number of sieve cells produced, if activity of phloem mother cells is curtailed in proportion to that of xylem mother cells. The rate of tissue production on the phloem side of the cambium is far slower than on the xylem side during midseason in most species (Bannan, 1955; Newman, 1956; Wilson, 1964a,b; Zimmermann, 1964c), so that if summer drought affects mitosis in the cambial zone, this might well result in cessation of fiber-sclereid production. The effect of drought is not known, however, on the differentiation, or partial differentiation, of sieve cells late in the season in those species whose overwintering sieve cells must function in transport the following spring. It is possible that the intrinsic pattern of phloem development is sufficiently strong to overcome problems associated with water stress in late summer or early fall. Regarding the structure and properties of outer bark, no information is available on the effect of water stress on phellogen activity or the formation of cork tissue (Srivastava, 1964).

A. WOOD FORMATION

1. Cambial Meristem

There is practically no experimental evidence for a direct effect of water stress on the various stages of differentiation of derivative cells in the cambial region of trees. Yet living cambial tissues must be under moderate to severe water stress almost daily during the growing season because of the high tensile forces that develop in the adjacent mature xylem. Certainly rates of cell division and enlargement are reduced when internal water stress is severe enough to cause dehydration and shrinkage of the tissue containing mother cells and derivatives. Much circumstantial evidence suggests this is so from measurements and counts of cells in xylem tissue formed under conditions of environmental water deficits (Fig. 3).

Diurnal dimension changes of the bole indicate that the cambial region is under internal water stress for at least several hours each day during warm weather. The bole serves as a reservoir of water that is depleted when transpiration is rapid and refilled overnight by absorption from moist soil. Possibly most cell enlargement in the cambial meristem occurs during the night and early morning, and a healthy young tree produces diurnal spurts of growth that register on sensitive dendrograph traces as the difference between successive daily maxima so commonly reported (Kozlowski and Winget, 1964).

Mitotic activity by mother cells, however, has been shown not to conform to this diurnal pattern. Wilson (1966) found a late afternoon maximum rate of mitosis in the cambial region of pole-sized *Pinus strobus* during the spring period when overall tree growth was most rapid. He suggested that periods of rapid cell division reflect peak periods of translocation of mitotic regulating substances from leaves, but there could certainly be a time lag between the peaks in transport and in mitosis. Wilson noted diurnal fluctuation in mitoses confined to the portion of the main stem within the crown of the tree, suggesting that a peak in activity was damped out at greater distances from the leaves.

Studies of mitosis in the cambial region agree that the rate of cell division decreases sharply in midsummer (Bannan, 1962; Wilson, 1966). This decrease coincides in time with high internal water deficits, because the water reservoir of the stem may not be fully refilled overnight and the cambial region may remain continually under relatively severe water stress. There is overwhelming circumstantial evidence, as presented in the following sections, that this midsummer period of internal water stress does cause a direct reduction in the rate of cell division in the cambial meristem of trees. The physiological

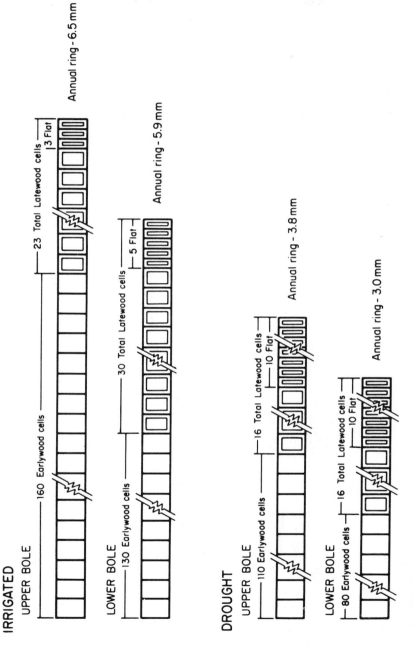

Fig. 3. Diagrams of single radial files of tracheids in the xylem rings of 20-year-old *Pinus resinosa* trees grown under irrigation and simulated drought. Redrawn from Zahner *et al.* (1964).

mechanisms and conditions that doubtless are responsible for this reduced growth are presented and discussed elsewhere (Zahner, 1963; Kramer, 1964; Kozlowski, 1964b).

The rapid production of new xylem derivatives is always associated with optimum soil moisture levels, rapid root and shoot development, and overall high vigor of the entire tree. At such times as many as 10 tracheids per week may be added to the radial file in *Abies concolor* and *Pinus strobus* (Wilson, 1963, 1964a), *Pinus resinosa* (Zahner and Oliver, 1962), and *Thuja occidentalis* (Bannan, 1955). Cell counts in conifer xylem show conclusively that the production of new tracheids slows and even ceases under drought conditions and may resume at a moderate rate when water stress is alleviated. The latter case is documented by Shepherd (1964), who measured a change in rate for field-grown *Pinus radiata* in Australia from less than 1 new cell per week under midseason drought up to 12 new cells per week following a soil water recharge.

2. Development of Xylem Tissues

Cell numbers, cell size, and cell wall thickness are the anatomical characteristics of xylem tissue most greatly influenced by water stress (Fig. 3). In the formation of the annual ring, all three of these features play an important role in determining the appearance and various properties of wood. The mature ring of all trees is the result of derivatives passing through successive stages of cell enlargement and wall thickening (Wilson *et al.* 1966), with each stage affected to some degree by water stress within the tree.

Normal development of the annual ring is apparently modified by summer drought in each of the three major zones of xylem differentiation, as analyzed specifically in *Pinus resinosa* by Whitmore and Zahner (1966; Fig. 4). Already discussed is the reduced rate of cell production in the mother cell zone, resulting in few cells and a narrow ring. Reduced also is the degree of radial enlargement of derivatives in the cell enlarging zone, resulting in radially flat cells (also tangentially narrow vessels in some angiosperms). Finally, in the maturing zone, the rate of secondary wall thickening is reduced also, but depending on the degree of water stress and the life span of derivatives, the net amount of wall thickening may range from less than normal to more than normal. Effects of drought on all three zones are both indirect through reduced crown and transport activities and direct through low water potential in the cambial meristem itself.

The development of angiosperm xylem has not been studied specifically in relation to water stress. Priestley *et al.* (1935), Priestley and Scott (1936), Wareing (1951), and others suggested that the large earlywood vessels in ring-porous species are not influenced greatly by the external environment of the tree, since these vessels are differentiated quickly and become mature

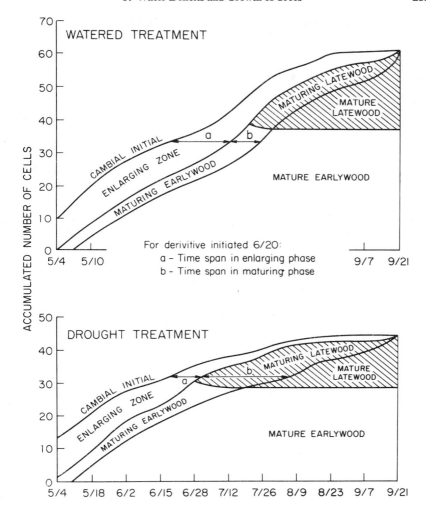

Fig. 4. Trends of seasonal development in the various zones of maturing tracheids in the xylem ring of 25-year-old *Pinus resinosa* trees. Lower curves: Drought imposed from June 1 to September 30. Upper curves: Trees watered when rain was less than 1 inch per week. Note that the time span of derivatives in the enlarging zone is reduced and in the maturing zone is increased by drought conditions. Redrawn from Whitmore and Zahner (1966).

water-conducting elements prior to foliage expansion in early spring. It is possible that the number and size of such vessels is at least indirectly influenced by water stress, however, controlled by the overall physiological vigor of the tree that is preconditioned by environmental stress and growth rates of previous growing seasons. Unpublished data of the author show that earlywood vessels in mature *Quercus rubra* trees on dry sites are 10–20% smaller in radial diameter than those from trees on moist sites. Savina (1956) also reported that vessel elements of *Quercus* were small in all dimensions when formed in trees under adverse water conditions.

Wood formation in the entire ring of diffuse porous angiosperms, and in the latewood portion of the xylem increment of ring-porous species, probably is affected by water stress in much the same manner as is annual ring formation in conifers (Priestley and Scott, 1936; Chalk, 1937). The all-tracheid wood of conifers has received considerable attention, both because of its relative simplicity and its great economic importance. For the purposes of this discussion, it is assumed that the effect of water stress on xylem differentiation is similar in many respects for both angiosperms and gymnosperms and that water stress *per se* may affect but does not determine the cell type. The first earlywood produced in each annual ring is quite different among species, but the maintenance of xylem differentiation throughout the growing season can be visualized as similar among species (Wilson *et al.*, 1966). Following initial earlywood, additional derivatives are produced in radial files from dividing mother cells and, whether they are destined to become vessels, tracheids, or fibers, they undergo the same phases of maturation under similar physiological controls.

Recent studies have described in detail the cell-by-cell development of xylem in conifers under varying conditions of water stress (Larson, 1963a; Shepherd, 1964; Whitmore and Zahner, 1966). Some of this work has been concerned with tracheid size in the radial dimensions, a characteristic of wood associated with the transition from earlywood to latewood. The time of the transition from wide to narrow tracheids in *Pinus* has been well correlated with the first period of soil moisture stress, whether it comes early or late in the growing season, both by circumstantial evidence (Chalk, 1951; Kraus and Spurr, 1961; Zahner, 1962) and direct measurement (Chalk, 1930; Zahner and Oliver, 1962; Shepherd, 1964; Zahner *et al.*, 1964; Whitmore and Zahner, 1966).

Oppenheimer (1945) and Larson (1963a, 1964) suggested that the earlywood–latewood transition in tracheid size is regulated by auxin levels and not directly by water stress. Larson (1963a) presents evidence that the effect of water stress on tracheid diameter in potted seedlings of *Pinus resinosa* subjected to drought is largely indirect through the reduced transport of auxin from shoots to xylem derivatives in the enlarging phase. However, Shepherd (1964) found a rapid and simultaneous reduction in the radial width of tracheids along the full length of boles of 30-year-old *Pinus radiata* undergoing severe drought (Fig. 5) and believed the response was too uniform to be consistent with a vertical auxin gradient. Shepherd suggested that a rapid decline in hydrostatic pressure within the tree as a result of dry soil can override the auxin factor in determining cell size. Kramer (1964) and Zahner (1963) also argue for this latter case.

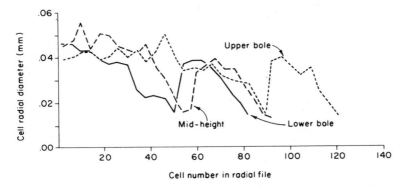

Fig. 5. The effect of severe summer drought on radial diameters of cells across radial files of tracheids in an annual sheath of wood at three positions along the main stem of *Pinus radiata*. Drought was alleviated by rainfall during the last part of the growing season, resulting in a false ring throughout the annual sheath. Redrawn from Shepherd (1964).

Experimental control of soil water levels in plantations of approximately 20-year-old *P. resinosa* has shown that the formation of narrow diameter tracheids can be hastened by more than 4 weeks in trees undergoing drought than in those irrigated all summer (Zahner *et al.*, 1964; Whitmore and Zahner, 1966). Unpublished data of F. W. Whitmore indicate that auxin levels in the cambial region of the trees under drought did not drop significantly below auxin levels of trees that were irrigated, yet the difference in diameter of tracheids at midseason was significant. Thus, low auxin levels need not be prerequisite for the failure of derivatives to enlarge fully, presenting further evidence that water stress can be a direct overriding factor in cell enlargement in cambial meristems of trees.

The role of water stress in thickening of secondary walls of xylem derivatives is at least severalfold, with each role complicated by both direct and indirect effects. The first requisite for deposition of new wall material is, of course, the presence of living cytoplasm within a maturing derivative. The length of time a new cell remains alive is obviously an initial governing factor in the wall thickening process. The role water stress, or lack of it, plays in maintaining the life span of a xylem derivative is not clear, but at first consideration the two seem anomalously related. In the spring, when water deficits are low, cytoplasm dies quickly and secondary wall thickening ceases abruptly. Later in the season, when water deficits are high, the cell remains alive for a longer period of time, often adding large amounts of cellulose. Figure 4 illustrates how summer drought conditions reduce the time span of a derivative in the enlarging phase and increase the time span in the maturing phase.

A current theory (Whitmore and Zahner, 1966) proposes that the life span of derivatives in the maturing phase of differentiation is related to competition, in the ecological sense, among individual cells in the various zones of the cambial meristem. When water stress is low and mother cells are actively producing new derivatives, older derivatives currently in the maturing zone are displaced at a rapid rate from the phloem, the source of carbohydrates and other substances, by the new derivatives entering the enlarging zone. As suggested by Shepherd (1964), the active cambial zone is the major assimilate sink, taking precedence over cells farther removed from the phloem, both by virtue of position and metabolic activity. Thus, derivatives in the maturing zone do not compete well for cell wall substrate and mature before exceptional secondary wall thickening occurs. As water stress develops in the tree later in the season, the zones of mother cell activity and derivative enlargement both become narrow in dimension and small in cell number. Their utilization of phloem substances is greatly reduced, and thus older derivatives in the maturing zone compete satisfactorily for cell wall substrate. Even though the rate of growth of the secondary wall may also be reduced by water stress, the alleviation of competition among cells is sufficient for considerable wall thickening to occur before maturity because derivatives remain alive for a much longer period of time.

That water stress in fact does have a direct effect on the growth of cell walls in xylem derivatives of trees has been shown by Whitmore and Zahner (1967). Tissues of developing xylem were excised from the main stems of *Pinus sylvestris* trees and incubated in solutions containing labeled glucose. Water stress induced with the osmotic agent polyethylene glycol caused significant lowering of glucose incorporation into tracheid cell walls. As water potential decreased from -3.1 atm, a level common in stems under favorable conditions, to -28.1 atm, a level common under conditions of environmental stress, incorporation of labeled glucose was reduced by over 50% (Fig. 6).

The net effect of water stress on secondary wall thickening in the intact tree growing under field conditions apparently varies considerably. Under moderate environmental water deficit, in which the rate of new cell production is merely slowed and derivatives are permitted a long life span, many exceptionally thick-walled xylem cells may be formed in the ring. At the extreme, under severe water stress which essentially stops production of derivatives by mother cells and also strongly reduces the rate of wall assimilation in the few derivatives remaining in the maturing phase, the final ring may contain only a few moderately thick-walled cells. The latter condition is caused by the combined direct inhibition of wall growth by water stress arising from tension in adjacent water-conducting xylem plus the indirect effect of decreased photosynthesis and phloem transport of growth substances to the cambial region.

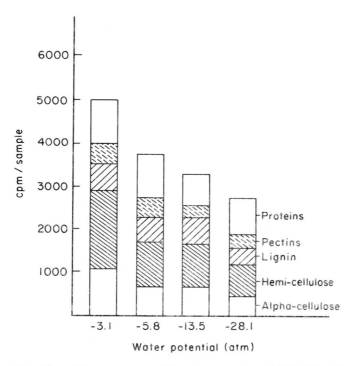

Fig. 6. The effect of low water potential on incorporation of ^{14}C-labeled glucose into cell wall constituents of differentiating xylem in 20-year-old *Pinus sylvestris* trees. Redrawn from Whitmore and Zahner (1967).

3. Wood Properties

The influence of environment on wood quality has received considerable attention since the work of R. Hartig and other German forest botanists at the end of the nineteenth century. An annotated bibliography on this subject has been published recently by Larson (1962). Studies of the effect of water stress have been limited almost exclusively to conifers, with emphasis on such properties as wood specific gravity and percentage of latewood in the annual ring. Subjective inferences concerning the relation between water in the tree's environment and wood structure have been made many times (e.g., Paul and Martz, 1931; Priestley and Scott, 1936; Chalk, 1951; Wellwood, 1952; Spurr and Hsuing, 1954; Savina, 1956; Larson, 1957; Goggins, 1961; Kennedy, 1961, Gilmore *et al.*, 1966), but few studies have as a stated objective the rela- tion between internal water deficits and properties of wood. Moreover, for convenience nearly all studies of annual ring anatomy in large trees until recently have been confined to breast height, although the effect of water on xylem development is quite different in the upper than in the lower stem (Smith and Wilsie, 1961; Zahner and Oliver, 1962). Intrinsic variation in wood pro- perties throughout the bole of a tree masks the true effect of environmental stress, and studies of wood formed in either young trees or at breast height in fully mature trees have resulted in conflicting conclusions (Richardson, 1961). The significant findings reviewed below have resulted only after strati- fication of the bole into annual ring sequences along various axes to remove intrinsic variation due to age and size.

In an analysis of entire annual sheaths of wood from ground level to the tops of 55-foot-tall trees of *Pinus taeda*, Smith and Wilsie (1961) showed conclusively that soil water deficits have a marked effect on several wood properties. Annual increments produced during years of moderately low summer water deficits were characterized by wide latewood zones at the base of the increment and narrow latewood zones at the apex (Fig. 7). Under

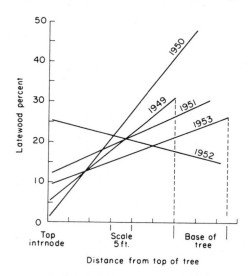

Fig. 7. Relationship between percentage of summerwood in *Pinus taeda* trees and distance from apex of stem for individual sheaths of annual increment over a 5-year period. 1950 was an exceptionally rainy growing season, with a calculated summer water deficit of only 1.0 inch; 1952 was a season of extreme drought, with a summer water deficit of 15.0 inches; deficits of other years shown fall proportionately between. Redrawn from Smith and Wilsie (1961).

drought conditions, the latewood zones were modified to uniformly narrow ones throughout the increment, even tending to be narrower at the base than at the apex. All wood properties associated with latewood showed this same pattern, that is, xylem anatomy in the lower bole was far more strongly influenced by water stress than that in the upper bole. Thus, during midsummer periods of water stress, wood formed in the lower stem contains far fewer total latewood cells, each with smaller radial dimensions and thinner secondary wall, yielding an annual ring with lower specific gravity, than that formed in the same stem position during years with more favorable moisture conditions. High in the crown, stem wood properties associated with latewood are not so drastically affected by drought, because of the intrinsic pattern of wood formation there. Lying close to the source of photosynthate and other crown-produced substances, the upper stem even under normal conditions produces few latewood cells, and these are formed late in the season. The effect of summer water stress in the upper stem is to telescope both earlywood and latewood proportionally, with the result that the percentage of latewood in the annual rings of the upper stem is little changed under natural year-to-year variation in moisture conditions. The lower stem, on the other hand, is affected in the extreme by reduced phloem transport under summer drought, and the annual ring formed there shows a sharp contrast between a normal earlywood band and the narrow latewood band resulting from summer water stress.

The wood formed by juvenile conifers is in some respects equivalent to that formed in the upper stem of old trees. In stems of young *P. taeda*, however, the width of the latewood band was found to be the same in trees subjected to both extreme drought and adequate watering (Zahner, 1962). In these young trees the percentage of latewood formed under drought conditions was significantly greater than that formed under irrigation, the difference resulting from a far greater quantity of earlywood formed under irrigation. Annual rings of irrigated trees were more than twice as wide as rings of trees under drought; since most of this difference was in the earlywood portion of the ring, it was concluded that the changeover to latewood was hastened by drought conditions. Thus, the equal bands of latewood for both treatments were each the result of separate mechanisms. Under drought, latewood was initiated early and few cells were formed during the remainder of the year. Under irrigation, the transition to latewood was delayed until near the end of the season when few additional cells were formed before normal fall dormancy.

Favorable water relations late in the growing season have been associated with increased width of the latewood band in several species of *Pinus* in the southern United States (Lodewick, 1930; Coile, 1936; Foil, 1961). These studies were made of wood in the lower boles, which position, for reasons of distance from the crown, apparently does not usually revert to the production of true earlywood when there is a return to favorable growing conditions late in the summer. It is probable that wood formation in the upper bole of such trees often does revert temporarily to earlywood production, and a false ring is the result. False rings are formed in lower boles in extreme cases of severe drought early in the season followed by high rainfall late in the season. Ladefoged (1952) and Shepherd (1964) described false rings in the lower boles of conifers, associated with severe summer drought later alleviated by the thorough recharge of soil moisture (Fig. 5).

A study of wood formed in the lower boles of 100-year-old *Pinus ponderosa* trees in dry climate of eastern Washington confirms the effect of summer water stress on wood properties associated with latewood in conifers (Howe, 1968). Alleviation of dry weather by midsummer irrigation over a 7-year period resulted in up to 40 % greater quantity of latewood than in trees not watered. The quite typical narrow band of latewood associated with ponderosa pine is therefore in part the result of environmental water stress and is not wholly an intrinsic species characteristic.

In other studies, wide bands of latewood in the lower boles of old trees have been artificially stimulated by summer irrigation (in 200-year-old *Pinus palustris*, by Paul and Martz, 1931), but not in young trees (20-year-old *Pinus resinosa* by Zahner *et al.*, 1964). This difference associated with age is probably the result of the vigor of crown meristems and of the distance from the crown to the position where increase in latewood was measured, as already mentioned. Only in lower boles of mature trees, apparently, does meristem activity slow sufficiently by midsummer for cell-wall assimilation to produce latewood tracheids in quantity when under irrigation treatment. As long as serious environmental water stress does not develop, the upper boles of conifers may continue to produce earlywood while the lower boles have already shifted to latewood production. Such conflicting responses associated with tree size illustrate why it is unsound to generalize from wood formed by tree seedlings under controlled environmental stress to wood formed by mature trees.

Latewood of all species of *Pinus* is not affected by summer water deficits. Foulger (1966) found no change in the quantity of latewood cells formed in *P. strobus* under varying conditions of water deficits. A thorough analysis of complete stems of 50-year-old trees revealed that the intrinsic characteristic of this species is to produce only a few (less than eight per radial file) latewood cells each year. This pattern held for all water stress conditions that occurred over 50 growing seasons. Even though total ring widths and total number of earlywood tracheids varied widely with water deficits, resulting in considerable variation in the percentage of latewood, total numbers of latewood cells remained constant. A distinction must be made, therefore, between the pattern of the annual rings formed by species like *P. strobus* and that of the other conifers discussed above. In this case, the percentage of latewood was highest during years of greatest water deficit.

In ring-porous *Quercus*, Savina (1956) reports the opposite type of early-wood–latewood relationship with water deficits. As defined earlier, "early-wood" in ring-porous angiosperms is more or less a constant in each annual increment, and therefore it is the latewood portion that is either extended by favorable growing conditions or curtailed by drought. Savina thus found that the percentage of latewood was lowest in trees most subject to greatest water deficits. Chalk (1930) and Priestley and Scott (1936) discuss this aspect of late-wood production in ring-porous woods.

Wood specific gravity is an important economic property closely related to the percentage of latewood in all tree species. Within the same tree, it is well established by most of the studies cited above that year-to-year variation in both features is correlated with water deficits. However, the variation in specific gravity from site to site may well indicate an anomoly. It was shown above that in several species of conifers the annual increment within the same tree, particularly in the lower bole, has a lower specific gravity when produced during drought years than when formed under favorable conditions (Smith and Wilsie, 1961). A study by Gilmore *et al.* (1966) confirms that environmental conditions related to the availability of moisture during the summer account for most of the variation in wood specific gravity of *Pinus taeda*, with lower specific gravities on drier sites. Several other studies have indicated that conifers on drier sites, however, contain wood of higher specific gravities than that grown on moister sites [in *Pseudotsuga menziesii* by Paul (1950), and by Wellwood (1952); in *Pinus banksiana* by Wilde *et al.* (1951); in *P. taeda* by Zobel and Rhodes (1955), and by Hamilton and Harris (1965); in *P. elliottii* by Larson (1957); in *P. resinosa* by Jayne (1958)]. In such cases the effect year after year of earlier transition to latewood on dry sites results in a high proportion of tracheids with small radial dimension and thus proportionally low lumen volume per unit volume of wood. The overall volume of earlywood produced on dry sites is not as great as that on moist sites, with the result that annual bands of latewood in dry-site trees are spaced close together with a net high specific gravity. These and other studies confirm that tracheid size is smaller and more compact on dry than on moist sites (Kienholz, 1931; Hamilton and Harris, 1965).

Kennedy (1961) found no correlation between site and latewood percentage of four annual rings in *Pseudotsuga menziesii*. He gives several clear examples of wide variations in earlywood among the 4 years, correlating relative amounts of earlywood and latewood with rainfall and temperature combinations before and after the initiation of latewood. For example, the annual ring with the largest percentage of latewood was formed under conditions of high internal moisture stress in May, resulting in a small amount of earlywood, so that only a moderate amount of latewood resulted in a high percentage of latewood. In Kennedy's study, such strong effects of individual seasons masked the underlying, more subtle effect of site that could not be manifested in only 4 years.

Wood specific gravity in diffuse porous angiosperms probably has the same complex relation with site and environmental water deficits as in conifers. In ring-porous species, however, the density of wood is lighter on dry sites that produce narrow growth rings than on moist sites that produce wide rings, because all ːings formed on dry sites contain proportionally more vessels (Priestley and Scott, 1936; Hale, 1959). Two groups of papers by R. Hartig (1894–1895) and H. Burger (1947–1950), annotated by Larson (1962), discuss further the relation between site and wood density in a variety of angiosperms and gymnosperms.

Tracheid and fiber lengths as affected by environment have received some attention, again because this property influences wood utilization. Hamilton and Harris (1965) report shorter tracheids in *Pinus elliottii* on dry sites than on moist sites, an observation that confirms several earlier studies in conifers (annotated by Larson, 1962). In angiosperms, Mell (1910) with *Juglans*, Savina (1956) with *Quercus* and *Populus*, Kennedy and Smith (1959) with *Populus*, and others have shown that fibers and other wood elements are significantly shorter when formed in trees growing on dry than on moist sites. No reports are available that the cambial initials themselves are shorter in trees on dry sites than in those on moist sites, although they may well be. Certainly, the effect of low water potential in the cambial region must restrict full expansion at the tips of enlarging derivatives. This direct water stress factor on tip growth of cambial derivatives must greatly overcompensate for the usually accepted explanation that narrow growth rings in fact contain longer elements because of fewer pseudotransverse divisions by the cambium. [The complex relationship between tracheid length and pseudotransverse cambial growth rate is explained by Larson (1963b), drawing heavily from many papers by Bannan.] Pseudotransverse divisions must have little effect on the average length of xylem elements as influenced by water stress because the frequency of such divisions is extremely low, less than 2 % of total divisions in fast-growing trees (Wilson, 1964a) and fewer in slow-growing trees. Thus, either cambial initials themselves are slightly shorter or normal tip growth is restricted, or probably both, in trees growing under water stress.

4. Integration of Terminal and Cambial Activities

As emphasized frequently in this review and by Larson (1963c, 1964), Kramer (1964), and Fritts (1966), the amount and type of growth produced by the cambial meristem is controlled by intrinsic characteristics and environmental conditions of the foliage and terminal meristems. The various effects of water stress on physiological processes of foliage are discussed elsewhere in this volume and by Kozlowski (1964b), and most of them have at least an indirect role in wood formation. Two results of water stress in foliage have been emphasized in the preceding review of cambial activity. First, certainly the quantity of photosynthate reaching the cambial region is of primary importance in wood formation (Zimmermann, 1964c; Kozlowski and Keller, 1966). Roberts (1964) has shown that both rates and amounts of transport of carbon-14-containing substances from photosynthetically assimilated $^{14}CO_2$ were reduced tenfold as leaf water deficits changed from 5 to 20% (by the Stocker method) in potted *Liriodendron tulipifera* trees. Second, certainly the water stress imposed directly by transpiring foliage on dividing and differentiating secondary xylem is of serious consequence, as indicated by the pattern of shrinkage in tree boles (Kozlowski and Winget, 1964; Worrall, 1966; Kozlowski, 1967). Unpublished data of the author show that more than two thirds of diurnal shrinkage of stems in four genera occurs in the living tissue of the cambial region as the mature xylem comes under transpirational-induced tension. Worrall (1966) measured an inverse linear relationship ($r = 0.98$) between the amount of stem shrinkage and the xylem sap tension in two tree species.

Thus, both the supplies of substrate for new xylem cells and the immediate tissue environment in which cells develop are regulated by water relations of leaves. When shoot growth and leaf development are curtailed by water stress, as reviewed in the first part of this chapter, it follows indirectly that there will be some effect on wood formation. The effects are both immediate and long lasting. Perhaps needles of an evergreen conifer, for example, fail to reach normal length because of a summer water deficit. The net amount of photosynthate reaching the cambial region is drastically reduced during the period of stress and is probably reduced at least below the normal potential for several subsequent growing seasons.

While shoots are expanding early in the season, there is a large drain on carbohydrates (Ziegler, 1964; Kozlowski and Keller, 1966), which doubtless has its effect on cambial activity. Larson (1964) and many others suggested that the availability of photosynthate for xylem cell-wall thickening increases substantially after shoot growth slows, and this in turn partially regulates the production of thick-walled latewood in the annual ring.

Internal competition for current and stored photosynthate between newly developing shoots and new xylem formation varies by intrinsic patterns of the meristems, and thus this effect of water stress in one species may not be similar to that in another. In ring-porous *Fraxinus* and *Quercus*, for example, the initial vessels must develop from stored substrate, because they are essentially formed prior to leaf flushing. As shoots flush, there must be a heavy drain on stored foods in the terminal direction, a direct competition with currently developing xylem. But in these genera the shoot flush is complete within a very short span of the growing season, and most of the remainder of the annual ring may develop relatively free of competition from the shoots. Phipps (1961) measured a distinct slowing of radial expansion in *Quercus* immediately following the expansion of earlywood vessels. Priestley and Scott (1936) reported that no latewood fibers formed in ring-porous species until late spring, and it is possible that the pause in cambial activity between differentiation of earlywood vessels and the production of fiberous latewood is related to intense competition between terminal and cambial meristems at this time.

At the other extreme, *Liriodendron* and *Betula* produce new foliage continuously throughout most of the growing season, which implies that there exists internal competition for stored foods and current photosynthate during the entire development of the annual ring. Thus, the stored food–current photosynthate relationship should be different in these two genera than in *Fraxinus* and *Quercus*. The role of foliar water stress on the formation of wood, therefore, even when viewed broadly is quite complex, and mechanisms certainly take different forms in different angiosperms. Intrinsic hormone relations between foliage and ring-porous vs. diffuse porous xylem formation are also complex (e.g., Digby and Wareing, 1966), and synthesis and transport of hormonal substances are unquestionably affected by water stress. Virtually nothing is known, however, from experimental evidence of the effect of water stress on the interactions between shoot and cambial activities in angiosperms.

Knowledge of the mechanisms of water stress is fairly advanced, in gross terms, for integration between shoot and wood formation in the simple all-tracheid conifers. Fritts et al. (1965) and Fritts (1966) present a biological model (Fig. 8), e.g., for the interactions between internal and external water conditions and between wood formation and shoot and needle production in *Pinus edulis, P. ponderosa*, and in several other conifers. Their model is based on field measurements of factors contributing to the formation of a narrow annual ring: beginning with low precipitation and high temperatures, and working through low soil moisture, rapid evaporation, increased water stress in the tree (not measured), decreased needle and shoot production, reduced photosynthesis, decreased concentrations of hormones (not measured), shorter period of cambial activity, radially narrow cells, less assimilation of cell parts, and finally the net production of a narrow ring. At least parts of such a scheme are oversimplified, however, because Whitmore and Zahner (1966) established that lack of expanding shoots *per se* do not contribute to the production of a narrow annual ring in *Pinus resinosa*.

In seedling *P. resinosa*, Larson (1963a) associated water stress with a decrease in auxin content of the new shoots. There is no evidence that this effect reaches the lower bole of trees beyond the seedling stage. Whitmore and

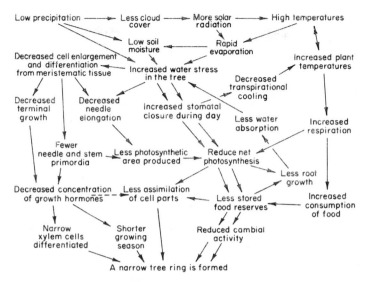

Fig. 8. Diagram of Fritts' (1966) proposed model for the effect of hot, dry weather on the formation of a narrow growth ring.

Zahner (1966) reported that extractable auxin in the cambial region of 20-year-old *P. resinosa* trees remained high in the main stem throughout a drought period that severely curtailed shoot and needle elongation and radial growth (Fig. 9). They found auxin content to be highly variable during the period of most active growth and becoming more uniform as shoot activity ceased, but found no significant difference in auxin levels between trees under drought and those irrigated.

Correlations probably exist between leaf development and differentiation of phloem tissue, as suggested by descriptive evidence but not yet supported by experimentation. Phloem transport theory insinuates that new secondary sieve cells in the stem be continuous longitudinally into new leaves. Thus, if primary growth is curtailed by water stress in a species that normally produces recurrent flushes throughout the growing season, then new secondary phloem tissue probably is also indirectly curtailed in proportion. Moreover, the amount of secondary phloem tissue that differentiates into functional transporting cells one year in other species may be regulated indirectly by the level of water stress of the previous year, if the number of primordia for new leaves are preformed and regulated by the vigor of the tree in the previous year. If fewer than normal sieve cells are produced currently because previous year drought affects the number of current leaves, then perhaps more than normal sclerified cells are differentiated in the same current phloem ring if water stress is not limiting to continued cambial activity.

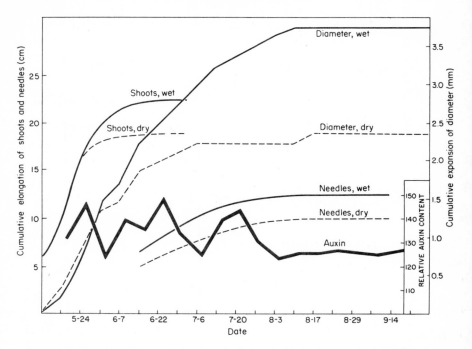

Fig. 9. Seasonal patterns of auxin levels in the vascular cambium, diameter growth, and shoot and needle elongation in *Pinus resinosa* trees under imposed drought (dry) and irrigation (wet) for one complete growing season. Auxin is expressed as a percentage of the final length of controls in a straight growth bioassay of first internode *Avena* segments. The plotted line is the average of upper and lower boles of four trees in each of the dry and wet treatments. There were no significant differences in auxin between positions in bole or between treatments. Unpublished data of F. W. Whitmore.

B. SOIL MOISTURE AND GROSS RADIAL EXPANSION

MacDougal (1921, 1923) and other early workers suggested that water deficits on some sites are responsible for more within-season variation in cambial growth than any other factor. More recently, Friesner and Walden (1946), Warrack and Joergensen (1950), Byram and Doolittle (1950), Dils and Day (1952), Buell *et al.* (1961), and others also concluded that midseason growth slackens and bole shrinkage occurs widely in many species after short periods without rain recharges. Kozlowski and Winget (1964), moreover, measured wide variability among species and positions on the stem of the same tree in the magnitude and timing of radial increment and decrement in relation to soil moisture recharges and depletion. Ring-porous *Quercus*, for example, showed less amplitude than nonporous *Pinus*, although both were consistent and in the same direction. They noted that the upper bole of *Populus* rehydrated far more rapidly than the lower bole following sudden rainfalls, probably reflecting absorption through foliage as well as through roots.

Studies of day-to-day variation in radial expansion are conducted with several types of dendrometers attached to the outside of the lower bole [Bormann and Kozlowski (1962) reviewed traditional dial gages and vernier bands; Fritts and Fritts (1955) devised a mechanically recording dendrograph; Impens and Schalck (1965) present a sensitive electric dendrograph], in much the same manner as MacDougal had done 30 years before, and in most cases soil moisture is estimated indirectly through daily records of precipitation. Several studies over the past 15 years have virtually proved the responsive relationship between soil water and girth changes through simultaneous measures of daily or weekly soil moisture and diameter increments and decrements. The more significant of these reports are reviewed below.

The first accurate weekly records of growth actually related to measured soil moisture regimes in the field were reported by Boggess (1953, 1956) for *Pinus echinata* and *Quercus alba*. Without exception, over a 4-year period cumulative basal area expansion in both species slowed when approximately half the available water was depleted from the surface 3 feet of soil. Expansion of trees ceased when three fourths of the available water was depleted, and bole shrinkage occurred under drier conditions. Such field studies of trees in established forests that are fully occupying their sites do not permit more than a qualitative estimate of the internal water relations of the trees, from even the most accurate measures of soil water. However, it is safe to assume that reduced radial growth and eventual shrinkage of the bole are definitely brought on by water stresses, because the tree responses in such studies occur simultaneously with the rapid depletion of soil water regardless of when during the growing season water is scarce (Kozlowski, 1965, 1967). In Boggess' examples, in some years both tree species ceased radial expansion in late June, whereas in a year with high levels of soil water recharge, rapid basal area increment was measured until mid-August (Fig. 10). In oak, for example, approximately twice the basal area increment was recorded for a wet year as for a dry one. During one season, the pine responded to a late-season soil water recharge and resumed rapid radial growth following a 3-week period of bole shrinkage. The oak, however, on a nearby site, failed to respond past midseason, except for immediate swelling of the bole during the rain recharge. This species difference reported by Boggess may be related to permanent desiccation damage in foliage or roots of the oak but not of the more drought-resistant pine.

McClurkin (1958), with *Pinus echinata*, and Fritts (1958, 1960), with *Fagus grandifolia* and two species of *Quercus*, used multiple regression techniques to test relationships between daily levels of environmental factors, including soil moisture, and daily radial enlargement in the lower boles of trees. Both investigators found the expected direct correlation of reduced growth with low levels of soil water availability, but in no case was a high proportion of the daily growth variation accounted for by the water factor. Their methods utilized radial enlargement on single days as the dependent variable, which was correlated with elapsed days of the growing season and with numerous

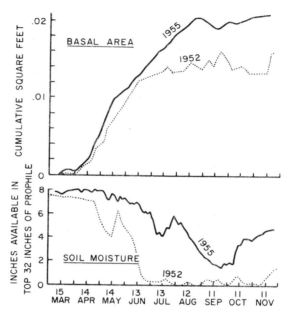

Fig. 10. Basal area growth per tree of *Pinus echinata* and trends of available soil moisture for relatively wet (1955) and dry (1952) growing seasons. Note that growth rate slowed in mid-June of 1952 but not until mid-August of 1955, each year when available soil water had been depleted to about 2 inches. Redrawn from Boggess (1956).

independent environmental variables. Since most environmental factors, as well as tree growth, are highly correlated with season, the effect of elapsed days alone absorbs much of the effect of such variables as soil moisture. For example, radial expansion on a day in midsummer may be restricted primarily because of a lack of available water, yet the full impact of the water factor is lost in the regression analysis because the number of elapsed days, as an independent variable, absorbs much of the effect of soil moisture. Consequently, the day of the season *per se* appears to be the causal factor, when in fact low levels of water availability at that season have resulted in reduced growth. Neither does the method permit the accumulated effect of soil moisture stress of previous days to be evaluated with the current day's growth. Thus, it is not surprising to find that the effects of independent variables, operating within seasonal sequences of days, are not highly associated with growth for a given day. For examples, McClurkin (1958) found that not more than 15% of the daily variation in radial enlargement was associated with moisture changes in the root zone of the pine plantations studied, and Fritts (1960) found that maximum daily temperatures were directly correlated with the daily growth of all species studied, even though high temperatures in midsummer contribute to internal water stress.

Radial growth in the lower bole of 16 species or northern hardwoods and conifers was reported to be closely related to soil moisture by Fraser (1956, 1962) in Canada. Several growing seasons out of the 10 years studied by Fraser were considered dry in June and July, confirmed by the detailed measurement of soil moisture. The radial increment of all species, including ring-porous and diffuse-porous angiosperms as well as gymnosperms on all sites, slowed following periods of rapid soil moisture depletion, with trees ceasing enlargement several weeks earlier than in more normal years. *Picea glauca* and *Betula lutea*, for example, ceased radial growth early in July on dry sites, and in late July on moist sites, whereas both species normally grow until late summer. Radial growth of most species studied was about half as great during dry years as during wet ones. Wet years were characterized by the absence of prolonged periods of soil moisture depletion between rain recharges.

Phipps (1961) recorded precise changes of radial increment and decrement in several species of angiosperms over a 5-year period in Ohio. High air temperatures during periods of low soil moisture in midsummer were consistently associated with temporary, and sometimes permanent, cessation of growth in all species. Soil moisture early in the season was never low enough to limit growth, and rain recharges late in the season extended the period of growth. Phipps' measurements of radial expansion of trees and soil moisture summarize the typical response of trees growing in fully stocked stands on upland sites in temperate climates; water stresses are not serious prior to midsummer, at which time it is normal for absorption by roots to lag far behind transpiration, and the resulting dehydration of tissues in the crowns and stems causes important limitations in growth below the potential for that time of year. If the water stress is alleviated by late-season rains, radial growth usually resumes if the mid-season water deficit has not been severe.

The effect of elimination or reduction of competition by silvicultural thinning has been studied specifically from the stand point of soil moisture and radial growth, particularly in pine plantations. It is well documented that thinning and wide spacing of trees in a forest stand result in less rapid depletion of water and thus higher levels of soil moisture throughout the growing season (Zahner, 1959; Della-Bianca and Dils, 1960; Zahner and Whitmore, 1960; McClurkin, 1961; DeVries and Wilde, 1962; Harms, 1962, Bay, 1963;

Bay and Boelter, 1963; Bassett, 1964a, Barrett and Youngberg, 1965). There is no question from these studies that water stresses that normally occur in trees in fully stocked, unthinned stands are alleviated by thinning. Radial growth occurs in residual trees at faster rates for longer periods in thinned

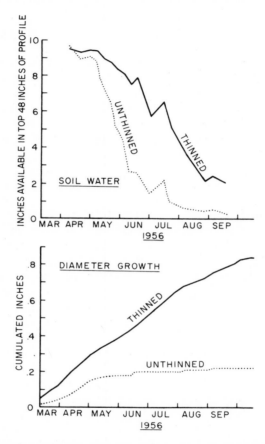

Fig. 11. Trends of soil water depletion and diameter increment per tree for average dominant *Pinus taeda* trees during one growing season, thinned plots and unthinned plots. Redrawn from Zahner and Whitmore (1960).

over unthinned stands, at least in part because of the greater availability of soil moisture throughout the growing season. For example, at the time radial expansion was ceasing in stems of young *Pinus taeda* that had not been released of competition, growth was occurring at a rapid rate in stems left widely spaced following a heavy thinning (Zahner and Whitmore, 1960).

Available soil moisture was nearly exhausted by midsummer under closely spaced trees, while at the same time water was not depleted halfway to the wilting point under widely spaced trees (Fig. 11). Total 5-year radial growth of individual trees under normal competitive conditions of full stocking was only 40% of that of residual trees in thinned stands. Bassett (1964a) also found that thinning in *Pinus taeda* stands alleviated midsummer water stress and that residual trees on heavily thinned plots grew throughout the season. Trees remaining on lightly thinned plots ceased radial expansion when soil water was depleted below 50% of that available and when calculated potential evapotranspiration was high.

C. RAINFALL–TREE RING CORRELATIONS AND OTHER INDIRECT MEASURES

A limited review of selected recent literature will illustrate that tree ring widths are readily correlated with the intensity of summer droughts. Studhalter *et al.* (1963) review in detail and Agerter and Glock (1965) annotate numerous attempts to correlate weather records with annual rings from stem analysis. Tryon *et al.* (1957) and Fritts (1958) present brief and pertinent reviews. As stem analyses, measures of drought, and analytical procedures become increasingly exacting and precise, they permit as much as 90% of the annual variation in ring width and basal area increment to be attributed to fluctuations in water stress.

The 100% correlation of exceptionally narrow individual rings with extremely dry growing seasons is common (for example, Lodewick, 1930; Glock, 1955; Phipps, 1964). It is now well accepted that drought years leave their record in the growth rings of trees.

Especially significant are studies such as that of Fritts (1962) that yield quantitative relationships between environmental water stress and tree rings. Of the 50 outer rings in the lower bole of mature trees of *Quercus alba* and *Fagus grandifolia*, up to 80 and 70%, respectively, of the variation in ring width was associated with rainfall and calculated water deficits. Fritts found, moreover, that periods of drought in both the previous and current years were equally effective in reducing the total width of the ring in diffuse porous *Fagus* but that the width of only the earlywood portion of the ring in *Quercus* was reduced by dry conditions of the previous year. Therefore, approximately one third of the annual variation in diameter growth of forest-grown beech was the result of water stress of the previous year (particularly August drought), one third of water stress of the current year, and approximately one third of other causes. In oak, about two thirds of diameter growth variation was the result of current year (June–July) water stress alone.

A 2-year effect of rainfall on the width of annual rings has been reported in *Quercus* by Bogue (1905), Robbins (1921), and Tryon and True (1958) and in *Fagus* by Diller (1935), all stressing inverse correlation between the severity of summer droughts and growth. Tryon and True found that a drought in 1949 severely restricted the growth of *Q. coccinea* the following year, 1950, which had more than normal rainfall during the growing season. A series of droughts every second year, 1951, 1953, and 1955, resulted in a pronounced decline in annual ring width over an 8-year period.

In conifers, up to 90% of variation in width of annual rings has been attributed to water stress in semiarid climates (Douglass, 1919; Fritts *et al.*, 1965) and up to 80% in humid temperate climates (Zahner and Donnelly, 1967). Climatic conditions in southwestern Colorado that contribute to narrow rings in mature *Pseudotsuga menziesii* trees were found by Fritts *et al.* (1965) to be a hot and dry previous summer and a dry current winter, spring, or summer. The unusual finding that winter droughts were so closely correlated with narrow rings of the subsequent growing season was explained by the dependence of annual recharge of soil moisture on water from snow melt in the area studied.

Zahner and Donnelly (1967) concluded that correlation coefficients of 0.80–0.90 are the upper limits for relating single environmental factors, such as water, to annual variations in radial growth of individual trees (Fig. 12). They reported that 14% of the variation in ring width of young *Pinus resinosa* in Michigan over a 10-year period was associated with moisture stress conditions of the previous season (July–September) and 68% was associated with moisture conditions of the current season (May–September). Accumulated daily water deficits, calculated from both rainfall and temperature records, for these critical periods were somewhat more highly correlated (inversely) with radial growth than were rainfall data *per se*.

Annual basal area growth of *Pinus monticola* in Idaho was closely related to moisture stress variables of both previous and current years (Zahner and Stage, 1966). Calculations of daily water deficits, from temperature and rainfall records over 40 growing seasons, combined with daily trends of other weather variables, were treated as functions of time over two growing seasons for each annual increment of basal area. The coefficients of the time trends, as independent variables in regressions of growth, accounted for 78% of the total variation in basal area increment.

The only near-perfect correlations reported between annual tree growth and indirect estimates of water stress were found by Bassett (1964b). He utilized a highly realistic approach to calculating the magnitude of growth permitted each day by prevailing soil moisture conditions. Indices of daily growth potential were calculated, based on availability of soil moisture, by estimating soil moisture stress for each day of the growing season over a 21-year period. The accumulated number of days, and portion of days, during which growth could occur gave a correlation coefficient of 0.97 with periodic basal area

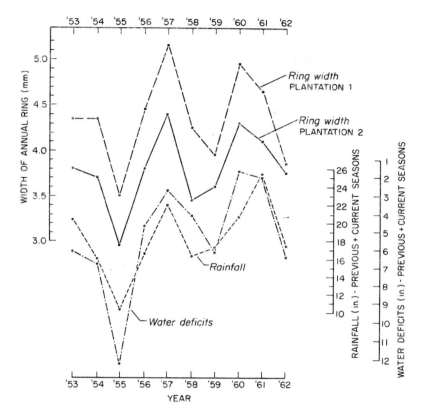

Fig. 12. Relation between widths of annual rings of *Pinus resinosa* trees in two plantations and rainfall and calculated water deficits over a 10-year period. Rainfall plotted for each year is the total of July 1–August 31 for the previous season plus May 16–September 15 for the current season. Deficits plotted are the total of August 16–September 30 for the previous season plus June 1–September 30 for the current season. Redrawn from Zahner and Donnelly (1967).

increment (Fig. 13). This excellent correlation between essentially a single factor and tree growth was not with ring width variation but with per acre production of all trees in a 100-acre all-aged stand of *Pinus taeda* and *P. echinata* in Arkansas. When soil moisture was estimated as limiting tree growth to only 50 days (or accumulated fractions of days) per year, annual basal area growth was 1.2 square feet per acre. When growth could occur for 100 days per year, annual basal area increment was nearly 3 times as great. Bassett's study offers very significant quantitative proof that annual forest productivity is strongly curtailed by normal summer water deficits. The higher correlations of growth and water deficits reported by Bassett over other reports in the literature probably result from his measure of the average response of the entire forest, which smooths out the excessive variation of single trees. Most correlations between tree growth and soil moisture or weather factors are limited to a small number of trees, usually dominants that reflect far greater annual variation than that of the average stand (Kozlowski and Peterson, 1962; Winget and Kozlowski, 1965).

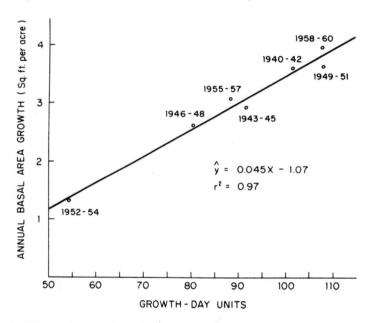

Fig. 13. Relation between annual basal area growth for a stand of Southern pines (*Pinus taeda* and *P. echinata*) and "growth-day" units per season. Growth units are the accumulated number of days and portions of days during which it was calculated that soil moisture availability permitted radial growth. Redrawn from Bassett (1964b).

Irrigation has been employed in several field studies of forest tree growth as a means of indicating the magnitude of radial growth that might occur if normal, incipient environmental water deficits were alleviated. A few studies are cited here, not to illustrate the effect of additional water so much as to emphasize that water stress is a common annual occurrence in many areas. Artificial watering in the field, however, inadvertently adds nutrients that are present in the irrigation water, and this effect is rarely recognized as an additional factor. Thus, it should be kept in mind that although water *per se* might be the factor of primary interest, it is usually confounded with a nutrient effect.

It is perhaps not surprising to find that irrigation of *Pinus ponderosa* during midsummer in the semiarid western United States resulted in increased radial growth (Mosher, 1960; Mace and Wagle, 1964). In Arizona, trees not receiving additional water grew only half as much as those given a supplement of 2 inches per week for the entire season. This irrigation was the equivalent of 3 times the normal rainfall on the sites studied. Mace and Wagle (1964) found that foliage water content of trees receiving water was not significantly greater than that of control trees, probably the result of the insensitivity of their method and the wide variation from tree to tree in such gross field studies. As discussed earlier, small decreases in leaf water content result in large decreases in leaf water potential, so that control trees presumably were under greater stress and produced less growth than irrigated trees.

Even in the humid climate of the eastern United States irrigation results in large increases in the radial growth of species normally considered to grow well on moist sites. Stout (1959) reported significant increases in the basal area of several species of hardwoods following summer irrigation on upland sites, and Broadfoot (1964) measured up to nearly double the radial growth in six bottomland hardwood species following irrigation as before irrigation. The latter response is especially interesting, because alluvial flood plain sites generally are interpreted as providing adequate moisture for optimum growth. Kraus and Bengtson (1960) mention other tests that indicate that the irrigation of forest trees almost always is followed by increased growth. Such reports emphasize indirectly that under normal field conditions most trees even on moist sites undergo periods of water stress that restrict maximum growth nearly every year.

IV. WATER DEFICITS AND ROOT DEVELOPMENT

It is almost certain that water stress in root meristems and adjacent tissues has similar adverse effects on apical and cambial growth as in analogous, above-ground stem parts. However, roots of trees beyond the seedling stage are far removed from the source of photosynthates and growth regulators and thus are perhaps subject to more frequent deficiencies in these substances when internal water stress causes reduction in phloem transport. On the other hand, root tissues are probably never under as severe water stress as crown tissues because of the time lag in the build-up of water tension between the transpiring leaf and the absorbing root. Roots are last to suffer the effect of stretched water columns and are first to recover turgor overnight.

The development of roots from an anatomical viewpoint has been described in detail for many tree species [e.g., Wilcox (1954, 1962) in *Abies* and *Libocedrus*; Wilson (1964b) in *Acer*], but not specifically from the standpoint of internal water relations. Root elongation and radial expansion must be reduced through restricted cell division, cell enlargement, and tissue differentiation just as are shoot extenstion and cambial activity in stems and branches. It is assumed that the direct effect of water stress on the primary growth of tree roots is the same as or similar to that reported for roots of herbaceous plants (e.g., Gingrich and Russell, 1957) and reviewed elsewhere in this volume.

Although root structure is similar among widely separate genera of trees [e.g., Laitakari (1929) in *Pinus*; Esau (1943) in *Pyrus*; Wilcox (1954) in *Abies*; Wilson (1964b) in *Acer*; Bogar and Smith (1965) in *Pseudotsuga*], the development of mycorrhizae is diverse among species. The morphology of mycorrhizal short roots is not known to be affected greatly by environmental stress other than nutrient deficiency (Harley, 1959), and gross forms are genetically controlled, varying from simple branches in *Fagus* and beads in *Acer* to complex dichotomy in *Pinus*. However, Worley and Hacskaylo (1959) altered the mycorrhizal association of *Pinus virginiana* Mill. by controlling the relative amount of available soil moisture. The proportion of infection by a species of black ectotropic fungus increased up to 100% in roots of first-year seedlings at low levels of soil moisture, whereas in roots of seedlings watered daily, species of white fungi replaced the black ones almost completely. The relative functions of the two types are not known. All forms have the effect of increasing root surface area, and it is generally assumed that a better balance between roots and shoots results from most mycorrhizal infections. Evidently, trees grow well without abundant mycorrhizae in nutrient culture and in highly fertile soil, but they depend greatly on the copious development of the fungi in soils of low fertility (Hatch, 1937; Kramer, 1949; Harley, 1959). Infertile soils are also generally those that store less available water and are more subject to effects of drought, but there is no experimental evidence on the role of mycorrhizae in supplying additional water and thus reducing water stress in trees on dry sites.

It is probable that the cambium in roots is under the same or similar intrinsic controls from the crown as the cambium in stems, as already discussed. Obviously, as with interrelations between terminal growth and cambial growth of the bole, any reductions in crown vigor, in production of photosynthate, or in rates of transport from foliage will have a relative effect on the maintenance of secondary growth in roots.

Secondary growth in roots, however, is more erratic than in stems. Annual rings are very often discontinuous, missing, or include false rings [e.g., Brown (1915) with *Pinus strobus*; Wilson (1964b) with *Acer rubrum*]. In stems these discontinuous rings occur most frequently in suppressed trees (Larson, 1956; Farrar, 1961), but in roots they occur also in vigorous dominants. The effect of environmental stress on such erratic and uncertain development in roots is not known, although it is probable that seasonal water deficits play at least an indirect role. Internal controls by growth regulators from above ground are unquestionably involved, and the effect of water stress on the production and transport of these substances is discussed elsewhere. Also, physical injury, including drought injury, to lateral roots often results in complex patterns or adventitious roots developing continually at and near the base of the stem and along all older roots (e.g., Lyford and Wilson, 1964), affecting the annual ring patterns of secondary growth.

A. SEASONAL AND DIURNAL ROOT ELONGATION

Both internal and external controls of primary root growth in trees are well recognized. Richardson (1958) and others have shown that the initiation of apical growth in the spring and the maintenance of root elongation in some species are dependent on hormones transported from shoots. Even though water stress at this time of year occasionally has an effect on translocation processes in the stem—for example, shoots may be desiccated by winds during a period when the soil is cold—it is doubtful that initiation of root elongation is affected because other factors, such as soil temperature, are limiting.

Much remains to be learned about the intrinsic controls of roots by shoots and, consequently, about the effects of water stress on these intrinsic controls. The levels and patterns of hormone regulation are probably quite different among species, because some require cold hardening to break root dormancy and others do not. In *Acer saccharinum*, for example, following cold periods roots begin elongation prior to leaf flushing, but they do not maintain growth unless shoots develop immediately (Richardson, 1958). In warm climates, on the other hand, roots elongate in most species of trees throughout the winter period, whether shoots are active or not, whether trees are evergreen or not (Kramer, 1949; Leshem, 1965).

It is well established that the period of root elongation in trees may be shortened by low levels of soil moisture during the growing season (Stevens, 1931; Turner, 1936; Reed, 1939; Kaufman, 1945; Kramer, 1949; Fielding, 1955; Leshem, 1965). As with other parts of the tree, maintenance of root growth in spring is rarely affected by water deficits, but summer water stress causes slowing and cessation of expanding root tissues. Fielding (1955) found that roots of *Pinus radiata* ceased elongation completely during warm, dry periods and that this apical growth in roots was highly correlated with that in shoots. Terminal meristems at both locations responded together to periods of water deficits. Further, Fielding correlated diurnal stem shrinkage with the cessation of elongation in both terminal meristems. Roots and shoots elongated during the night after stem swelling indicated rehydration.

Thus, alleviation of internal water stress permits roots to grow for brief periods each day as long as sufficient soil water is available for absorption.

During dry periods in mid-season, when turgor is not recovered overnight, roots cease elongation altogether, as do shoots. Late in the season, if soil moisture recharges occur, roots may resume elongation in some species of trees even though shoots have initiated bud formation and do not resume extension growth. Although a lack of correlation between intrinsic patterns of root and shoot growth is common (e.g., Reed, 1939; Wilcox, 1962; Leshem, 1965), water stress obviously overrides intrinsic factors and the net result is that elongation rates of roots and shoots are correlated at least during those periods when both meristems are growing.

The rates of root elongation plotted over time often result in seasonal M-shaped curves, because root growth of trees usually increases again near the end of the growing season following the midsummer low (Turner, 1936; Kaufman, 1945; Kramer, 1949; Leshem, 1965). The explanations of Kramer (1958), Kozlowski (1964b), Gates (1965), Fritts (1966), and others indicate that midday stress conditions that curtail photosynthesis and other processes in midsummer are alleviated by shorter day length and cooler temperatures in early fall. Thus, the diurnal amplitude of the daily growth curve changes from spring to summer to fall, and is controlled above all else by internal water stress. In many species of trees, roots have the intrinsic potential to elongate from the time of initiation in spring to the time of dormancy in winter. Thus, terminal meristems of tree roots exhibit a seasonal pattern of growth inversely related to the pattern of water stress in the tree, which often results in the M-shaped growth curve so commonly reported.

It is assumed that internal water stress affects mitosis and the enlargement of new cells in root meristems exactly as in shoot meristems. The substrate for new tissue development in roots, however, must come from stored foods, at least during the early part of the growing season, while in shoots much substrate may come from current photosynthate in the newly developing leaves (reviewed by Kozlowski and Keller, 1966). Thus, root growth must be affected by all factors that govern the quantity and type of food stored over winter, of which water stress is probably most important. Environmnetal water deficits of the previous season certainly affect the amount of food stored in roots for current-season utilization. There are apparently few or no observations, however, on the 2-year effect of environmental stress on root growth as there are for shoot growth. Per unit of tissue, roots accumulate far greater quantities of stored carbohydrate than do stems (Kozlowski and Keller, 1966); thus, it is possible that luxury amounts are available for assimilation during normal years and that droughts of previous years do not impose as severe restrictions on current root growth as on growth in above-ground parts. It is more likely, however, that the mobilization of foods stored in the roots is competitive between stem and root meristems and, thus, when current stem growth suffers from previous year droughts, current root growth does also.

B. Soil Moisture and Root Development

This section is concerned with effects of external soil water stress on root development. Among others, Kozlowski (1949) has shown that large root systems develop in tree seedlings when grown in soil maintained close to field capacity, in contrast to sparse root development in soils allowed to dry almost to permanent wilting before rewatering. It is impossible in such studies to distinguish between several separate effects of deficient soil moisture on root growth. Some studies have emphasized that soil strength increases sharply as the soil dries, and the resulting physical resistance to penetration by a root tip is considered limiting independent of the deficiency of water for absorption (Barley, 1963; Taylor and Gardner, 1963). Therefore, the failure of roots to grow into dry soil is probably more the result of physical impedance than soil water stress *per se*. The external resistance to root tips by drying soil is a factor that should not affect secondary growth seriously. Thus, it is possible that cambial growth in roots continues slowly after elongation ceases during periods of rapid soil water depletion. Nevertheless, the net effect of increased soil strength is that soil water is not available for absorption through elongating roots, and root tissues dehydrate by continued evaporation from aboveground parts.

Kramer and Bullock (1966) emphasize that tree roots can absorb large quantities of water through cracks and lenticels of suberized portions of root systems, and thus many trees are not dependent on actively growing new root tips. This may be true for trees growing in saturated soils with a good water contact over the root surface or seasonally on better drained sites for short periods following rain recharges. However, as soil surrounding suberized roots dires below field capacity most water contacts must break at the root surface and only limited absorption must take this route for most of the growing season.

Excess water and poor aeration of the soil (reviewed by Kramer, 1949) have at least as great an influence on gross root systems of trees as do deficient soil water or dense, impermeable substrata. Soils with limited gas exhange, either because of high bulk density or excess water, support trees with superficial, shallow root systems. During dry periods, when water in the root zone is exhausted quickly, trees on such sites often undergo severe drought damage and may develop symptoms of disease decline [e.g., Campbell and Copeland (1954), littleleaf disease in *Pinus echinata*; Leaphart and Copeland (1957), pole blight disease in *Pinus monticola*; Toole and Broadfoot (1959), dieback in *Liquidambar styraciflua*].

Several studies of tree seedlings planted in modified environments indicate that deficient soil water affects both the general morphology of roots and the root/shoot ratio. Bilan (1960) found that lateral roots of *Pinus taeda* seedlings penetrated a sandy loam soil deeply when the soil surface was exposed to evaporation that dried the surface layers. In contrast, lateral roots under mulched or shaded soil surfaces grew at and near the surface. The relative effects of temperature vs. water in Bilan's study could not be separated. The shading–mulching treatment that conserved soil water most also significantly decreased the dry-weight root/shoot ratio. In general, large tree seedlings with large shoots and relatively large roots result when soil moisture conditions are optimum, but a higher ratio of roots to shoots is obtained when there is a limited supply of water (Kramer, 1949; Steinbrenner and Rediske, 1964; Ledig and Perry, 1966). There are not consistent reports, however, on this effect of soil moisture on root/shoot ratios. Strothmann (1967) found that large, healthy seedlings of *Pinus resinosa* resulted when released of root competition from surrounding vegetation, and the ratio of roots to shoots was not different in the large seedlings than in the small seedlings under severe competition.

Steinbrenner and Rediske (1964) found that root systems of first-year seedlings of *Pinus ponderosa* reacted to soil moisture stress differently than those of seedling *Pseudotsuga menziesii*. *Pseudotsuga* roots were concentrated near the surface when the soil was kept well watered, but they penetrated deeply when surface soil moisture was not optimum. Thus, roots of this species grew where water was readily available, independent of a genetic pattern. On the other hand, there was deep, vertical penetration of the *Pinus* roots regardless of soil moisture treatment, indicating the strong intrinsic development of a tap root in this species.

Roots of mature trees are known to develop profusely in zones of the soil profile that contain an adequate supply of available water. Wilde (1958) and Gary (1963) presented examples of the gross morphology of tree roots developed in the capillary fringe above ground water. A proliferation of small roots is common at the end of large sinker roots that encounter ground water. In well-aerated soil profiles, tree roots are known to proliferate at the contact above any textural change that results in an increased supply of available water (Brown and Lacate, 1961). The high concentrations of roots of individual trees beneath the stems of neighboring trees in a forest have been attributed to the convergence of stem flow water that results in a greater recharge of soil moisture beneath stems than in the area between trees (Preston, 1942; Stout, 1956). Coile (1937) suggested that, among other factors, frequent recharges of the surface soil by summer rainstorms contribute to far greater concentration of roots near the soil surface than deeper in the profile.

There is extensive literature on the gross development of root systems of forest and orchard trees on various sites. Little of this is related to water stress except indirectly by inference. For example, Horton (1958) and Brown and Lacate (1961) determined by excavation the general morphology of root systems of three species of *Pinus* in a range of soil textures and topographic situations reflecting various water regimes. The adaptability of the vertical root system in this genus to moisture availability was clearly indicated by deep, well-developed roots on sites with optimum water relations and by shallow, poorly developed sinker roots on both dry sites and peat bogs. Bannan (1940) also found by excavation that roots of several eastern Canadian conifers were shallow and platelike when grown on dry sand and developed no more deeply there than in swamps with saturated soil and poor aeration. Most such field studies show similar logical relationships between the extent of root systems and the extremes of soil moisture, with sparse, shallow root development on dry sites. Reviews that refer to this subject include Laitakari (1927), Büsgen and Münch (1931), Stevens (1931), Biswell (1935), and Dunning (1949).

V. WATER DEFICITS AND REPRODUCTIVE GROWTH

A. SEXUAL REPRODUCTION IN TREES

The effect of water deficits on flowering and fruiting has been studied much more intensively in orchard trees than in forest trees. Because of the wide variation in response of reproductive growth among species in different climates, there is no general agreement even for fruit trees on the effects of water stress or the physiological mechanisms altered by water stress. For example, Hartmann and Panetsos (1961) found that continuous irrigation was necessary for the greatest production of flowers and subsequent fruiting in *Olea*, whereas Alvim (1960) reported that brief drought periods between irrigation were essential to stimulate abundant flowering in *Coffea*. In forest trees it is even more difficult to establish the overall effects of water stress on reproductive growth.

Every season of the year is represented by some species of tree in some intrinsic tage of sexual reproduction. Although in many species flower bud primordia are generally formed in the late season one year and flowers bloom in the early season the next year, differentiation and maturation of fruit and seed in these same trees are highly variable in time. On a given site at the same time adjacent trees of different species may be in various stages of reproductive growth, from initiation of flower buds, to full inflorescence and pollination, to maturation of embryos and fruits, and to dispersal of seeds. Moreover, many species, especially conifers, require several seasons to complete all stages of reproductive growth, and each year's flower and seed crop is in its own separate phase of development at a given time. Thus, the effects of environmental stress superimposed at various seasons certainly result in almost endless variation in the physiological processes and in conditions that regulate the many stages of the anatomical development of flowers and fruits. Nitrogen metabolism, carbohydrate metabolism, and various hormonal controls are known to form the basis for the regulation of the reproductive development in trees (Kramer and Kozlowski, 1960). Internal water stress can have a significant effect on any of these processes and conditions at several times during the year, and thus can certainly affect, at a particular time, almost any stage of reproductive development in trees.

Unlike vegative growth, therefore, which for all species in temperate climates is initiated in early spring and develops in a more or less predictable pattern over the growing season, reproductive growth is so variable among species that it is unwise to extrapolate generally from one pattern to another. What is learned about the effect of summer environmental water stress on seed production in *Pseudotsuga*, e.g., which requires more than two seasons to complete all reproductive stages, will probably not be specifically applicable to the same processes in *Salix*, which flowers in late winter and matures fruits within 2 months, or in *Liriodendron*, which flowers in late spring following full leaf and does not begin to mature fruits until autumn. When the effect of water stress on seed maturation, for example, is understood for one species, the information gained probably will not hold for many other species, simply because of the variable periodicity of development. Each group of trees with the same intrinsic pattern of reproductive growth will have to be evaluated separately.

It is likely that any stage of sexual reproduction in trees that occurs during mid-season is subject to the effects of severe transpirationally induced water stress. Thus, in some species, midsummer drought conditions may result in fewer seeds; in others, in smaller, less viable seed; and in still others, in the reduction of both numbers and sizes of seed, some perhaps for 2 successive years. As with shoot growth, there is certainly at least a 2-year effect of water deficits on flowering in temperate and cool climates. Most species of both forest and horticultural trees bloom in early spring either just prior to or along with leaf flushing. Thus, the previous season's weather conditions, especially summer and fall droughts, must be grossly correlated with the potential for flowering from overwintering flower buds. For example, Wenger (1957) reported that seed production in stands of *Pinus taeda* was positively correlated with the rainfall of the season 2 years preceding that of cone maturation. It was the formation of flower bud primordia that was most sensitive to water stress for the period of years studied by Wenger. It is probable that differentiating seed in conelets of this species 1 year prior to maturation are also affected by drought, but such an effect may be manifested largely in germination properties of the embryo (not studied by Wenger) and not in the total numbers of cones produced.

The net effect of weather, however, on success or failure of early spring flowering in temperate climates is usually determined by temperature conditions before and during the appearance of inflorescences (Kramer and Kozlowski, 1960). More often than not adverse sequences of warm and freezing temperatures in early spring completely negate the effects of high or low water deficits of the previous season. Part of the failure to correlate seed production with environmental water conditions is also associated with the fact that one stage of reproductive growth can be adversely affected by high rainfall at one time in the season, which may mask positive effects of high rainfall at another time. For example, extended rainy periods during flowering may decrease pollen dispersal, whereas a short time later in the season such a rain could benefit embryo development. Another factor of reproductive growth that obscures the effects of conditions of previous seasons is the characteristic bearing of fruit in abundance only in alternate years (Davis, 1957). It is therefore a rare sequence of environmental events that permits the effect of water stress in one year to be fully expressed in subsequent years.

At present, there is little direct evidence on the detailed effects of water stress on any stage of flower and fruit production in any species of forest trees. There is much general speculation, based on the knowledge that conditions promoting high rates of photosynthesis and carbohydrate accumulation are conducive to the production of large fruit and seed crops (Kramer and Kozlowski, 1960). Thus, soil moisture availability certainly affects, at least indirectly, the reproductive processes of trees. Trees on moist sites produce more seed than those on dry sites, when other factors are equal. Open-grown and dominant trees in a forest stand usually are the most prolific seeders, presumably resulting from high photosynthetic activity in their large crowns. It is well documented that releasing *Pinus* species from competition by thinning surrounding stems promotes large increases in cone production (Allen, 1953; Wenger, 1954). Wenger (1957) suggested that the increase in flowering and fruiting resulting from release and wide spacing in *P. taeda* may be attributed more to the increased availability of soil water than to other factors such as improved light conditions. This seems logical, since an increase in cone crops is usually noted the second year following the release of pine seed trees, indicating that an increased production of flower bud primordia occurs shortly following release, before a build-up in new foliage can respond to the increased light.

In orchard trees, the stage of fruit enlargement has been studied intensively for many decades in relation to soil moisture and irrigation practices [e.g., in *Citrus* by Bartholomew (1926), Furr and Taylor (1939), Rokach (1953); in *Pyrus* by Lewis *et al.* (1935), Ryall and Aldrich (1944); in *Malus* by Boynton (1937); in *Phoenix* palm by Moore and Aldrich (1938); and especially by Hendrickson and Veihmeyer (1931) in *Vitis*, (1942) in *Pyrus*, (1950) in *Prunus*]. No attempt will be made to review the work with horticultural species, but the firm conclusions are that the rate of enlargement by fleshy fruits is strongly reduced following the rapid depletion of soil moisture, and the final size and quality of individual fruits are strongly regulated by the amount of water available during enlargement. Few of these studies present quantitative data on internal water potential, but some involve careful measurements of soil moisture stress in relation to fruit enlargement. These studies also show that the quality of fruits, with respect to sugars, starch, and content of other organic compounds, was altered by low levels of soil moisture at various stages of fruit development. For example, marked reduction in the growth of *Pyrus* fruits was associated with soil moisture depletion below 70% of the available capacity (Lewis *et al.*, 1935), and well-watered *Pyrus* trees produced fruits that were smoother in texture, higher in sugar content, and lower in acids than were fruits on trees growing under normal summer soil water deficits (Ryall and Aldrich, 1944).

Diurnal contraction and expansion of fleshy fruits are the obvious result of midday transpirational water stress and the restoration of turgor overnight, respectively, and have been measured many times in orchard trees (e.g., Bartholomew, 1926; Magness *et al.*, 1935; Schroeder and Wieland, 1956). Kozlowski (1964b, 1965, 1968), reviewing this subject, states that diurnal water deficits in fruits are known to become more serious as the soil is depleted of water and that fruits attached to trees in midsummer often fail to regain full weight overnight during periods of extended drought. *Citrus* fruits supply water to leaves for transpiration during dry periods and thus prevent wilting.

Most species of forest trees bear fruits that are less fleshy than those of most horticultural species, and their fruits are not such large reservoirs from which adjacent transpiring leaves pull water. Nevertheless, MacDougal (1924) measured diurnal dimension changes in the fruits of developing *Juglans* that were correlated with diurnal changes in water stress, emphasizing that even highly lignified fruits are subject to periods of low water potential when the rest of the tree is under stress.

B. SEED GERMINATION AND SEEDLING ESTABLISHMENT

Since seed size affects the viability and germination of the seed and also survival of the seedling (Korstian, 1927; U.S. Dept. Agr., 1948; Squillace and Bingham, 1958), water deficits in the mother tree during the period of embryo maturation and accumulation of stored foods in the seed have a detrimental effect on the subsequent establishment of seedlings. Thus, although emphasis is usually placed on environmental moisture conditions at the time of seed germination, the success of seedling establishment depends largely on conditions under which the seed was produced on the parent tree. A germinating seedling from a physically small seed with a low proportion of stored food usually has a poorer chance of surviving the rigors of a current water deficit than one of the same species from a large seed. The small seed lacks the capacity for the early establishment of an effective root system before it depends entirely on producing its own photosynthate.

Following dissemination, seeds are often subject to dehydration, and severe water deficits prior to germination possibly are detrimental to the subsequent development of seedlings. Although there is a large variation among species in the susceptibility of seed coats to water loss, many seeds in natural field situations probably undergo periods of some evaporative water loss, with effects on the balance between soluble and insoluble foods and on the general metabolism of the embryo. Little published information is available on physiological changes in tree seeds that are induced by water stress or on their effects on subsequent germination. However, mild water stress probably is desirable in the seed during its normal dormant period, because the premature uptake of environmental water may initiate chemical changes leading to untimely germination. Successful germination, in any case, depends eventually on the absorption of water and, in most species, this amounts to a severalfold increase in water content (Kramer and Kozlowski, 1960). Thus, seeds in unfavorable microsites do not germinate during dry weather.

The survival of germinating embryos depends on the absorption of water almost immediately by the elongating root apex. This early, critical period in the development of tree seedlings has received much attention from the aspect of forest ecology (e.g., Toumey, 1929; Holch, 1931; Kozlowski, 1949). Considerable variation may be found among species in the rate and extent of root development in germinating seedlings, and there is general agreement that intrinsic rooting morphology accounts for the survival or death of certain species under conditions of water deficits.

That water is at least as important as light or other factors in the establishment of tree seedlings under a forest canopy was established early by the classical trenching studies of German foresters and later in North America by Korstian and Coile (1938), and others. The relative importance of water vs. light in the survival and growth of forest seedling reproduction will not be reviewed here, and the reader is referred to Oosting and Kramer (1946), Kozlowski (1949), Ferrell (1953), and Gatherum et al. (1963) for thorough discussions of this subject. Water availability is evidently more critical than light immediately following germination for successful seedling establishment; subsequently, light becomes the more important factor for the survival and growth of established seedlings (Strothmann, 1967).

REFERENCES

Abbe, L. B., and Crafts, A. S. (1939). Phloem of white pine and other species. *Botan. Gaz.* **100**, 695.

Agerter, S. R., and Glock, W. S. (1965). "An Annotated Bibliography of Tree Growth and Growth Rings, 1950–1962." Univ. of Arizona Press, Tucson, Arizona.

Alfieri, F. J., and Evert, R. F. (1965). Seasonal phloem development in *Pinus strobus*. *Am. J. Botany* **52**, 626.

Allen, R. M. (1953). Release and fertilization stimulate longleaf pine cone crop. *J. Forestry* **51**, 827.

Alvim, P. deT. (1960). Moisture stress as a requirement for flowering of coffee. *Science* **132**, 354.

Artschwager, E. (1950). The time factor in the differentiation of secondary xylem and phloem in pecan. *Am. J. Botany* **37**, 15.

Bannan, M. W. (1940). The root systems of northern Ontario conifers growing in sand. *Am. J. Botany* **27**, 108.

Bannan, M. W. (1955). The vascular cambium and radial growth in *Thuja occidentalis* L. *Can. J. Botany* **33**, 113.

Bannan, M. W. (1962). "The vascular cambium and tree ring development." *In* "Tree Growth" (T. T. Kozlowski, ed.), pp. 3–21. Ronald Press, New York.

Barley, K. P. (1963). Influence of soil strength on growth of roots. *Soil Sci.* **96**, 175.

Barrett, J. W., and Youngberg, C. T. (1965). Effect of tree spacing and understory vegetation on water use in a pumice soil. *Soil Sci. Soc. Am. Proc.* **29**, 472.

Bartholomew, E. T. (1926). Internal decline of lemons. III. Water deficit in lemon fruits caused by excessive leaf evaporation. *Am. J. Botany* **13**, 102.

Bassett, J. R. (1964a). Diameter growth of loblolly pine trees as affected by soil moisture availability. *U.S. Dept. Agr. Forest Serv. Res. Paper* **S0–9**.

Bassett, J. R. (1964b). Tree growth as affected by soil moisture availability. *Soil Sci. Soc. Am. Proc.* **28**, 436.

Bay, R. R. (1963). Soil moisture and radial increment in two density levels of red pine. *U.S. Dept. Agr. Forest Serv. Res. Note* **LS–30**.

Bay, R. R., and Boelter, D. H. (1963). Soil moisture trends in thinned red pine stands in northern Minnesota. *U.S. Dept. Agr. Forest Serv. Res. Note* **LS–29**.

Bell, D. B. (1957). The relationship between height growth in conifers and the weather. *J. Oxford Univ. Forestry Soc.* [4] **5**.

Bilan, M. V. (1960). Root development of loblolly pine seedlings in modified environments. *Stephen F. Austin State Coll. Forestry Bull.* **4**.

Biswell, H. H. (1935). Effects of environment upon the root habits of certain deciduous forest trees. *Botan. Gaz.* **96**, 676.

Bogar, G. D., and Smith, F. H. (1965). Anatomy of seedling roots of *Pseudotsuga menziesii*. *Am. J. Botany* **52**, 720.

Boggess, W. R. (1953). Diameter growth of shortleaf pine and white oak during a dry season. *Univ. Illinois Agr. Expt. Sta. Forestry Note* **37**.

Boggess, W. R. (1956). Weekly diameter growth of shortleaf pine and white oak as related to soil moisture. *Proc. Soc. Am. Foresters* **1956**, 83.

Bogue, E. E. (1905). Annual rings of tree growth. *Missouri Weather Rev.* **33**, 250.

Bormann, F. H., and Kozlowski, T. T. (1962). Measurements of tree ring growth with dial gauge dendrometers and vernier tree ring bands. *Ecology* **43**, 289.

Boynton, D. (1937). Soil moisture and fruit growth in an orchard situated on shallow soil in the Hudson Valley, New York, in 1937. *Proc. Am. Soc. Hort. Sci.* **34**, 169.

Brix, H. (1962). The effect of plant water stress on the rates of photosynthesis and respiration in tomato plants and loblolly pine seedlings. *Physiol. Plantarum* **15**, 10.

Broadfoot, W. M. (1964). Hardwoods respond to irrigation. *J. Forestry* **62**, 579.

Brown, H. P. (1915). Growth studies in forest trees. II. *Pinus strobus* L. *Botan. Gaz.* **59**, 197.

Brown, W. G. E., and Lacate, D. S. (1961). Rooting habits of white and red pine. *Can. Dept. Forestry, Forest Res. Note* **108**.

Buell, M. F., Buell, H. F., Small, J. A., and Monk, C. D. (1961). Drought effect on radial growth of trees in the William L. Hutchinson Memorial Forest (New Jersey). *Bull. Torrey Botan. Club* **88**, 176.

Büsgen, M., and Münch, E. (1931). "The Structure and Life of Forest Trees," 3rd ed. (transl. by T. Thomson). Wiley, New York.

Byram, G. M., and Doolittle, W. T. (1950). A year of growth for a shortleaf pine. *Ecology* **31**, 27.

Campbell, W. A., and Copeland, O. L. (1954). Little leaf disease of shortleaf and loblolly pines. *U.S. Dept. Agr. Circ.* **940**.

Carmean, W. H. (1961). Soil survey refinements needed for accurate classification of black oak site quality in southeastern Ohio. *Soil Sci. Soc. Am. Proc.* **25**, 394.

Carvell, K. L. (1956). Summer shoots cause permanent damage to red pine. *J. Forestry* **54**, 271.

Chalk, L. (1930). The formation of spring and summer wood in ash and in Douglas fir. *Oxford Forestry Mem.* **10**.

Chalk, L. (1937). A note on the meaning of the terms early wood and late wood. *Proc. Leeds Phil. Lit. Soc.* **3**, 324.

Chalk, L. (1951). Water and the growth of wood of Douglas fir. *Quart. J. Forestry* **45**, 237.

Clemments, J. R. (1966). Growth responses of red pine seedlings to watering treatments applied in two different years. *Petawawa Forest Expt. Sta., Can. Dept. Forestry, File Rept.* **P–400**.

Cockerham, G. (1930). Some observations on cambial activity and seasonal starch content in sycamore. *Proc. Leeds Phil. Lit. Soc.* **2**, 64.

Coile, T. S. (1936). The effect of rainfall and temperature on the annual radial growth of pine in southern U.S. *Ecol. Monographs* **6**, 534.

Coile, T. S. (1937). Distribution of forest tree roots in North Carolina Piedmont soils. *J. Forestry* **35**, 247.

Coile, T. S. (1952). Soil and the growth of forests. *Advan. Agron.* **4**, 329.

Craib, I. J. (1929). Some aspects of soil moisture in the forest. *Yale Univ. School Forestry Bull.* **25**.

Daubenmire, R., and Deters, M. E. (1947). Comparative studies of growth in deciduous and evergreen trees. *Botan. Gaz.* **109**, 1.

Davis, L. D. (1957). Flowering and alternate bearing. *Proc. Am. Soc. Hort. Sci.* **70**, 545.

Della-Bianca, L., and Dils, R. E. (1960). Some effects of stand density in a red pine plantation on soil moisture, soil temperature, and radial growth. *J. Forestry* **58**, 373.

DeVries, M. L., and Wilde, S. A. (1962). The effect of the density of red pine stands on moisture supply in sandy soils. *Netherlands J. Agr. Sci.* **10**, 235.

Digby, J., and Wareing, P. F. (1966). The relationship between endogenous hormone levels in the plant and seasonal aspects of cambial activity. *Ann. Botany* **30**, 607.

Diller, O. D. (1935). The relation of temperature and precipitation to the growth of beech in northern Indiana. *Ecology* **16**, 72.

Dils, R. E., and Day, M. W. (1952). The effect of precipitation and temperature upon radial growth of red pine. *Am. Midland Naturalist* **48**, 730.

Doolittle, W. T. (1957). Site index of scarlet and black oaks in relation to southern Appalachian soil and topography. *Forest Sci.* **3**, 114.

Douglass, A. E. (1919). Climatic cycles and tree growth; A study of the annual rings of trees in relation to climate and solar activity. *Carnegie Inst. Wash. Publ.* **289**.

Duff, G. H., and Nolan, N. J. (1958). Growth and morphogenesis in Canadian forest species. III. The time scale of morphogenesis at the stem apex of *Pinus resinosa* Ait. *Can. J. Botany* **36**, 687.

Dunning, D. (1949). A selected list of references on *Roots of Forest Trees. U.S. Dept. Agr. Forest Serv. Rocky Mt. Forest Range Expt. Sta. Res. Note* **52**.

Esau, K. (1943). Vascular differentiation in the pear root. *Hilgardia* **15**, 299.

Evert, R. F. (1963). The cambium and seasonal development of the phloem in *Pyrus malus. Am. J. Botany* **50**, 149.

Farrar, J. L. (1961). Longitudinal variation in the thickness of the annual ring. *Forestry Chron.* **37**, 323.

Ferrell, W. K. (1953). Effect of environmental conditions on survival and growth of forest tree seedlings under field conditions in the Piedmont region of North Carolina. *Ecol. Monographs* **24**, 667.

Fielding, J. M. (1955). The seasonal and daily elongation of shoots of Monterey pine and the daily elongation of roots. *Commonwealth Australia Forestry Timber Bur. Leaflet* **75**.

Foil, R. R. (1961). Late season soil moisture shows significant effect on width of slash pine summerwood. *Louisiana State Univ. Forestry Note* **44**.

Foulger, A. N. (1966). Variation in certain wood properties of eastern white pine. *Ohio Agr. Res. Develop. Center Res. Bull.* **985**.

Fraser, D. A. (1952). Initiation of cambial activity in some forest trees in Ontario. *Ecology* **33**, 259.

Fraser, D. A. (1956). Ecological studies of forest trees at Chalk River, Ontario, Canada. II. Ecological conditions and radical increment. *Ecology* **37**, 777.

Fraser, D. A. (1962). Tree growth in relation to soil moisture. *In* "Tree Growth" (T. T. Kozlowski, ed.), pp. 183–204. Ronald Press, New York.

Friesner, R. C., and Walden, G. (1946). A five year dendrometer record of two trees of *Pinus strobus. Butler Univ. Botan. Studies* **8**, 1.

Fritts, H. C. (1958). An analysis of radial growth of beech in a Central Ohio forest during 1954–1955. *Ecology* **39**, 705.

Fritts, H. C. (1960). Multiple regression analysis of radial growth in individual trees. *Forest Sci.* **6**, 334.

Fritts, H. C. (1962). The relation of growth ring widths in American beech and white oak to variations in climate. *Tree Ring Bull.* **25**, 2.

Fritts, H. C. (1966). Growth rings of trees: Their correlation with climate. *Science* **154**, 973.

Fritts, H. C., and Fritts, E. C. (1955). A new dendrograph for recording radial changes of a tree. *Forest Sci.* **1**, 271.

Fritts, H. C., Smith, D. G., and Stokes, M. A. (1965). The biological model for paleoclimatic interpretation of Mesa Verde tree-ring series. *Am. Antiquity* **31**, 101.

Furr, J. R., and Taylor, C. A. (1939). Growth of lemon fruits in relation to moisture content of the soil. *U.S. Dept. Agr. Tech. Bull.* **640**.

Gaertner, E. E. (1963). Water relations of forest trees. *In* "The Water Relations of Plants" (A. J. Rutter and F. H. Whitehead, eds.), Brit. Ecol. Soc. Symp., pp. 366–378. Blackwell, Oxford.

Gary, H. H. (1963). Root distribution of five-stamen tamarisk, seepwillow, and arrowweed. *Forest Sci.* **9**, 311.

Gates, D. M. (1965). Energy, plants and ecology. *Ecology* **46**, 1.

Gatherum, G. E., McComb, A. L., and Lewis, W. E. (1963). Effects of light and soil moisture on forest tree seedling establishment. *Iowa Agr. Expt. Sta. Res. Bull.* **513**, 776.

Gibbs, R. D. (1958). Patterns in the seasonal water content of trees. *In* "Physiology of Forest Trees" (K. V. Thimann, ed.), pp. 43–69. Ronald Press, New York.

Gilmore, A. R., Boyce, S. G., and Ryker, R. A. (1966). The relationship of specific gravity of loblolly pine to environmental factors in southern Illinois. *Forest Sci.* **12**, 399.

Gingrich, J. R., and Russell, M. B. (1957). A comparison of effects of soil moisture tension and osmotic stress on root growth. *Soil Sci.* **84**, 185.

Glock, W. S. (1955). Tree Growth. II. Growth rings and climate. *Botan. Rev.* **21**, 73.

Goggins, J. F. (1961). The interplay of environment and heredity as factors controlling wood properties in conifers. *N. Carolina State Univ. School Forestry Tech. Rept.* **11**.

Haasis, F. W. (1932). Seasonal shrinkage of Monterey pine and redwood trees. *Plant Physiol.* **7**, 285.

Hale, J. D. (1959). Physical and anatomical characteristics of hardwoods. *Tappi* **42**, 670.

Hamilton, J. R., and Harris, J. B. (1965). Influence of site on specific gravity and dimensions of tracheids in clones of *Pinus elliottii* Engelm. and *Pinus taeda* L. *Tappi* **48**, 330.

Hammel, H. T. (1967). Freezing of xylem sap without cavitation. *Plant Physiol.* **42**, 55.

Harley, J. L. (1959). "The Biology of Mycorrhiza." Leonard Hill, London.

Harms, W. R. (1962). Spacing-environmental relationships in a slash pine plantation. *U.S. Dept. Agr. Forest Serv. Southeast. Forest Expt. Sta. Paper* **150**.

Harris, E. H. M. (1955). The effect of rainfall on the late wood of Scots pine and other conifers in East Anglia. *Forestry* **28**, 136.

Hartman, H. T., and Panetsos, C. (1961). Effect of soil moisture deficiency during floral development on fruitfulness in the olive. *Proc. Am. Soc. Hort. Sci.* **78**, 209.

Hatch, A. B. (1937). The physical basis of mycotrophy in the genus *Pinus. Black Rock Forest Bull.* **6**.

Helms, J. A. (1965). Diurnal and seasonal patterns of net assimilation in Douglas-fir as influenced by environment. *Ecology* **46**, 698.

Hendrickson, A. H., and Veihmeyer, F. J. (1931). Irrigation experiments with grapes. *Proc. Am. Soc. Hort. Sci.* **28**, 151.

Hendrickson, A. H., and Veihmeyer, F. J. (1942). Readily available soil moisture and the size of fruit. *Proc. Am. Soc. Hort. Sci.* **40**, 13.

Hendrickson, A. H., and Veihmeyer, F. J. (1950). Irrigation experiments with apricots. *Proc. Am. Soc. Hort. Sci.* **55**, 1.

Holch, A. E. (1931). Development of roots and shoots of certain deciduous tree seedlings in different forest sites. *Ecology* **12**, 259.

Horton, K. W. (1958). Rooting habits of lodgepole pine. *Can. Dept. Forestry, Forest Res. Tech. Note* **61**.

Hosner, J. F., and Leaf, A. L. (1962). The effect of soil saturation on the dry weight, ash content, and nutrient absorption of various bottomland tree species. *Soil Sci. Soc. Am. Proc.* **26**, 401.

Howe, J. P. (1968). The influence of irrigation on wood formed in ponderosa pine. *Forest Prod. J.* **18**, 85.

Hunt, F. M. (1951). Effects of flooded soil on growth of pine seedlings. *Plant Physiol.* **26**, 363.

Husch, B. (1959). Height growth of white pine in relation to selected environmental factors on four sites in northeastern New Hampshire. *New Hampshire Agr. Expt. Sta. Tech. Bull.* **100**.

Hustich, I. (1948). The Scotch pine in northernmost Finland and its dependence on the climate in the last decades. *Acta Botan. Fennica* **42**, 4.

Hygen, G. (1965). Water stresses in conifers during winter. *In* "Water Stress in Plants" (B. Slavik, ed.), *Proc. Symp. Prague*, 1963, pp. 89–98. Junk, The Hague.

Impens, I., and Schalck, J. (1965). A very sensitive electric dendrograph for recording radial changes of a tree. *Ecology* **46**, 183.

Jarvis, P. G., and Jarvis, M. S. (1963). The water relations of tree seedlings. I. Growth and water use in relation to soil water potential. *Physiol. Plantarum* **16**, 215.

Jayne, B. A. (1958). Effect of site and spacing on the specific gravity of wood of plantation-grown red pine. *Tappi* **41**, 162.

Kaufman, C. M. (1945). Root growth of jack pine on several sites in the Cloquet Forest, Minn. *Ecology* **26**, 10.

Kennedy, R. W. (1961). Variation and periodicity of summer-wood in some second-growth Douglas-fir. *Tappi* **44**, 161.

Kennedy, R. W., and Smith, J. H. G. (1959). The effects of some genetic and environmental factors on wood quality in poplar. *Pulp Paper Mag. Can.* **60**, T35.

Kenworthy, A. L. (1949). Soil moisture and growth of apple trees. *Proc. Am. Soc. Hort. Sci.* **54**, 29.

Kienholz, R. (1931). Effect of environmental factors on the wood structure of lodgepole pine, *Pinus cortorta* Loudon, *Ecology* **12**, 354.

Kirkwood, J. E. (1914). The influence of the preceding seasons on the growth of yellow pine. *Torreya* **14**, 115.

Korstian, C. F. (1927). Factors controlling germination and early survival in oaks. *Yale Univ. School Forestry Bull.* **19**, 7.

Korstian, C. F., and Coile, T. S. (1938). Plant competition in forest stands. *Duke Univ. School Forestry Bull.* **3**.

Kozlowski, T. T. (1949). Light and water in relation to growth and competition of Piedmont forest tree species. *Ecol. Monographs* **19**, 207.

Kozlowski, T. T. (1958). Water relations and growth of trees. *J. Forestry* **56**, 498.

Kozlowski, T. T. (ed.) (1962). "Tree Growth." Ronald Press, New York.

Kozlowski, T. T. (1964a). Shoot growth in woody plants. *Botan. Rev.* **30**, 335.

Kozlowski, T. T. (1964b). "Water Metabolism in Plants." Harper, New York.

Kozlowski, T. T. (1965). Expansion and contraction of plants. *Advancing Frontiers Plant Sci.* **10**, 63.

Kozlowski, T. T. (1967). Diurnal variations in stem diameters of small trees. *Botan. Gaz.* **128**, 60.

Kozlowski, T. T. (1968). Diurnal changes in diameters of fruits and stems of Montmorency cherry. *J. Hort. Sci.* **43**, 1.

Kozlowski, T. T., and Clausen, J. J. (1966). Shoot growth characteristics of heterophyllous woody plants. *Can. J. Botany* **44**, 827.

Kozlowski, T. T., and Keller, T. (1966). Food relations of woody plants. *Botan. Rev.* **32**, 293.

Kozlowski, T. T., and Peterson, T. A. (1962). Seasonal growth of dominant, intermediate, and suppressed red pine trees. *Botan. Gaz.* **124**, 146.

Kozlowski, T. T., and Winget, C. H. (1964). Diurnal and seasonal variation in radii of tree stems. *Ecology* **45**, 149.

Kramer, P. J. (1943). Amount and duration of growth of various species of tree seedlings. *Plant Physiol.* **18**, 239.

Kramer, P. J. (1949). "Plant and Soil Water Relationships." McGraw-Hill, New York.

Kramer, P. J. (1958). Photosynthesis of trees as affected by their environment. *In* "The Physiology of Forest Trees" (K. V. Thimann, ed.), pp. 157–186. Ronald Press, New York.

Kramer, P. J. (1962). The role of water in tree growth. *In* "Tree Growth" (T. T. Kozlowski, ed.), pp. 171–182. Ronald Press, New York.

Kramer, P. J. (1964). The role of water in wood formation. *In* "The Formation of Wood in Forest Trees" (M. H. Zimmermann, ed.), pp. 519–532. Academic, Press New York.

Kramer, P. J., and Bullock, H. C. (1966). Seasonal variation in the proportion of suberized and unsuberized roots of trees in relation to the absorption of water. *Am. J. Botany* **53**, 200.

Kramer, P. J., and Kozlowski, T. T. (1960). "Physiology of Trees." McGraw-Hill, New York.

Kraus, J. F., and Bengtson, G. W. (1960). The use of irrigation in forestry. *In* "Southern Forest Soils," *8th Ann. Forestry Symp. Louisiana State Univ. 1959.* pp. 96–105. Louisiana State Univ. Press, Baton Rouge, Louisiana.

Kraus, J. F., and Spurr, S. H. (1961). Relationship of soil moisture to the springwood-summerwood transition in southern Michigan red pine. *J. Forestry* **59**, 510.

Ladefoged, K. (1952). The periodicity of wood formation. *Kgl. Danske Videnskab. Selskab Biol. Skrifter* **7**, 1.

Laitakari, E. (1927). The root system of pine (*Pinus sylvestris*): a morphological investigation. *Acta Forest. Fennica* **33** (Engl. summary, p. 307).

Larcher, W. (1965). The influence of water stress on the relationship between CO_2 uptake and transpiration. *In* "Water Stresses in Plants" (B. Slavik, ed.), pp. 184–194. Junk, The Hague.

Larson, P. R. (1956). Discontinuous growth rings in suppressed slash pine. *Tropical Woods* **104**, 80.

Larson, P. R. (1957). Effect of environment on the percentage of summerwood and specific gravity of slash pine. *Yale Univ. School Forestry Bull.* **63**.

Larson, P. R. (ed.) (1962). "The Influence of Environment and Genetics on Pulpwood Quality," *Tappi Monograph Ser.* **24**. Tech. Assoc. Pulp Paper Ind., New York.

Larson, P. R. (1963a). The indirect effect of drought on tracheid diameter in red pine. *Forest Sci.* **9**, 52.

Larson, P. R. (1963b). Microscopic wood characteristics and their variations with tree growth. Paper Presented to the Working Group on Wood Quality, International Union of Forestry Research Organizations.

Larson, P. R. (1963c). Stem form development in forest trees. *Forest Sci. Monographs* **5**.

Larson, P. R. (1964). Some indirect effects of environment on wood formation. *In* "The Formation of Wood in Forest Trees" (M. H. Zimmermann, ed.), pp. 345–365. Academic Press, New York.

Leaphart, C. D., and Copeland, O. L., Jr. (1957). Root and soil relationships associated with pole blight disease of western white pine. *Soil. Sci. Soc. Am. Proc.* **21**, 551.

Ledig, F. T., and Perry T. O. (1966). Physiological genetics of the shoot-root ratio. *Soc. Am. Foresters Ann. Meeting Proc.* **1965**, 39.

Lemon, E. R. (1965). Energy conversion and water use efficiency in plants. *In* "Plant Environment and Efficient Water Use," pp. 28–48. Am. Soc. Agron., Madison, Wisconsin.

Leshem, B. (1965). The annual activity of intermediary roots of the aleppo pine. *Forest Sci.* **11**, 291.

Lewis, M. R., Work, R. A., and Aldrich, W. W. (1935). Influence of different quantities of moisture in a heavy soil on rate of growth of pears. *Plant Physiol.* **10**, 309.

Lodewick, J. E. (1928). Seasonal activity in the cambium of some northeastern trees. *Syracuse Univ. Coll. Forestry Tech. Publ.* **23**.

Lodewick, J. E. (1930). The effect of certain climatic factors on the diameter growth of longleaf pine in western Florida. *J. Agr. Res.* **41**, 349.

Lotan, J. E., and Zahner, R. (1963). Shoot and needle responses of 20-year-old red pine to current soil moisture regimes. *Forest Sci.* **9**, 497.

Lyford, W. H., and Wilson, B. F. (1964). Development of the root system of *Acer rubrum* L. *Harvard Forest Paper* **10**.

McClurkin, D. C. (1958). Soil moisture content and shortleaf pine radial growth in north Mississippi. *Forest Sci.* **4**, 232.

McClurkin, D. C. (1961). Soil moisture trends following thinning in shortleaf pine. *Soil Sci. Soc. Am. Proc.* **25**, 135.

McClurkin, D. C. (1963). Soil-site predictions for white oak in north Mississippi and west Tennessee. *Forest Sci.* **9**, 108.

McDermott, R. (1954). Effects of saturated soil on seedling growth of some bottomland hardwood species. *Ecology* **35**, 36.

MacDougal, D. T. (1921). Growth in trees. *Carnegie Inst. Wash. Publ.* **307**.

MacDougal, D. T. (1923). Records of tree growth. *Geograph. Rev.* **13**, 661.

MacDougal, D. T. (1924). Growth in trees and massive organs in plants. *Carnegie Inst. Wash. Publ.* **350**, 50.

Mace, A. C., and Wagle, R. F. (1964). A measure of the effect of soil and atmospheric moisture on the growth of ponderosa pine. *Forest Sci.* **10**, 454.

Magness, J. R., Degman, E. S., and Furr, J. R. (1935). Soil moisture and irrigation experiments in eastern apple orchards. *U.S. Dept. Agr. Tech. Bull.* **491**.

Mell, C. D. (1910). Determination of quality of locality by fiber length of wood. *Forestry Quart.* **8**, 419.

Miller, L. N. (1965). Changes in radiosensitivity of pine seedlings subjected to water stress during chronic gamma irradiation. *Health Phys.* **11**, 1653.

Mitchell, A. F. (1965). The growth in early life of the leading shoot of some conifers. *Forestry* **38**, 121.

Moore, D. C., and Aldrich, W. W. (1938). Leaf and fruit growth of the date in relation to moisture in a saline soil. *Proc. Am. Soc. Hort. Sci.* **36**, 216.

Mosher, M. M. (1960). Irrigation and fertilization of ponderosa pine. *Wash. State Univ. Agr. Expt. Sta. Circ.* **365**.

Motley, J. A. (1949). Correlation of elongation in white and red pine with rainfall. *Butler Univ. Botan. Studies* **9**, 1.

Mueller-Dombois, D. (1964). Effect of depth to water table on height growth of tree seedlings in a greenhouse. *Forest Sci.* **10**, 306.

Newman, I. V. (1956). Pattern in meristems of vascular plants. I. Cell partition in living apices and in the cambial zone in relation to the concepts of initial cells and apical cells. *Phytomorphology* **6**, 1.

Oosting, H. J., and Kramer, P. J. (1946). Water and light in relation to pine reproduction. *Ecology* **27**, 47.

Oppenheimer, H. R. (1945). Cambial wood production in stems of *Pinus halepensis*. *Palestine J. Botany* **5**, 22.

Owston, P. W. (1968). Late summer second shoots in *Pinus strobus*. *Forest Sci.* **14**, 66.

Parker, J. (1952). Desiccation in conifer leaves: anatomical changes and determination of the lethal level. *Botan. Gaz.* **114**, 189.

Paul, B. H (1950). Wood quality in relation to site quality of second growth Douglas-fir. *J. Forestry* **48**, 175.

Paul, B. H., and Martz, R. O. (1931). Controlling the proportion of summerwood in longleaf pine. *J. Forestry* **29**, 784.

Pearson, G. A. (1918). The relation between spring precipitation and height growth of western yellow pine saplings in Arizona. *J. Forestry* **16**, 677.

Pessin, L. J. (1938). Effect of soil moisture on the rate of growth of longleaf and slash pine seedlings. *Plant Physiol.* **13**, 179.

Pharis, R. P. (1966). Comparative drought resistance of five conifers and foliage moisture content as a viability index. *Ecology* **47**, 211.

Phipps, R. L. (1961). Analysis of five years dendrometer data obtained within three deciduous forest communities of Neotoma. *Ohio Agr. Expt. Sta. Res. Circ.* **105**.

Phipps, R. L. (1964). Ring analysis of selected trees at Neotoma. Part X of Bioclimatic and Soils Investigations in Forest Environments. *Rept. to U.S. Atomic Energy Comm., Ohio, Agr. Expt. Sta., Ohio State Univ. Contract No.* **AT(11-1-552)**.

Preston, R. J. Jr. (1942). The growth and development of the root systems of juvenile lodgepole pine. *Ecol. Monographs* **12**, 449.

Priestley, J. H. (1930). Studies in the physiology of cambial activity. *New Phytologist* **29**, 316.

Priestley, J. H., and Scott, L. I. (1936). A note upon summer wood production in the tree. *Proc. Leeds Phil. Lit. Soc.* **3**, 235.

Priestley, J. H., Scott, L. I., and Malins, M. E. (1935). Vessel development in the angiosperm. *Proc. Leeds. Phil. Lit. Soc.* **3**, 42.

Ralston, C. W. (1964). Estimation of forest site productivity. *Intern. Rev. Forestry Res.* **1**, 171.

Reed, J. F. (1939). Root and shoot growth of shortleaf and loblolly pines in relation to certain environmental conditions. *Duke Univ. School Forestry Bull.* **4**.

Richardson, S. D. (1958). Bud dormancy and root development in *Acer saccharinum. In* "The Physiology of Forest Trees" (K. V. Thimann, ed.), pp. 409–425. Ronald Press, New York.

Richardson, S. D. (1961). A biological basis for sampling in studies of wood properties. *Tappi* **44**, 170.

Robbins, W. J. (1921). Precipitation and the growth of oaks at Columbia, Missouri. *Missouri Univ. Agr. Expt. Sta. Res. Bull.* **44**, 3.

Roberts, B. R. (1964). Effects of water stress on the translocation of photosynthetically assimilated carbon-14 in yellow poplar. *In* "The Formation of Wood in Forest Trees" (M. H. Zimmermann, ed.), pp. 273–288. Academic Press, New York.

Roeser, J., Jr. (1940). The water requirements of Rocky Mountain conifers. *J. Forestry* **38**, 24.

Rokach, A. (1953). Water transfer from fruits to leaves in the Shamouti orange tree and related topics. *Palestine J. Botany* **8**, 146.

Romberger, J. A. (1963). Meristems, growth, and development in woody plants. *U.S. Dept. Agr. Tech. Bull.* **1293**.

Ryall, A. L., and Aldrich, W. W. (1944). The effect of water deficits in the tree upon maturity, composition, and storage quality of Bosc pears. *J. Agr. Res.* **68**, 121.

Sacher, J. A. (1954). Structure and seasonal activity of the shoot apices of *Pinus lambertiana* and *Pinus ponderosa. Am. J. Botany* **41**, 749.

Sands, K., and Rutter, A. J. (1959). Studies in the growth of young plants of *Pinus sylvestris*. II. The relation of growth to moisture tension. *Ann. Botany (London) [N.S.]* **23**, 269.

Savina, A. V. (1956). The physiological justification for the thinning of forests. Translated from Russian by the Israel Program for Scientific Translations for the National Science Foundation and the *U.S. Dept. Agr. Office Tech. Serv. Publ. No.* **OTS60-51130**.

Scholander, P. F., Hammel, H. T., Bradstreet, E. D., and Hemmingsen, E. A. (1965). Sap pressure in vascular plants. *Science* **148**, 339.

Schroeder, C. A., and Wieland, P. A. (1956). Diurnal fluctuations in size in various parts of the avocado tree and fruit. *Proc. Am. Soc. Hort. Sci.* **68**, 253.

Schulman, E. (1958). Bristlecone pine, oldest known living thing. *Natl. Geographic* **113**, 355.

Shepherd, K. R. (1964). Some observations on the effect of drought on the growth of *Pinus radiata* D. Don. *Australian Forestry* **28**, 7.

Small, J. A., and Monk, C. D. (1959). Winter changes in tree radii and temperature. *Forest Sci.* **5**, 229.

Smith, D. M., and Wilsie, M. C. (1961). Some anatomical responses of loblolly pine to soil-water deficiencies. *Tappi* **44**, 179.

Spurr, S. H. (1964). "Forest Ecology." Ronald Press, New York.

Spurr, S. H., and Hsuing, W. (1954). Growth rate and specific gravity in conifers. *J. Forestry* **52**, 191.

Squillace, A. E., and Bingham, R. T. (1958). Selective fertilization in *Pinus monticola* Dougl. *Silvae Genet.* **7**, 188.

Srivastava, L. M. (1964). Anatomy, chemistry, and physiology of bark. *Intern. Rev. Forestry Res.* **1**, 203.

Stanhill, G. (1957). The effect of differences in soil moisture status on plant growth: A review and analysis of soil moisture regime experiments. *Soil Sci.* **84**, 205.

Steinbrenner, E. C., and Rediske, J. H. (1964). Growth of ponderosa pine and Douglas-fir in a controlled environment. *Weyerhaeuser Forestry Paper No.* **1**.

Stevens, C. L. (1931). Root growth of white pine (*Pinus strobus* L.). *Yale School Forestry Bull.* **32**.

Stone, E. C. (1957a). Dew as an ecological factor. I. A review of the literature. *Ecology* **38**, 407.

Stone, E. C. (1957b). Dew as an ecological factor. II. The effect of artificial dew on the survival of *Pinus ponderosa* and associated species. *Ecology* **38**, 414.

Stout, B. B. (1956). Studies of the root systems of deciduous trees. *Black Rock Forest Bull.* **15**.

Stout, B. B. (1959). Supplemental irrigation of 75-yr.-old hardwoods. *Black Rock Forest Paper* **25**.

Stransky, J. J. (1963). Needle moisture as a mortality index for southern pine seedlings. *Botan. Gaz.* **124**, 178.

Stransky, J. J., and Wilson, D. R. (1964). Terminal elongation of loblolly and shortleaf pine seedlings under soil moisture stress. *Soil Sci. Soc. Am. Proc.* **28**, 439.

Strothmann, R. O. (1967). The influence of light and moisture on the growth of red pine seedlings in Minnesota. *Forest Sci.* **13**, 182.

Studhalter, R. A., Glock, W. S., and Agerter, S. R. (1963). Tree growth—Some historical chapters in the study of diameter growth. *Botan. Rev.* **29**, 245.

Taylor, H. M., and Gardner, H. R. (1963). Penetration of cotton seedling taproots as influenced by bulk density, moisture content and strength of soil. *Soil Sci.* **96**, 153.

Tepper, H. B. (1963). Leader growth of young pitch and shortleaf pines. *Forest Sci.* **9**, 344.

Thames, J. L. (1963). Needle variation in loblolly pine from four geographic seed sources. *Ecology* **44**, 168.

Toole, E. R., and Broadfoot, W. M. (1959). Sweetgum blight as related to alluvial soils of the Mississippi River floodplain. *Forest Sci.* **5**, 2.

Toumey, J. W. (1929). Initial root habit in American trees and its bearing on regeneration. *Proc. Intern. Congr. Plant Sci., 1st Congr., Ithaca 1926*, p. 713.

Trimble, G. R., and Weitzman, S. (1956). Site index studies of upland oaks in the northern Appalachians. *Forest Sci.* **2**, 162.

Tryon, E. H., and True, R. P. (1958). Recent reductions in annual radial increments in dying scarlet oaks related to rainfall deficiencies. *Forest Sci.* **4**, 219.

Tryon, E. H., Cantrell, J. O., and Carvell, K. L. (1957). Effect of precipitation and temperature on increment of yellow-poplar. *Forest Sci.* **3**, 32.

Tucker, C. M., and Evert, E. F. (1964). Phloem development in *Acer negundo*. *Am. J. Botany* **51**, 672.

Turner, L. M. (1936). Root growth of seedlings of *Pinus echinata* and *Pinus taeda*. *J. Agr. Res.* **53**, 145.

U.S. Dept. Agr. (1948). Woody Plant Seed Manual. *Misc. Publ.* **654**.

Wadleigh, C. H., and Gaugh, H. G. (1948). Rate of leaf elongation as affected by the intensity of the total soil moisture stress. *Plant Physiol.* **23**, 485.

Walker, L. C. (1962). The effects of water and fertilizer on loblolly and slash pine seedlings. *Soil Sci. Soc. Am. Proc.* **26**, 197.

Wareing, P. F. (1951). Growth studies in woody species. IV. The initiation of cambial activity in ring porous species. *Physiol. Plantarum* **4**, 546.

Wareing, P. F. (1958). The physiology of cambial activity. *J. Inst. Wood. Sci.* **1**, 34.

Warrack, G., and Joergensen, C. (1950). Precision measurement of radial growth and daily radial fluctuations in Douglas fir. *Forestry Chron.* **26**, 52.

Wellwood, R. W. (1952). The effect of several variables on the specific gravity of second growth Douglas-fir. *Forestry Chron.* **28**, 34.

Wenger, K. F. (1952). Effect of moisture supply and soil texture on the growth of sweetgum and pine seedlings. *J. Forestry* **50**, 862.

Wenger, K. F. (1954). Stimulation of loblolly pine seed trees by preharvest release. *J. Forestry* **52**, 115.

Wenger, K. F. (1957). Annual variation in the seed crops of loblolly pine. *J. Forestry* **55**, 567.

Went, F. W. (1942). Some physiological factors in the aging of a tree. *Natl. Shade Tree Conf. Proc.* **18**, 330.

White, D. P. (1958). Available water: the key to forest site evaluation. *Proc. 1st North Am. Forest Soils Conf., Michigan State Univ., East Lansing, Michigan*, pp. 6–11.

Whitmore, F. W., and Zahner, R. (1966). Development of the xylem ring in stems of young red pine trees. *Forest Sci.* **12**, 198.

Whitmore, F. W., and Zahner, R. (1967). Evidence for a direct effect of water stress in the metabolism of cell wall in *Pinus*. *Forest Sci.* **13**, 397.

Wilcox, H. (1954). Primary organization of primary and dormant roots of noble fir, *Abies procera*. *Am. J. Botany* **41**, 812.

Wilcox, H. (1962). Growth studies of the root of incense cedar, *Libocedrus decurrens*. I. The origin and development of primary tissues. *Am. J. Botany* **49**, 221.

Wilde, S. A. (1958). "Forest Soils." Ronald Press, New York.

Wilde, S. A., Paul, B. H., and Mikola, P. (1951). Yield and quality of jack pine pulpwood produced on different types of sandy soils in Wisconsin. *J. Forestry* **49**, 878.

Wilson, B. F. (1963). Increase in cell wall surface area during enlargement of cambial derivatives in *Abies concolor*. *Am. J. Botany* **50**, 95.

Wilson, B. F. (1964a). A model for cell production by the cambium of conifers. *In* "The Formation of Wood in Forest Trees" (M. H. Zimmermann, ed.), pp. 19–36. Academic Press, New York.

Wilson, B. F. (1964b). Structure and growth of woody roots of *Acer rubrum* L. *Harvard Forest Paper* **11**.

Wilson, B. F. (1966). Mitotic activity in the cambial zone of *Pinus strobus* L. *Am. J. Botany* **53**, 364.

Wilson, B. F., Wodzicki, T. J., and Zahner, R. (1966). Differentiation of cambial derivatives: proposed terminology. *Forest Sci.* **12**, 438.

Winget, C. H., and Kozlowski, T. T. (1964). Winter shrinkage in stems of forest trees. *J. Forestry* **62**, 335.

Winget, C. H., and Kozlowski, T. T. (1965). Seasonal basal area growth as an expression of competition in northern hardwoods. *Ecology* **46**, 786.

Worley, J. F., and Hacskaylo, E. (1959). The effect of available soil moisture on the mycorrhizal association of Virginia pine. *Forest Sci.* **5**, 267.

Worrall, J. (1966). A method of correcting dendrometer measures of tree diameter for variations induced by moisture stress changes. *Forest Sci.* **12**, 427.

Zahner, R. (1959). Soil moisture utilization by southern forests. *8th Ann. Forestry Symp., Louisiana State Univ.* pp. 25–30.

Zahner, R. (1962). Terminal growth and wood formation by juvenile loblolly pine under two soil moisture regimes. *Forest Sci.* **8**, 345.

Zahner, R. (1963). Internal moisture stress and wood formation in conifers. *Forest Prod. J.* **13**, 240.

Zahner, R., and Donnelly, J. R. (1967). Refining correlations of rainfall and radial growth in young red pine. *Ecology* **48**, 525.

Zahner, R., and Oliver, W. W. (1962). The influence of thinning and pruning on the date of summerwood initiation in red and jack pines. *Forest Sci.* **8**, 51.

Zahner, R., and Stage, A. R. (1966). A procedure for calculating daily moisture stress and its utility in regressions of tree growth on weather. *Ecology* **47**, 64.

Zahner, R., and Whitmore, F. W. (1960). Early growth of radically thinned loblolly pine. *J. Forestry* **58**, 628.

Zahner, R., Lotan, J. E., and Baughman, W. D. (1964). Earlywood-latewood features of red pine grown under simulated drought and irrigation. *Forest Sci.* **10**, 361.

Ziegler, H. (1964). Storage, mobilization, and distribution of reserve material in trees. *In* " The Formation of Wood in Forest Trees " (M. H. Zimmermann, ed.), pp. 303–320. Academic Press, New York.

Zimmermann, M. H. (ed.) (1964a). "Wood Formation in Forest Trees." Academic Press, New York.

Zimmermann, M. H. (1964b). Sap movements in trees. *Biorheology* **2**, 15.

Zimmermann, M. H. (1964c). The relation of transport to growth in dicotyledonous trees. *In* " The Formation of Wood in Forest Trees " (M. H. Zimmermann, ed.), pp. 289–301. Academic Press, New York.

Zobel, B. J., and Rhodes, R. R. (1955). Relationship of wood specific gravity in loblolly pine to growth and environmental factors. *Texas Forest Serv. Tech. Rept.* **11**.

CHAPTER 6

WATER DEFICITS IN VASCULAR DISEASE

P. W. Talboys

EAST MALLING RESEARCH STATION, KENT, ENGLAND

INTRODUCTION

Vascular mycoses and bacterioses are induced by a miscellaneous group of organisms, some of which occur in an exceedingly wide range of host species. Nevertheless, the resulting disease syndromes have much in common. This convergence of syndromes is generally interpreted in terms of a common mechanism of disease induction, viz, the development of water stress. This chapter considers the extent to which the symptoms of vascular disease are consistent with the effects of water stress, how water stress can be induced, and the mechanisms that may ameliorate the effects of water stress.

A dominant feature of vascular disease etiology is that once the pathogen has entered the vascular system, either directly through wounds or indirectly after penetrating intact or damaged extravascular tissues in the root, it generally remains within the nonliving conducting tissues of the xylem until the disease has reached an advanced stage. As the diseased plant becomes moribund, the pathogen invades xylem parenchyma, medullary rays, phloem, and cortex. This feature distinguishes vascular pathogens, such as certain *Verticillium* species and *Fusarium* species, from wood-rotting fungi that, although

255

predominantly xylem-colonizing organisms, can kill the living parenchyma and ray cells and then colonize them.

Thus, although the pathogenic activities of the "silver leaf" pathogen, *Stereum purpureum*, have certain features in common with those of vascular pathogens, and will be discussed in this connection, the organism is more appropriately grouped with other related "wood-rotting" fungi because of its capacity to invade ray and parenchyma cells.

Evidence of early growth in xylem parenchyma and ray cells by pathogens causing oak wilt (*Ceratocystis* [= *Endoconidiophora*] *fagacearum*), persimmon wilt (*Cephalosporium diospyri*), and Dutch elm disease (*Ceratocystis* [= *Ceratostomella*] *ulmi*) has been given by Wilson (1961, 1963, 1965) and suggests that possibly these pathogens have more in common with *Stereum* than with the *Verticillium* and *Fusarium* species.

A study of the role of water stress in vascular pathogenesis is thus concerned mainly with activities of the pathogen from the time it enters a vascular system until symptoms are fully developed, and with the consequences of its activity in terms of its effects on the water relations of the host plant. First, however, the outward and visible manifestations of vascular pathogenesis will be considered.

II. THE SYNDROME IN VASCULAR DISEASE

It would be reasonable to expect that specific patterns of disease development would be associated with particular pathogens and would be modified in detail by particular host species. However, vascular disease syndromes seem to be broadly divisible into two main groups based *not* on taxonomic relationships of the host or parasite but on the *growth form* of the host, viz, whether it is an herbaceous plant or one with a woody perennial shoot system. The two groups of syndromes have much in common, and the differences between them relate, at least in part, to the fact that in plants with perennial shoot systems there is a greater possibility of one year's disease development being influenced by the host/pathogen interactions in previous years.

A. PATTERNS OF DISEASE IN HERBACEOUS PLANTS

The basic pattern of disease development is exemplified by verticillium wilts of tomato (Butler and Jones, 1949), hop (Keyworth, 1942), lupin and sunflower (Griffiths and Isaac, 1963), cotton (Presley, 1950), and potato (Robinson *et al.*, 1957) and fusarium wilts of tomato (Walker, 1957) and banana (Wardlaw, 1961). In these diseases foliar chlorosis, leading to necrosis and defoliation, develops first on the oldest leaves and spreads upward in plants with an elongated axis or inward in rosette plants. Leaf epinasty occurs in some wilt diseases and flaccidity of leaves may precede or accompany chlorosis but does not invariably do so. Vein clearing has been reported as a characteristic early sympton in several vascular diseases, including fusarium wilts of tomato (Foster, 1946), cotton (Subba-Rao, 1954), and bean (*Phaseolus* species) (Thomas, 1966).

In annual herbaceous plants, defoliation is followed by necrosis of the stem and roots, but in perennial herbs the underground parts may survive and develop new shoots, often symptomless, in the following year. In this respect they differ from plants with perennial shoot systems, in which the effects of one year's disease tend to be apparent the following year.

In early stages of a disease, or in a mild disease, symptoms may develop only on one side of a plant as, for example, in peppermint verticillium wilt (Nelson, 1950) and hemp fusarium wilt (Noviello and Snyder, 1962). Unilateral symptoms may even be evident in individual leaves, in which only a small sector may be chlorotic or necrotic while the rest of the leaf is apparently normal; this occurs in mild attacks of verticillium wilt in hops (see Fig. 1).

These unilateral patterns are a consequence of the manner in which the pathogen colonizes the vascular system. Movement from the initial infection site occurs partly by mycelial growth and partly by the distribution of propagules in the transpiration stream (Banfield, 1941; Sewell and Wilson, 1964). The major movement is therefore upward from vessel to vessel in catenary succession. Because vascular pathogens do not at the outset invade living xylem parenchyma or ray cells, and often do not readily enter thick-walled fibers, their lateral spread tends to be limited to growth through pits to adjacent vessels. Hence, distribution of the pathogen depends on the number and location of entries to the vascular system and on the structure of the host xylem. Infection of a few roots on one side of a plant thus may lead to a unilateral symptom pattern.

Extensive invasion of a vascular system from numerous infection loci results in relatively uniform symptoms over the whole of each leaf. Such symptoms appear to be of two main types. In the first there is initially an overall yellow mottling, sometimes preceded by vein clearing. Necrosis may occur in irregular patches or may be initiated at the tip or margins, later spreading uniformly across the lamina toward the petiole, although defoliation may occur before necrosis becomes extensive. In the second type, chlorosis is intensified in interveinal areas and necrosis, initially marginal, also spreads along interveinal areas, leaving the veins green until a relatively late stage. The result is a striking pattern of dark brown, yellow, and green that, in hop verticillium wilt, has been termed "tiger striping" (Fig. 2). A similar pattern occurs in raspberry verticillium wilt (Harris, 1925). Development of this pattern tends to be relatively rapid, and eventually the whole leaf becomes necrotic and in-rolled and is abscissed.

A number of so-called "wilt diseases" such as hop verticillium wilt are

Fig. 1. Hop leaf showing sectorial chlorosis and necrosis characteristic of the mild wilt syndrome. (*Verticillium albo-atrum*.)

inappropriately named because neither leaves nor stems become flaccid; instead, the leaves become stiff and brittle with progressive desiccation.

Nevertheless, in some other diseases, e.g., tomato fusarium wilt at high temperatures, leaves collapse and wither in a manner consistent with the term "wilt." Similar rapid collapse of leaves occasionally occurs in diseases normally characterized by desiccation without flaccidity; such atypical symptoms usually result from exceptionally severe infections or exceptionally adverse environmental conditions and are often described as "quick wilt." When leaves are killed rapidly while they are still green they generally do not abscise and they remain attached to the plant for a long time.

It is notable that the syndrome in oil palm fusarium wilt corresponds with that of an herbaceous plant even though there is a perennial shoot system. This similarity is, presumably, associated with the fact that the vascular system in most monocotyldons remains in a condition comparable with that in a

Fig. 2. Marginal and interveinal necrosis characteristic of acute syndrome in hop verticillium wilt.

young dicotyledonous plant with discrete vascular bundles. There is no secondary xylem "skeleton" supporting renewed cambial growth each year. Wilt disease syndromes in monocotyledons appear to differ from those in dicotyledons in that abscission of the leaves does not occur in the former.

The symptoms described so far are the ones that immediately direct attention to the presence of the pathogen. In vascular pathogenesis, however, the onset of chlorotic or necrotic symptoms is often preceded by reduced plant growth resulting in shorter internodes and a tendency to rosette formation in terminal shoots. Reduced growth is not customarily regarded as a typical feature of wilt diseases. Nevertheless, it has been reported, e.g., in diseases caused by *Verticillium albo-atrum* in tomato (Selman and Pegg, 1957) and peppermint (Nelson, 1950*), by *Fusarium oxysporum* in oil palm (Prender-

* In this, and in many other American publications, microsclerotial isolates of *Verticillium* are included in the species *V. albo-atrum*. Elsewhere they are considered as a separate species, *V. dahliae* (Isaac, 1949).

gast, 1963), by *Fusarium oxysporum* f. *pisi* in pea (Linford, 1931b), and by *Fusarium oxysporum* in French bean (Thomas, 1966). Growth impairment without subsequent acute foliar chlorosis or necrosis may be a more frequent consequence of vascular infection than has generally been recognized (Woolliams, 1966).

B. Disease Patterns in Plants with Perennial Shoot Systems

In dicotyledonous trees and shrubs there are a number of variants of the syndrome resulting from vascular infection, though the same species may show different syndromes under different circumstances.

In the most acute form there is a sudden collapse of leaves, often starting with the youngest ones in the outermost parts of the crown. The leaves may die while still green or they may show various degrees of yellowing, bronzing, and necrosis. Defoliation commonly occurs, though it is sometimes delayed when there is rapid death of the leaves and normal abscission does not take place. Some or all branches of the tree or shrub may be affected initially, but leaf symptoms eventually spread throughout the crown. The whole plant is killed within a year of infection.

This acute syndrome is exemplified by oak wilt (*Ceratocystis fagacearum*) in red and black oaks (Henry *et al.*, 1944; Kuntz and Riker, 1956), Dutch elm disease (*Ceratocystis ulmi*) in American elms (Walter *et al.*, 1943; Peace, 1962), mimosa wilt (*Fusarium perniciosum*; Hepting, 1939), some cases of wilt and die-back of white elm caused by a *Cephalosporium* species (Goss and Frink, 1934), and persimmon wilt (Crandall and Baker, 1950) incited by *Cephalosporium diospyri*.

Acute symptoms do not, however, invariably lead to the rapid death of the whole plant. There may be a progressive extension of the disease throughout the crown of the plant over a number of years, as in Dutch elm disease in English elms (Peace, 1962), in oak wilt in white and bur oaks (Kuntz and Riker, 1956), in cephalosporium wilt of white elm (Goss and Frink, 1934), and in sapstreak disease of sugar maple, caused by *Endoconidiophora virescens* (Hepting, 1944). This pattern of disease development may be conveniently termed a subacute syndrome.

In a further variant of the acute syndrome the severe foliar symptoms are transient and are limited in extent. Only one or a few branches are affected initially, and thereafter the disease progresses no further and the plant apparently recovers from the disease. This transient syndrome occurs in some cases of oak wilt in white and bur oaks (Kuntz and Riker, 1956) and of Dutch elm disease in English elms. It also occurs in the verticillium wilt diseases of avocado (Zentmyer, 1949), olive (Wilhelm and Taylor, 1965) and various other fruit tree species (Parker, 1959; Blodgett, 1965). Intermittent die-back and renewed growth from below the affected shoots have been reported for verticillium wilt of yellow poplar (tulip tree) by Waterman (1956).

The variants of the vascular disease syndrome in woody plants described above have all been characterized by acute foliar symptoms in the year of infection. In other diseases the syndrome is a mild or chronic one in which leaves show yellowing commonly on one side of a tree or on a few branches, followed by premature defoliation of the affected parts. Such symptoms often develop in acropetal succession along the branch.

The symptoms may progress from year to year, with leaves becoming smaller and paler, growth being retarded, and some die-back of shoots and branches taking place, though new shoots are produced each year and the plant as a whole survives. Frequently, however, symptoms are transient, occurring in one year and followed apparently by complete recovery of the plant. In other cases the symptoms are intermittent, appearing in some years and not in others, apparently in relation to environmental conditions.

These variants of the mild syndrome occur in the verticillium wilt diseases of many woody trees and shrubs [see, e.g., Parker (1959) and Blodgett (1965) for verticillium wilts in fruit tree species and Peace (1962) for verticillium wilts of various trees and shrubs].

On a much smaller scale, strawberry wilt (*Verticillium dahliae*) has certain features in common with the vascular diseases of tree species, e.g., either progressive reduction in leaf size from year to year or apparent recovery by the replacement of diseased axes.

In raspberry wilt (*V. dahliae*) the biennial shoots often show acropetal foliar symptoms with marginal and interveinal necrosis in the first year, as in herbaceous plants, and either failure of buds to open or yellowing and collapse of leaves in the second year (Harris, 1925).

A further variant of the mild syndrome is one in which no obvious foliar symptoms occur in the year in which infection takes place, but in the following year the buds on some shoots fail to open. Die-back from the tip of the shoot also occurs. This disease pattern has been observed in quince infected with *V. dahliae* (Talboys, unpublished) and may be of fairly common occurrence in that host, although the damage has probably been attributed to other disorders of uncertain etiology such as "die-back" or "winter injury."

As in herbaceous plants, the presence of the pathogen may cause a reduction in growth, either without the development of other symptoms or as a prelude to more acute foliar symptoms or during recovery after an acute phase of disease. Thus, Bewley (1922) showed that inoculation with *Verticillium* led to dwarfing in elm and sycamore seedlings. Dwarfing is a feature of strawberry wilt caused by *V. dahliae* (Talboys *et al.*, 1961), and work by Talboys and Bennett (Anonymous, 1963) showed that shoot growth of Conference pear on quince rootstocks was significantly reduced by soil inoculation with

V. dahliae even though no other symptoms were evident and few plants showed vascular infection above soil level.

Apparent recovery of plants from vascular diseases may be associated with the death of the pathogen within the host tissues, as has been reported in olive verticillium wilt (Wilhelm and Taylor, 1965) and apricot verticillium wilt (Taylor, 1963) or with "burial" of the pathogen after the production of new xylem tissues. In some diseases the pathogen appears to be restricted to the wood formed in the year of infection; it does not spread from one annual ring to the next. Thus, Fig. 3 (p. 240) shows the limitation of *V. dahliae* in a plum tree. In such cases renewed symptoms in successive years apparently result from reinfection from an external source, except where infection of the late-formed xylem exerts an indirect effect in the following year by killing or inhibiting adjacent cambium, shoot apices, or buds, as in raspberry wilt (Harrris, 1925) or quince wilt (Talboys, unpublished).

III. SYMPTOMS OF WATER DEFICIT IN PLANTS:
THE DROUGHT SYNDROME

The manifestations of vascular disease are reproduced in many respects in noninfected plants by water stress resulting from suboptimal soil moisture or excessive water loss. Thus, sudden and acute water stress is expressed by the wilting and collapse of leaves and shoots, death of leaves while still green, and failure of the abscission mechanism. Death of the whole plant may subsequently occur, but die-back of shoots and branches may constitute an intermediate stage in woody perennials. Less acute water stress may cause marginal necrosis and defoliation.

Prolonged subacute stress results in mottled chlorosis of leaves. These do not become flaccid and are eventually shed by normal abscission. Development of drought symptoms in herbaceous plants is commonly acropetal, the oldest leaves (i.e., the *lowest* in plants with an elongated axis and the *outermost* in rosette plants) being affected first. This sequence is not generally evident in woody perennials.

Nonlethal water stress also leads to reductions in leaf size, shoot extension, and cambial growth. Each of these water stress phenomena has its counterpart in at least one of the vascular disease syndromes outlined above. However, certain characteristics of vascular diseases apparently are not reproduced in nonpathogenic water stress, notably one-sided development of symptoms, and the strikingly prominent striped patterns of interveinal necrosis characteristic of certain wilt diseases, which may sometimes simulate mineral deficiency symptoms rather than those of water stress.

IV. MECHANISMS OF PATHOGENIC WATER STRESS

The association of symptom development with disturbances in water relationships in intact infected plants has been demonstrated for several vascular diseases (Linford, 1931a; Harris, 1940; Ludwig, 1952; Scheffer and Walker, 1953; Threlfall, 1959; Beckman *et al.*, 1962). The *mechanisms* of such disturbances have been ascribed both to interruption of the transpiration stream between the root surface and the terminal tracheids that form the ultimate foliar elements of the vascular system and to excessive water loss, which could result from failure of stomatal control or loss of the water-retaining properties of the mesophyll.

A. INDUCTION OF PATHOGENIC WATER DEFICIENCY: OCCLUSION THEORY

1. Changes in Physiology of Extravascular Root Tissues

Soil-borne vascular pathogens, notably *Verticillium* species and *Fusarium* species, are apparently able to enter the host by direct penetration of epidermal cells and to proceed by inter-cellular and intracellular growth through the cortex and nonsuberized endodermal cells to enter the vascular system of the root (Talboys, 1958a, 1964).

Although these wilt diseases develop from initial root infections, root damage is not usually a conspicuous result and is generally evident only in the terminal stages of the disease, when the pathogen passes outward from the vascular elements as the surrounding tissues become moribund, although Orton (1902) considered root damage to be significant in cow pea wilt. Even the loci of initial root penetration are usually difficult to detect. Nevertheless, the presence of the pathogen in contact with epidermal and cortical cells can lead to structural alterations in cell walls, which become thickened and impregnated with ligninlike substances. Cortical infection may also cause endodermal suberization to occur earlier than in noninfected roots (Talboys, 1958a). In species in which hypodermis suberization is a normal feature of root development, premature suberization may be induced by the presence of a pathogenic organism in or on the epidermis. There seems to be little direct evidence that such changes in root-tissue physiology resulting from invasion by vascular pathogens have any influence either on water and nutrient uptake and transport via the protoplasts or on mass flow through free space. The possibility of such effects needs investigation, because they might be specially significant when infective units of the pathogen are present in very large numbers in the soil. Obviously, cells that are severely damaged or killed and colonized by a pathogen lose their osmotic properties, but in vascular diseases caused by *Verticillium* species and *Fusarium* species such cells seem to be relatively few in number, and the effects are unlikely to be as dramatic as those reported in a tobacco root rot (Jenkins, 1948) in which root destruction caused acute symptoms of drought and mineral deficiency. More immediately significant is the work of Sadasivan and Saraswathi-Devi (1957), who showed marked differences in the balance of various inorganic constituents of cotton plants depending on the presence or absence of the wilt pathogen *Fusarium oxysporum* f. *vasinfectum*. The changes associated with the presence of the pathogen were attributed to derangement of the selective absorption mechanisms in the roots as a result of the action of fungal toxins. The activities of fungal metabolities, including their effects on cell permeability, are considered in some detail in Section IV,B,4.

2. Occlusion of Vascular Tissues

Vessels may be occluded by the pathogen itself, by metabolites of the pathogen, by products of host-tissue degradation, or by pathogen-induced host responses that may simultaneously contribute to host resistance.

a. Obstruction by the Pathogen. The intense vascular colonization often evident in terminal stages of vascular wilts, especially in herbaceous plants, has led some observers to conclude that symptom development is a consequence of water shortage induced by blockage of the vessels by masses of mycelium or bacteria.

This appears to have been first postulated by Stewart (1897) for a maize bacterial wilt (*Phytomonas stewarti*) and by Smith (1899), who, referring to fusarium wilt of watermelon, stated: "The gross symptoms ... are those of a plant transpiring freely, and insufficiently supplied with water, although at the same time there is an abundance of moisture in the soil.... The water ducts are clogged [by mycelium] to such an extent that they cannot function." Smith noted that wilting was acute in hot, dry weather and tended to be reduced by the onset of cool, moist conditions. Leaves showing early symptoms of wilt recovered turgor when they were removed from the plant and placed with their petioles in water.

Experimental evidence of reduced flow through vascular tissues of cabbage stems infected with *Fusarium oxysporum* f. *conglutinans* was obtained by Melhus *et al.* (1924). The rate of flow in diseased stems was only about 12% of that in healthy ones, although Gilman (1916), working with the same disease, had shown that only about 17% of the stem vessels contained mycelium.

Many other workers, from Hutchinson (1913) onward have asserted that the amounts of bacteria or mycelium present in vascular diseases are insufficient to account for the whole of the disease syndrome (Brandes, 1919; Rosen, 1926; Gottlieb, 1944; Ludwig, 1952; Waggoner and Dimond, 1954; Talboys, 1958c).

This issue was investigated by Ludwig (1952) in studies of tomato fusarium wilt and by Waggoner and Dimond (1954), working with a model system have ing fluid flow characteristics similar to those of vessels of infected tomato plants.

Data from earlier studies on the movement of radioactive phosphorus in the vascular system (Dimond and Waggoner, 1953c) together with determinations of density and viscosity of xylem sap and the mean radius of vessels in a given length of stem permitted determination of the Reynolds number (Re) for tomato vessels;

$$Re = 2dvr/u \tag{1}$$

where d = density of the fluid, v = its velocity, u = its viscosity, and r = the radius of the vessel.

The value obtained for the Reynolds number (Re = 0.10) indicated that water movement in the vessel would have the characteristics of laminar rather than turbulent flow. Obstruction in the vessels, therefore, would not interfere with water movement by increasing turbulence but would merely increase resistance by frictional drag.

Resistance to flow in such a system may be expressed in terms of the drag coefficient C:

$$C = (4pr)/dv^2l \tag{2}$$

where p = pressure drop along a length l of a tube of radius r, d = density, and v = velocity of the fluid.

The rate of laminar flow in narrow tubes is expressed in terms of Poiseuille's law:

$$V = (pr^4)/(8lu) \tag{3}$$

where V = flow rate (in volume measure per second) and p, r, l, and u are as defined above.

The validity of this law as applied to flow in the vascular system of the tomato has been demonstrated in experiments by Ludwig (1952), in which he determined the effects of varying the length of the system, the cross-sectional area of the conducting tissue, the applied pressure, and the viscosity of the fluid. All of these caused the rate of flow to vary as expected within limits of experimental error.

Poiseuille's law may also be expressed in terms of linear flow per unit time:

$$V = (pr^2)/(8lu) \tag{4}$$

where V = velocity (in linear units per second). From the above equations

Waggoner and Dimond (1954) showed that for a smooth cylindrical tube

$$C = 64/\mathrm{Re} \qquad (5)$$

Experimental determinations of C and Re in a model system 6 cm long × 0.249 cm in radius, with glycerol–water mixtures representing vascular fluids, showed that at a range of Re from 0.04 to 1.70 the relationship shown in Eq. (5) was valid. Addition of "perforation plates" with apertures one half the tube diameter at the ends of the test section increased the value of the constant approximately twofold, i.e., for a given Re, the drag coefficient was doubled. The addition of model "hyphae" one tenth the diameter of the "vessel" caused further increases in C, ranging from times 1.7 for two "hyphae" to 6.0 for 10 "hyphae," the latter resulting in a reduction in flow of 83% at the same pressure or requiring a sixfold increase of pressure to maintain the same flow rate.

Thus, in determinations of flow rates in stem segments by methods such as those of Melhus *et al.* (1924) and Ludwig (1952), the presence of 10 hyphae in all the vessels would be expected to reduce flow rate by about 83%. A somewhat smaller reduction would be expected in intact plants under tensions induced by transpiration. Homeostasis in the plant's hydraulic system is such that a reduced rate of flow can be partially compensated by increased tension derived from increased OP. However, infection as extensive as 10 hyphae per vessel over the *whole* xylem cross section rarely occurs at the time symptoms are induced in vascular wilts. In Ludwig's experiment, for example, only 16% of the stem vessels in cross sections contained mycelium. Applying Poiseuille's law, taking into account the reduced size of a diseased plant and assuming that vessels containing mycelium were completely occluded, he estimated that at a given pressure the flow rate in a diseased stem should be approximately half of that in a healthy one. However, his measurements showed that water flow in the diseased stem was only 6% of that in a healthy one. In this instance, as in the case of cabbage infected with *Fusarium oxysporum* f. *conglutinans* (Melhus *et al.*, 1924), a reduction in water flow occurred that could not be ascribed wholly to vascular occlusion by mycelium. The point must be made, however, that in some vascular diseases appreciable numbers of vessels may be much more intensively colonized than would be represented by Waggoner and Dimond's model. An example is shown in another connection in Fig. 17, in which many hyphal branches are orientated across the line of flow in a vessel. Furthermore, the density of mycelial masses generally is greater in the lower parts of a catena of infected vessels than in upper parts.

Nevertheless, it is generally accepted that occlusion by the *organism*, as distinct from its products, is not a major factor in most vascular diseases, and a number of other occluding agents have been recognized; these may be metabolites produced by the pathogen or host products formed as a result of other activities of the pathogen.

b. Obstruction by Pathogen Metabolites. One major group of metabolites, the polysaccharides, has been implicated in wilt induction in a number of diseases including Dutch elm disease (Dimond, 1947), oak wilt (White, 1955), maple verticillium wilt (Caroselli, 1954), and bacterial wilts caused by *Pseudomonas solanacearum* (Husain and Kelman, 1958a). They have generally been considered as "toxins," but Hodgson *et al.* (1949) showed that polysaccharides probably obstructed vessels to various degrees, depending on their molecular weight. Polysaccharides forming opalescent colloidal "solutions" in water caused flaccidity of tomato stems and petioles. More soluble ones of low molecular weight, giving clear solutions in water, caused wilting of leaflets.

It was inferred that high molecular weight polysaccharides plugged vessels near the bases of cuttings, whereas those of low molecular weight plugged only the small terminal elements of the vascular system. Husain and Kelman (1958b) attributed the wilt-inducing action of polysaccharides to increased viscosity of vascular fluids rather than direct plugging of the vessels.

Polysaccharide production *in vitro* has been reported for a number of vascular pathogens, including *Fusarium solani* f. *eumartii* (Thomas, 1949), *Verticillium albo-atrum* (Porter and Green, 1952; Caroselli, 1954; Le Tourneau, 1957), *Cephalosporium gramineum* (Spalding, 1960), *Ceratocystis ulmi* (Dimond, 1947), and *Ceratocystis fagacearum* (White, 1955).

Husain and Kelman (1958a) found that a virulent strain of *Pseudomonas solanacearum* produced a polysaccharide slime that apparently increased the viscosity of the vascular stream and interfered with water movement in tomato vessels. Avirulent or low-virulence strains did not produce slime.

Feder and Ark (1951) and Leach *et al.* (1957) showed that polysaccharides produced by certain nonwilt-inducing phytopathogenic bacteria could also induce wilting in sunflower and tomato shoots. Thus, fungal and bacterial polysaccharides supplied in solution to cut shoots of plants consistently induce flaccidity that in some cases develops initially in stems and petioles; in other cases the leaf laminae collapse first. Leaves gradually become desiccated, often without marked change in color. The symptoms are entirely consistent with the failure of water transport in the vascular system, but they do not generally simulate the whole syndrome induced by the pathogen in the host. This does not necessarily exclude them as significant factors in disease development. Applying a relatively concentrated solution to the whole vascular system of a shoot could not be expected to reproduce exactly the effect of the same material supplied in low but increasing concentrations to a small proportion of the vascular cylinder. The role of polysaccharides in synergistic action with toxins will be considered later.

c. Obstruction by Host Degradation Products. Bewley (1922) noted the occurrence of hyaline gumlike deposits, later becoming brown tinted, in vessels of tomato plants infected with *Verticillium*. He suggested that the deposits might be products of the pathogen or host breakdown products resulting from the action of fungal enzymes. The deposits were evident in water-mounted hand sections but were not visible in permanent preparations. Similar observations were reported for *Fusarium*-infected tomato by Ludwig (1952), who suggested that the deposits might be hydrophilic pectic colloids that contracted to a thin film lining the vessel wall when dehydrated during the preparation of permanent mounts.

Gothoskar *et al.* (1953) found that *F. oxysporum* f. *lycopersici* produced a pectin methylesterase (PME) but very little polygalacturonase (PG) *in vitro*. Exo-enzyme preparations induced typical wilt symptoms in tomato shoots, and PME was proposed as the chief factor causing wilt and vascular discoloration. Similar results were obtained by Winstead and Walker (1954) for a number of *formae* of *F. oxysporum* and *F. solani*.

Waggoner and Dimond (1955) showed that both PME and PG were produced in the presence of pectin, and PME of fungal origin was detected in the vascular system of diseased plants. They suggested that PME and PG macerated xylem cells and that debris and calcium uronide gels lodged in vessels and occluded them. As a result of maceration, host phenolic glycosides were released and hydrolyzed either by fungal enzymes or those released by the host. Oxidation of the phenolic compounds would cause characteristic browning, as postulated by Dimond *et al.* (1954).

Combined action of PME and a pectin depolymerase was shown by Pierson *et al.* (1955) to induce granular vascular plugs, which stained with ruthenium red, in tomato stems. Similar plugs were found in stems invaded with *Fusarium*. Preparations containing only PME did not cause plugging.

Production of pectolytic enzymes has been found in a number of other vascular pathogens, including *Fusarium oxysporum* f. *conglutinans* (Heitefuss *et al.*, 1960), *F. oxysporum* f. *lini* (Trione, 1960), *F. solani* f. *phaseoli* (Bateman, 1966), *Ceratocystis fagacearum* (White, 1954; Fergus and Wharton, 1957), *Ceratocystis ulmi* (Beckman, 1956), *Cephalosporium gramineum* (Spalding *et al.*, 1961), *Pseudomonas solanacearum* (Husain and Kelman, 1957), *Verticillium albo-atrum* (Talboys, 1958b; Deese and Stahmann, 1962; McIntyre, 1964), *V. dahliae* (Kamal and Wood, 1956), and *Verticillium* species (Matta, 1963).

Most of the major vascular pathogens thus have been shown capable of producing pectolytic enzymes, but agreement has not been reached on whether these enzymes are a significant factor in the development of vascular occlusions. Thus, McDonnell (1958) found that a strain of *F. oxysporum* f. *lycopersici* that failed to produce pectolytic enzymes was pathogenic to tomato and induced the normal syndromes. Kamal and Wood (1955) reported that *Verticillium* filtrates lacking protopectinase were toxic to cotton shoots, as were pectinase-containing filtrates before and after boiling. On the other hand, Leal and Villanueva (1962) found that nonpathogenic species of *Verticillium* failed to produce pectic enzymes.

In plum silver-leaf disease the " silvering " symptom results from the entry of air into spaces formed by the separation of palisade cells from one another and from the epidermis. Brooks and Brenchley (1931) showed that cell separation was induced by a thermolabile metabolite of the pathogen *Stereum purpureum*. It has been suggested, e.g., by Wood (1960) that this was an effect of a translocated pectolytic enzyme. A direct effect of pectic enzymes on leaf tissues does not seem to have been proposed as a pathogenic mechanism in vascular wilt diseases, although, as Wood (1960) has pointed out, leaf tissues appear to provide more suitable substrates for pectolytic action than the inner surfaces of lignified xylem elements.

The possible role of cellulases in causing vascular occlusion by products of tissue disintegration has been studied less extensively than that of pectinases, although there is a potentiality for releasing substances of large molecular weight into the lumina of vessels, as with the pectinases. There is commonly, however, a limiting factor to cellulase production; the presence of other carbohydrates tends to supress cellulase formation by the pathogen. No such limitation operates for pectinase production. Cellulase is produced only in the presence of cellulose, but pectinases can be adaptive or constitutive, depending upon the organism and the enzyme involved.

Nevertheless, cellulase production *in vitro* has been reported for several vascular pathogens, including *Ceratocystis ulmi* (Beckman, 1956), *Fusarium oxysporum* f. *lycopersici* (Husain and Dimond, 1958), *Verticillium albo-atrum* (Talboys, 1958b), and *Pseudomonas solanacearum* (Husain and Kelman, 1957). In the latter there was evidence of *in vivo* cellulose degradation. Purified cellulase from *Myrothecium verrucaria* induced wilting in tobacco shoots (Husain and Kelman, 1958b).

A tissue change caused by a nonenzymic metabolite, fusaric acid, has been reported by Beckman (1964), who found that the toxin induced swelling of perforation plates and primary cell walls in vessels to form a gel. Such gel plugs resembled those induced by *Fusarium oxysporum* f. *cubense* and other fungi (Beckman and Halmos, 1962). The gel appeared to be mainly an esterified acid polymer, probably containing pectic materials (Beckman, 1964).

*d. Obstruction by Host Reactions: Gummosis and Tylosis.** Invasion of vascular tissues by parasitic organisms frequently leads to development of vascular

---

* The term *tylosis* is used here in accordance with the convention that the suffix *-osis* denotes a process or condition. Hence, *tylosis* is the process of tylose (plural, tyloses) formation or the condition resulting from the formation of tyloses.

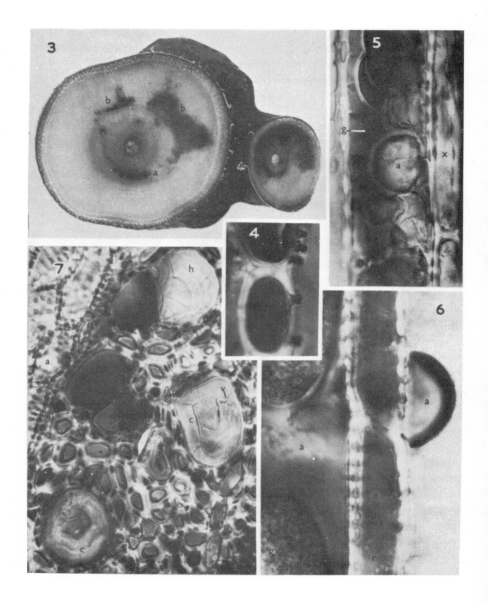

occlusions as a result of changes in the metabolic processes of the living xylem parenchyma and ray cells. As Talboys (1958c, 1964) noted, such responses commonly correspond with wound reactions and generally result in the production of gums or tyloses or both. Chattaway (1949) has pointed out that these are commonly produced as wound responses and during heartwood development and that the specific response is related to the dimensions of the pits connecting vessel elements and ray cells. Species with large pits tend to form tyloses, whereas in those with small pits gummosis is predominant; occasionally both gums and tyloses are produced.

Gums formed as a result of metabolic activity of xylem parenchyma and ray cells must be distinguished from the gels and gums resulting from enzymic breakdown of host tissues. Host-synthesized gums are polysaccharides based commonly on glucuronic acid with associated hexose and pentose sugars (Norman, 1937; Hirst, 1949).

On hydrolysis, the gums formed in the bark, fruits, and stones of plum, for example, have been shown to yield glucuronic acid, galactose, mannose, arabinose, xylose, and rhamnose (Hough and Pridham, 1959). Analytical studies of gums in other Rosaceae include those of cherry (Jones, 1939), damson (*Prunus insititia*; Hirst and Jones, 1938); peach (Jones, 1950) and almond (Brown *et al.*, 1948).

Plum gum was shown to contain certain phenolic substances, as well as some peroxidase and slight phenolase activity. The brown coloring that develops in the gums is probably a result of oxidation of the phenolic compounds.

The biogenesis of gums is uncertain. Some workers concluded that they are formed from pectic constituents of cell walls, whereas others have suggested that starch is the starting material. There seems little doubt that gums are

---

Fig. 3. Vascular discoloration in Brompton plum infected with *Verticillium dahliae*. Note that the main area of infection in the second-year wood of the main stem (a) did not extend into the third year's wood, where new infections (b, b) resulted in obvious symptoms for the first time, with defoliation of the lateral branches on that side of the trunk.

Fig. 4. Early stage of gum secretion from ray cells into vessels of plum infected with *Stereum purpureum* (longitudinal section).

Fig. 5. Longitudinal section of vessel of plum infected with *V. dahliae*, showing granular (g) and spherical amorphous (a) gum masses secreted from adjacent xylem parenchyma cells (x).

Fig. 6. Hemispherical and meniscuslike masses of amorphous "basiphil" gum (a) and granular gum (g) apparently produced by ray cells (r) as a result of *Stereum* infection. Longitudinal section.

Fig. 7. Transverse section of root of raspberry infected with *V. dahliae*, showing hemispherical (h), amorphous (a), and concentric (c) occlusions. A single hypha (v) is visible.

synthesized in the living xylem parenchyma or ray cells and are exuded either fully formed or as non polymerized precursors into vessels via pits (Figs. 4 and 5).

Gum formation in plum may be stimulated by mechanical and insect damage, by bacterial pathogens, e.g., *Pseudomonas mors-prunorum* and by fungal pathogens including *Stereum purpureum* (Brooks and Moore, 1926), and *Verticillium dahliae* (Talboys, unpublished). Gum formation has also been noted in vascular tissues of oak invaded by *Ceratocystis fagacearum* (Struckmeyer *et al.*, 1954), by *V. dahliae* in maple and elm (Van der Meer, 1926), by *Cephalosporium diospyri* in persimmon (Crandall and Baker, 1950), by *Fusarium perniciosum* in mimosa (Hepting, 1939), by *V. dahliae* and *V. albo-atrum* in quince and strawberry (Talboys, unpublished), and by *V. albo-atrum* in cherry (Van der Meer, 1925).

Although gummosis may be extensive in plum silver leaf disease (*Stereum purpureum*), there appears to have been no assertion that the resulting vascular occlusion is the cause of the silvering symptom, which has been attributed to the action of transported pectic enzymes (Wood, 1960, from evidence of Brooks and Brenchley, 1931) or phytolysin (Naef-Roth, *et al.*, 1963).

The virus causing Pierce's disease of grape and also causing dwarf disease of alfalfa induces vascular gummosis in both hosts; a similar response is induced by phony disease virus in peach (Esau, 1948).

A number of workers, including Brooks and Storey (1923) and Esau (1948) noted that the gums produced as a response to pathogenesis show varying staining characteristics, with some taking up cellulose-staining dyes and others lignin-staining dyes. Bartholomew (1928) and Esau (1948) inferred that variations in dye uptake reflect changes in properties of substances intermediate between carbohydrates, such as starch, and the final gum product, which generally reacts with phloroglucinol and hydrochloric acid to give a scarlet color different from that given by lignified cell walls.

Observations by Talboys (unpublished) on pathogen-induced gummosis in plum suggest that the granular gum taking up acidic dyes has a limited capacity for movement in the vascular fluids. The gum staining with basic dyes appears to form brittle resinlike masses with less tendency to flow, initially appearing as hemispherical globules or as meniscuslike occlusions across the vessels (Fig. 6). Both forms give positive reactions with phloroglucinol-HCl, but the granular form gives a less intense reaction in tests for phenolic substances. It is suggested that the hard "basiphil" gum results from terminal stages of damage to the cells producing it, with the result that relatively large quantities of phenolic substances are released. Products of phenolic oxidation commonly stain with basic dyes, and differences in the quantities of such oxidized material would account for changes in staining properties of the gum. Confluent meniscuslike occlusions can readily simulate the appearance of tyloses; Chattaway (1949) has also noted that gum masses may simulate tyloses.

Gum formation in raspberry roots infected with *V. dahliae* sometimes takes the unusual form of concentric accretions, deposited in layers on the vessel wall and around occasional hyphae in the vessels (Fig. 7).

Tyloses appear to provide an alternative to gums as a host-response to damage to the vascular system and, like gums, result from metabolic activity of xylem parenchyma and ray cells. Distension of the pit-closing membrane (Fig. 8) and synthesis of new cell wall material lead to the formation of spherical or ovate bodies that may completely occlude a vessel (Figs. 9, 10, and 11). Several tyloses may be initiated from a single parenchyma or ray cell, but the nucleus of the cell moves into one tylose, which often enlarges conspicuously at the expense of the others. Secondary thickening may occur in the cell walls, and masses of starch grains may accumulate. Tyloses sometimes are so numerous and tightly packed in a vessel that they present the appearance of parenchymatous tissue. Chattaway (1949) discussed the role of tyloses in heartwood formation and as a wound response. Tyloses have been shown to occlude vessels in disease caused by *Fusarium oxysporum* f. *cubense* in banana (Beckman, 1964; Beckman *et al.*, 1961), *F. oxysporum* f. *batatas* in sweet potato (McClure, 1950), *Ceratocystis fagacearum* in oak (Struckmeyer *et al.*, 1954), *Verticillium albo-atrum* in hop (Talboys, 1958a), *V. dahliae* in *Robinia pseudoacacia* (Talboys, unpublished), *V. albo-atrum* in tomato (Blackhurst and Wood, 1963), *F. oxysporum* in oil palm (Kovachich, 1948), *V. albo-atrum* in cherry (Van der Meer, 1925) and in numerous other vascular diseases. Beckman (1966) has discussed tylose formation as part of a general host response limiting vascular invasion by a wide variety of microorganisms.

Tyloses have been regarded as the immediate cause of the visible symptoms of oak wilt. Thus, Struckmeyer *et al.* (1954) showed that in twigs showing incipient wilt symptoms, 50–75% of vessels in any cross section were blocked. The amount of blockage in diseased trees appeared to be sufficient to account for an 85% reduction in the movement of radioactive rubidium previously reported by Beckman *et al.* (1953). Gummosis appeared to contribute to vascular occlusion in severely diseased trees.

There appears to be no conclusive evidence on the mechanism by which tyloses and gums are induced by vascular pathogens. Beckman *et al.* (1953) suggested that the formation of tyloses in oak wilt resulted from weakening of pit membranes by the action of pectinases, but later work (Beckman, 1966) indicated participation of hormonelike substances in tylose induction. McClure (1950) attributed the formation of tyloses in sweet-potato wilt to growth stimulating metabolites of the pathogen, and Talboys (1958c) postulated a metabolite of *V. albo-atrum* that would stimulate tylosis at low concentrations and suppress it at high concentrations.

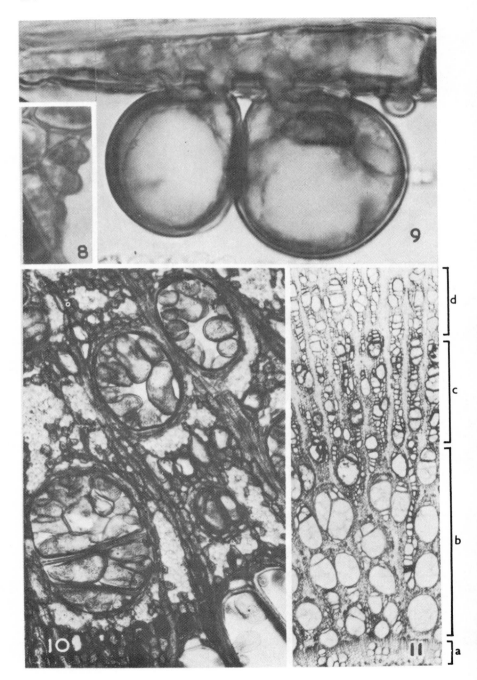

Pegg and Selman (1959) recognized IAA as a metabolite of *V. albo-atrum* and suggested that one of its growth-regulating effects might be the initiation of tyloses. Investigations by Mace and Solit (1966) confirmed this for banana fusarium wilt. However, the ubiquity of tylosis as a response to various physical and chemical stimuli suggests that it may be a consistent reaction to IAA or related hormones, whether these are released from host cells damaged by the inciting stimulus or produced by an invading pathogen. The complex role of growth regulators in plant diseases has been reviewed by Sequeira (1963).

Even less is known about mechanisms of gum induction in vascular diseases. However, the close parallelism between gum formation and the development of tyloses in response to various stimuli suggests a common inciting mechanism. Whatever the immediate incitant may be, the effect appears to be to stimulate synthetic activity of living cells in the xylem. When the pits between such cells and vessels are large, the synthetic activities result in the production of additional cell wall material and increased cell content. Under the constraint of *small* pits, synthetic activity appears to be directed into other channels, resulting in exudation of polyuronide carbohydrates, phenolic substances, etc. Such diversion seems to result in the accumulation of gums, tannins, etc. in reacting cells and finally causes their premature death.

e. Vascular Occlusion as a Water–Stress Mechanism. There is abundant evidence that occlusion of a given proportion of the vessel cross-sectional area at a point in the stem does not necessarily result in a corresponding reduction in water flow to the shoot system above that point. Blockage of vascular elements at one point also does not necessarily lead to water deficit in distal tissues normally supplied through those elements (Ludwig, 1952; Talboys, 1958c). There are two reasons for this: (1) A constant demand from the shoot system will result in an increased flow rate in intact vessels if some

---

Fig. 8. Initiation of tyloses from xylem parenchyma cells in hop.

Fig. 9. Two tyloses, formed from a single xylem parenchyma cell, in vessel of hop root infected with *V. albo-atrum*.

Fig. 10. Transverse section of hop root from moderately resistant variety, showing heavily tylosed vessels, with no obvious mycelium of the incitant, *V. albo-atrum*. Note also the distribution of xylem parenchyma, rays, and fibers that limit the lateral spread of the pathogen in the vascular cylinder.

Fig. 11. Transverse section of stem base of hop showing mild wilt syndrome: (a) primary xylem; (b) normal secondary xylem, heavily tylosed; (c) additional (hyperplastic) secondary xylem, also tylosed; (d) outer ring of hyperplastic xylem, not tylosed, providing for passage of water to the upper parts of the plant. From Talboys (1958c).

vessels are put out of action (2) Vascular tissues form a complex branching and anastomosing system that allows both lateral and longitudinal movement of water, so that water can move within limits from regions of relative plenty to regions of deficit. Studies of water movement around saw cuts in tree trunks, cited by Kozlowski (1964), provide clear evidence of this. Hence, very extensive occlusions may be necessary to induce a sufficient water deficit in the leaves to cause symptoms in vascular diseases. It follows also that limited occlusion, for example, occurring over a short distance in a single quadrant at the base of a stem, would not be expected to lead to water stress sufficient to induce disease symptoms in the upper parts of the shoot.

Dimond and Edgington (1960) have reported a mathematical study of the effects of vessel blockage in tomato, caused by *Fusarium*. The conductive cross-sectional areas of individual bundles were determined, and flow rates in stems and petioles were obtained from transpiration rates of the individual leaves. Application of Poiseuille's, Ohm's, and Kirchoff's laws enabled determination of flow rates in individual vascular bundles of healthy and diseased plants. The main stem bundles are large, include large vessels, and have a network of by-passes. A decrease of 50% in the effective radius of a large stem-bundle over a distance of two nodes can be compensated by a pressure increase of only 3%. In a petiole bundle, however, where there are no by-passes, a similar decrease in effective area would require a 1600% increase in pressure to maintain the same water supply to the leaf. The leaf therefore wilts while flow in the stem is relatively unimpeded. This was presumed to account for the acropetal sequence of symptom development as successive petioles are invaded and occluded. An extended study of pressure and flow relations in tomato xylem was reported by Dimond (1966).

Although Dimond's investigations appear to elucidate the tomato fusarium wilt syndrome they do not adequately account for syndromes in diseases characterized by clearly demarcated marginal, interveinal, or sectorial areas of chlorosis and necrosis such as occur, for example, in verticillium wilts of hops (Figs. 1 and 2) and raspberries. It is difficult to conceive a pattern of occlusion in the main vascular system or petioles that would be likely to arise regularly and in such a manner as to induce a water-deficit pattern of this kind. Such patterns of deficit might develop if occlusion took place in the terminal tracheids of the vascular system in the leaf. Blocking of such elements by a fungal metabolite was suggested by Talboys (1957) as one possible mechanism of induction of interveinal necrosis in the hop leaf by *V. albo-atrum*. More recent observations have shown the presence of granular deposits in the terminal tracheids of such leaves (Fig. 20). However, these deposits appear to resemble gums of host origin rather than fungal metabolites, and they are probably the *consequence* of damage to the surrounding mesophyll cells rather than the *cause*.

3. Damage to Leaf Cells Surrounding Terminal Vascular Elements

Comparatively little attention has been paid to processes blocking water movement from the leaf vascular system to the transpiring surfaces in the mesophyll, although Gäumann and Jaag (1950) and Gäumann (1958) suggested that polysaccharides of low molecular weight might "paralyze" water movement in the leaf by occluding the intermicellar spaces in the cell walls. Uptake of inulin solutions, for example, rapidly caused almost complete cessation of water movement into and out of detached tomato shoots

A different mechanism was suggested by Talboys (1958c), who found that foliar necrosis in hop verticillium wilt was associated with the development of brown granular masses filling the cells in contact with terminal tracheids (Fig. 20). Subsequently, the contents of the palisade and mesophyll cells more remote from the tracheids degenerated and collapsed. The appearance of the tissues suggested that toxin-induced dysfunction (see Section IV,B) of the cells adjacent to the tracheids had led to failure of water transport to the more remote cells and hence to their desiccation.

4. " Drought hardening" and Some Other Indications of Water Stress

Much evidence in support of an occlusion theory of wilt induction is based on the obvious resemblance of vascular disease syndromes to that of water deficiency, on observations of vascular obstruction in diseased plants, and on reduced flow rates of water in segments of infected stems relative to rates in healthy ones.

Evidence of occlusion has also been obtained in studies of transpiration changes in intact plants infected with vascular pathogens. Such studies (e.g. Harris, 1940; Ludwig, 1952; Scheffer and Walker, 1953; Threlfall, 1959) have generally shown that transpiration rates in infected plants either are similar to those of healthy plants or are slightly higher, until about the time that symptoms are first initiated. At that stage water loss from infected plants falls below that of healthy ones and continues to decline until the whole plant is severely diseased. The fact that water loss during or after initiation of symptoms rarely exceeds that of healthy plants has been taken as evidence that the symptoms result from water shortage rather than from excessive water loss, although Threlfall (1959) noted a transient prewilting increase in transpiration interrupting the general decline.

Transpiration measurements of a whole plant obviously represent mean values for a set of leaves in various stages of disease development and could conceal increased water loss in some leaves and reduced loss in others. However, Dimond and Waggoner (1953c) and Threlfall (1959) found that transpiration rates of individual leaves from diseased plants were reduced irrespective of the leaf position, i.e., irrespective of the symptom intensity.

However, the *degree* of reduction was greater in yellowed or wilted leaves than in symptomless ones (Dimond and Waggoner, 1953c). Noninfected plants subjected to suboptimal water regimes also showed reduced transpiration rates that were maintained in detached leaves even when water was readily available. Persistence of low transpiration rates in fully turgid leaves from both droughted and infected plants was associated with persistent stomatal closure. Such an association is characteristic of " drought hardening."

Therefore, it has been implied that since both vascular infection and water deficiency can induce " drought hardening," the result of infection is water deficiency. Other workers, e.g., Linford (1931b), reported that detached leaves from diseased plants wilted more slowly than leaves from healthy plants. In this connection it is notable that Blackhurst (1963) found that the verticillium wilt-resistant tomato " Loran Blood " was more drought resistant than wilt susceptible "Ailsa Craig"; True and Tryon (1956) reported that red oaks were more drought sensitive than white oaks; the difference corresponds with the relative sensitivity of the two groups to oak wilt.

The role of water shortage in the oak-wilt syndrome has been studied by Beckman *et al.* (1953). Movement of a rubidium radioisotope in the transpiration stream was drastically reduced 3–4 days before symptoms appeared in inoculated trees. During the period of low water conduction before permanent wilt set in, the leaves commonly wilted during the day and recovered at night. A further indication of water stress was reported by Kozlowski *et al.* (1962): Diurnal changes in stem radius, superimposed upon the sigmoid seasonal pattern of growth and related to changes in tissue hydration, were eliminated 3–5 days before symptoms developed in the foliage, but cambial activity resulting in radial growth continued until initiation of foliar symptoms. The loss of diurnal fluctuations evidently indicated a degree of vascular occlusion such that water deficits developing during the day could not be eliminated at night. Cambial damage was clearly not a casual factor in the etiology of the initial syndrome, although it would be significant in long-term prognosis (see Section V).

One of the most sophisticated investigations on the role of water shortage in wilt development was that of Beckman *et al.* (1962) on *Pseudomonas solanacearum* in banana. Transpiration from a limited area of a leaf was monitored continuously with a recording lithium chloride hygrometer, and CO_2 absorption was also determined continuously by infrared gas analysis. Several days before disease symptoms appeared in inoculated plants, marked periodic fluctuations in transpiration were recorded. Similar fluctuations occurred in droughted plants. Transpiration declined and the fluctuations were reduced in amplitude until, at the time symptoms appeared, water loss had been reduced to a very low rate.

The fluctuations in transpiration were attributed to alternate opening and closing of stomata. Guard cells apparently lost turgor, causing stomatal closure when transpiration loss substantially exceeded the impaired supply. Turgor of guard cells was restored, causing stomatal opening, when the balance had been restored. These cycles were completed in 20–30 minutes. Under high ambient humidity, the fluctuations were eliminated.

Simultaneous determinations of apparent photosynthesis (CO_2 uptake) and transpiration indicated that during the phase of water stress leading to sympton initiation, photosynthesis was limited by water stress, but removal of internal water stress by increasing ambient humidity resulted in full recovery of photosynthetic activity. These observations indicated that during development of water stress there had been no acute physiological damage that could have been attributed to a toxic action of fungal metabolites in the leaf. Hence, the data indicated that vascular occlusion was the immediate cause of symptom induction in banana bacterial wilt and that toxins played no significant role.

B. INDUCTION OF PATHOGENIC WATER LOSS: TOXIN THEORY

Although much evidence has been presented in support of the theory that vascular occlusion is the main mechanism of water stress induction and symptom development in vascular diseases, there are also strong indications that toxins contribute significantly to the etiology of some diseases, though whether they act through a water stress mechanism may in some cases be open to doubt.

1. Early Evidence for the Toxin Theory of Wilting

The concept of wilt induction through toxigenic activity of the pathogen was introduced by Hutchinson (1913). In studies of tobacco bacterial wilt incited by *Bacterium solanacearum*, he showed that wilting and necrosis could be induced by injecting the vascular system with an aqueous extract of alcohol-precipitated material from a bouillon culture of the pathogen. Furthermore, transverse sections of diseased stems showed that only a small proportion of the vessels contained bacterial masses. A substantially larger proportion could be put out of action by a transverse incision halfway across a healthy stem without causing any leaf symptoms. Hutchinson commented: "Wilting is very generally assumed to be due to the plugging of the water vessels by masses of bacteria and the purely mechanical interference with water supply which follows; ... my observations do not support this view, I am inclined on the contrary to attribute the wilting effect to the action of secreted toxins upon the cell protoplasm. ... " So far as wilting is connected with failure of water supply it seems reasonable to conclude that this failure may be due largely to the interference with osmotic pressure consequent on protoplasmic intoxication."

It is noteworthy, incidentally, that Hutchinson's toxin preparations were thermolabile and caused tissue disintegration by destroying the middle lamella. Pectolysis was not a familiar aspect of pathogenesis at that time, especially in vascular diseases.

In this as in many subsequent investigations, the toxin-induced syndrome did not exactly reproduce the normal pattern of disease development. There was no indication that unilateral injections gave unilateral symptoms comparable with those resulting from point inoculation on one side of the vascular cylinder, and wilting and necrosis showed basipetal progression in contrast to the acropetal progression characteristic of the disease.

Dimond and Waggoner (1953b) defined a *toxin* as "any compound produced by a microorganism which is toxic to plants." They further defined a *vivo-toxin* as a "substance produced in the infected host by the pathogen and/or its host which functions in the production of disease but is not itself the initial inciting agent of the disease." They proposed a modification of Koch's postulates to establish criteria for proving the pathogenic role of vivo-toxins, viz, (1) reproducible separation from the diseased plant, (2) purification, and (3) reproduction of at least part of the disease syndrome when the metabolite is introduced into a healthy plant. The conclusion that a vivo-toxin is involved in the etiology of a disease may nevertheless be drawn if only the first and last criteria are satisfied.

Wheeler and Luke (1963) redefined vivo-toxins as chemically defined substances isolated from diseased plants and shown to produce at least a part of the disease syndrome. They introduced the term "pathotoxin" to describe substances shown to play an important causal role in disease, e.g., in cases where a single substance applied at concentrations that could reasonably be expected to occur in or around the diseased plant would induce in a susceptible host plant all the symptoms of the disease, would show the same suscept specificity as the pathogen, and would be produced in amounts directly related to the ability of the pathogen to cause disease. None of the wilt toxins so far studied is referable to the group of pathotoxins, in which Wheeler and Luke (1963) included victorin, produced by *Helminthosporium victoriae*, and toxins from *Periconia circinata* and *Pseudomonas tabaci*.

Hutchinson (1913) demonstrated that the bacterial pathogen produced a toxin, but he did not show conclusively that this was a vivo-toxin implicated in the disease syndrome. Much subsequent work on the role of fungal metabolites in vascular pathogenesis has provided similar circumstantial evidence of toxin involvement, and the pattern of Hutchinson's work has been repeated for many vascular diseases.

Other early reports of toxin production by pathogens growing *in vitro* came from Haskell (1919) and Young and Bennett (1921) working with *Fusarium oxysporum* in potato; Goss (1924) with *F. eumartii* in potato; and Brandes (1919) with *F. oxysporum* f. *cubense* in banana.

Dowson (1922) found that the symptoms induced in Michaelmas daisies by metabolites produced *in vitro* by a species of *Cephalosporium* closely resembled those of plants infected with the organism.

2. *Production of Two-Component Toxins In Vitro*

An early attempt to identify the active substances in toxin preparations was made by Fahmy (1923), who showed that ammonia and oxalates were present in culture filtrates of *F. solani* but at concentrations too low to be probable wilt-inducing agents in broad-bean shoots. Rosen (1926) concluded that a volatile alkaline compound and an inorganic nitrite were involved in wilting of cotton seedlings by culture solutions from *F. oxysporum* f. *vasin-fectum*.

Observations on tobacco fusarium wilt (*F. oxysporum*) led Johnson (1921) to conclude that the leaf symptoms, characterized by yellowing of the lamina while still turgid, with subsequent collapse and necrosis, probably resulted from the action of toxins either secreted by the pathogen or formed as a result of its action on the host. Subsequent experiments by Wolf and Wolf (1948) suggested that *F. oxysporum* f. *nicotianae* produces, *in vitro*, two thermostable nonvolatile substances, one inducing wilting of leaves, the other inducing necrosis.

Two-component toxigenic systems have been reported on several occasions. White (1927) found that *F. oxysporum* f. *lycopersici* produced a non-dialyzable thermostable fraction with volatile and nonvolatile components, both of which could induce wilting in tomato shoots.

V. albo-atrum from lucerne (*Medicago sativa*) produced a thermostable toxin that induced general wilting and a thermolabile material that caused necrosis of the leaf blades (Kiessig and Haller-Kiessig, 1957).

Zentmyer (1942) reported thermostable metabolites of *Ceratocystis ulmi* that were toxic to elm shoots, and Dimond (1947) separated two components: One was an alcohol-insoluble polysaccharide that caused up-curling and marginal withering of the leaves; the other, an alcohol- and water-soluble but ether-insoluble substance, caused severe interveinal necrosis.

According to White (1955), the oak wilt pathogen, *Ceratocystis fagacearum*, also produces an alcohol-insoluble polysaccharide that causes drying and wilting of leaves and a soluble toxin causing necrosis.

Caroselli (1954) found that isolates of *Verticillium albo-atrum* causing a wilt desease of maple produced a stem-wilting polysaccharide and also thiourea, which caused wilting of leaves. Polysaccharides have also been implicated in the pathogenesis of *Fusarium solani* f. *eumartii* in tomato (Thomas,

1949) and *Cephalosporium gramineum* in wheat (Spalding, 1960). Production of polysaccharides by *Verticillium* was reported by Porter and Green (1952), but their role in pathogenesis was discounted by Green (1954). Le Tourneau (1957) found that *Verticillium albo-atrum* produced tri-, tetra-, and penta-saccharides in culture. Polysaccharides formed by nonwilt-inducing phytopathogenic bacteria were shown by Feder and Ark (1951) to be capable of inducing wilting in sunflower and tomato shoots. Reference has already been made in Section IV,A,b to the work of Hodgson *et al.* (1949), which indicated that polysaccharides caused wilting by occluding the vascular system. Polysaccharides of high molecular weight caused obstruction in the large vessels of the stem, and those of low molecular weight obstructed the small elements in petioles and leaves. Similar results were obtained with a range of "carbowaxes" of various molecular weights.

The possible significance to be attached to the prevalence of polysaccharides in two-component systems is discussed later.

3. Chemically Defined Wilt Toxins: Lycomarasmin, Ethylene, and Fusaric Acid

The first of the wilt-inducing toxins to be chemically defined, lycomarasmin, is produced by *Fusarium oxysporum* f. *lycopersici* (Clauson-Kaas *et al.*, 1944; Plattner and Clauson-Kaas, 1945).

Lycomarasmin is a dipeptide, N-[α-(α-hydroxy propionic acid)]-glycylasparagine (Woolley, 1948) with a molecular weight of 277. It induces wilting and desiccation in tomato shoots. The fact that the symptoms do not exactly simulate those of diseased plants does not necessarily exclude lycomarasmin as a significant agent in the pathogenesis of tomato wilt, but Dimond and Waggoner (1953a) discounted its importance because they concluded that lycomarasmin is a product of lysis in aging rather than of actively growing mycelium. However, Gäumann (1957) showed that lycomarasmin could be detected after 7 days *in vitro*, although more than 40 days were required for maximum production. Its production *in vivo* has not been demonstrated.

The production of ethylene by *Fusarium oxysporum* f. *lycopersici*, and participation of this compound in the induction of tomato fusarium wilt symptoms, were demonstrated by Dimond and Waggoner (1953d). They showed that ethylene was the cause of leaf epinasty. It did not induce defoliation, but it caused a slow yellowing of the oldest leaves.

The physiological activity of ethylene in plants is closely interrelated with that of auxins (Hall and Morgan, 1964). Auxin production has been demonstrated in a number of fungi (Gruen, 1959), including several vascular pathogens, and it is possible that symptom induction may be mediated by accelerated ethylene formation resulting from the action of fungal auxin.

A further chemically defined wilt toxin, fusaric acid (5-butylpicolinic acid), has been obtained from several species of Hypocreaceae. It was first isolated by Yabuta *et al.* (1934) from cultures of *Fusarium heterosporium*. Fusaric acid is formed *in vitro* during active mycelial growth (Gäumann, 1957) and its formation *in vivo* has been demonstrated in cotton fusarium wilt (Lakshminarayanan and Subramanian, 1955; Kalyanasundarum and Venkata Ram, 1956), watermelon fusarium wilt (Nishimura, 1957), tomato fusarium wilt (Kern and Kluepfel, 1956), and banana fusarium wilt (Page, 1959b).

Nevertheless, application of fusaric acid solutions to cut shoots results in symptoms that differ from those induced by invasion by the pathogen. Thus, in tomato (Gäumann, 1957) the first indication of injury is "furrowing" over the main vascular bundles, first of the stem and later of the petioles. Leaf injury, which occurs later, takes the form of interveinal necrotic spotting. However, as Gäumann points out, the concentrations applied to give positive reactions in a reasonably short time are substantially higher than those occurring in naturally infected tissues. Furthermore, it is not to be expected that a single toxin would necessarily reproduce the syndrome induced by a toxin complex. The action of a toxin needs to be considered in the context of its accompanying metabolites, which may or may not be actively toxic.

Indirect evidence of toxin production and activity *in vivo* has been presented for tomato fusarium wilt by Davis (1954) and Keyworth (1964). Both investigators examined disease development in composite plants prepared by grafting hosts of different resistance, with Davis using intergeneric grafts of various Solanaceous plants on susceptible tomato rootstocks and Keyworth using intervarietal grafts of wilt-resistant and susceptible tomatoes. Scions of genera and varieties normally resistant to fusarium wilt showed acute symptoms when infected via susceptible rootstocks. The symptoms developed before the pathogen entered the scions. There was no evidence that the symptoms were induced by vascular occlusion, and Keyworth's double-grafts, resistant stem/susceptible stem/susceptible rootstock, apparently excluded any possibility of occlusion, because the leaves on the intermediate "susceptible" stem section showed acute symptoms at least 7 days after those of the "resistant" sections immediately above them.

Direct evidence of toxin production was provided by Gottlieb (1943, 1944), who found that vascular fluids extracted from diseased tomato plants had toxic properties, whereas extracts from healthy plants exhibited no toxic effects.

There seems little doubt, therefore, for some diseases, that symptom development can be a consequence of the action of toxic substances. Attention will now be given to whether these substances operate by inducing water stress and, if so, how this is effected.

4. Action of Toxins on Water Relations of Plants

One of the early attempts to relate a wilt syndrome to the water economy of plants was that of Linford (1931a), who showed in limited experiments that during the development of fusarium wilt of peas (*F. orthoceras* f. *pisi*) transpiration of infected plants was *lower* than that of healthy ones *before* wilting became evident. As wilting occurred, there was an increased rate of water loss, but when all the leaves had wilted the transpiration rate diminished to a level lower than the original rate and continued progressively to decline.

As Beckman (1964) pointed out, Linford's results were not conclusive and could have been interpreted as supporting the vascular occlusion concept of wilt induction. However, Linford's conclusion was that wilting resulted *not* from a diminished water *supply* but from an excessive water *loss*. He suggested that this indicated the loss of normal powers of water retention by leaf protoplasts.

Metabolites from the pathogen grown *in vitro* caused marginal water soaking of leaflets, which *subsequently* lost turgor, collapsed, and became desiccated (Linford, 1931b). Stems and petioles collapsed later than leaflets. This sequence of events was the reverse of that in shoots inserted in concentrated sugar solutions, which caused reduced water uptake.

The impression gained by many workers, that wilting resulted from excessive water loss, was substantiated by Thatcher (1942), who examined the effects of culture filtrates of *Fusarium oxysporum* f. *lycopersici* on the permeability of tomato leaves to water. Cells adjacent to veins showed a slight reduction in permeability, but in palisade cells remote from the main veins, permeability was almost doubled by the culture filtrates. "The presence in the leaves of a metabolic product of a hadromycotic fungus causes an increase in permeability to water which is most pronounced where accumulation of the fungal irritant occurs most extensively. This increase in permeability promotes more rapid transpiration and the consequent additional demand for water upsets the delicate water balance of the plant, and, under conditions normally nearly limiting for maintenance of turgor, causes progressive wilting, which finally becomes permanent and leads to death of the mesophyll cells." Sivadjian and Kern (1958) later showed that the interveinal areas developed the highest transpiration rates in toxin-treated leaves.

Gottlieb (1944) demonstrated that toxic tracheal fluids from tomato stems infected with *F. oxysporum* f. *lycopersici* caused substantially increased permeability of pith cells: "Normally, there is a critical balance between the intake and the loss of water by the plant. A similar ratio must also exist in the leaf tissues, so that wilting can occur either because of increased transpiration or decrease in absorption of water. The toxins produced in the plant would tend to reduce that ratio below the critical level." Thatcher showed that wilting was prevented under high humidity, and recovery could also occur under these conditions. He pointed out that this was consistent with the diurnal pattern of wilting in some diseases.

In Gottlieb's experiments, also, wilting was prevented by high ambient humidity or by coating leaves with petroleum jelly. Plants that wilted at low humidity recovered when transferred to high humidity.

The most extensive studies of the action of fungal toxins on water relations of plants have been those of Gäumann's school at Zurich, working with a number of toxic metabolites including lycomarasmin (Gäumann and Jaag, 1946, 1947); fusaric acid (Gäumann, 1958), patulin (Gäumann and Jaag, 1947); lycomarasmic acid, baccatin, culmomarasmin, and skyrin (Gäumann et al., 1959); and a toxic metabolite and "slime substance" from *Bacterium solanacearum* (Kunz, 1953). Work with alternaric acid was simultaneously reported by Brian et al. (1952) and Gäumann et al. (1952a). Gäumann and Jaag (1950) made a comparative study of water relations in a toxigenic wilt induced by lycomarasmin and a physically induced wilt induced by a glucosan polysaccharide, inulin.

Not all of the toxins examined at Zurich are products of vascular pathogens, but studies of their effects on water relations throws light on possible mechanisms of vascular pathogenesis.

a. Patterns of Disturbance of Water Relations in Cut Shoots: "Leaf" and "Vascular" Toxins. The effect of these toxic substances on the water relations of shoots varies with their concentrations and such factors as light intensity, ambient temperature, and humidity. Nevertheless, two main patterns of disturbance in the water balance of detached tomato shoots can be recognized in the literature.

The first pattern, exemplified by lycomarasmin, is characterized by immediate and often drastic reduction in both water uptake and transpiration (termed by Gäumann the "shock phase"), followed by an equally rapid increase in transpiration rate to a peak exceeding the level that occurred before application of the toxin. Subsequenly, the water loss diminishes progressively to a very low level (Fig. 12). Water uptake continues to follow the general pattern of water loss after the "shock phase," but during or immediately after development of the transpiration peak, the rate of water uptake falls below that of loss and an adverse balance is generally maintained until absorption and transpiration approach equality at a very low level of water turnover. The fresh weight of the shoot falls at a rate determined by the magnitude of the deficit. Visible leaf damage, generally leading to interveinal necrosis with or without flaccidity, begins to appear immediately *before* or *during* the development of an unfavorable water balance. Symptoms are commonly initiated long before the water deficit in the leaf has approached what would be required to cause wilting in a healthy shoot deprived of water.

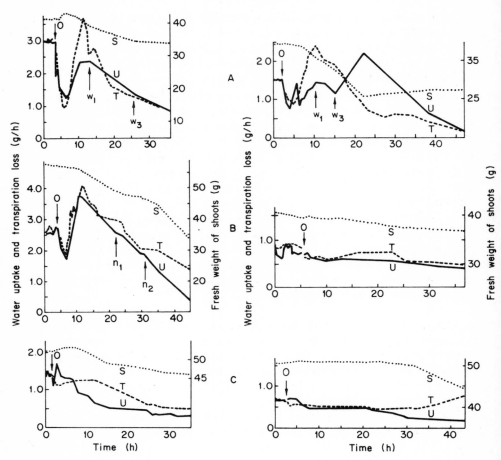

Fig. 12. Water relations in tomato shoots under treatment with lycomarasmin: A, 10^{-2} M solution; B, 10^{-3} M solution; C, 10^{-4} M solution; for each pair the left-hand diagram shows the response in light, the right-hand, the response in darkness. Adapted from Gäumann and Jaag (1947). O = time at which treatment was applied; S = shoot fresh weight; T = water loss by transpiration; U = water uptake; n_1 = first appearance of necrosis; n_2 = necrosis in all leaves; w_1 = first appearance of wilting in lower leaves; w_2 = all leaves wilted; w_3 = whole plant wilted; g = grooves in stem; r = start of inrolling of leaf edges.

This pattern of response was first shown for lycomarasmin (Gäumann and Jaag, 1947), and subsequently (Figs. 13 and 14) for the lycomarasmin-iron complex, for fusaric acid applied at pH values above 7.0, and for alternaric acid (Gäumann *et al.*, 1952a; Gäumann *et al.*, 1958; Brian *et al.*, 1952). With lycomarasmic acid the shock phase developed less rapidly, and the subsequent rise in water uptake was relatively slight (Gäumann *et al.*, 1959).

A lycomarasmic acid–iron complex and culmomarasmin (Gäumann *et al.*, 1959) both induced an immediate increase in transpiration and water uptake without an intervening shock phase (Fig. 13). The appearance of necrotic symptoms coincided with the beginning of a decline in water uptake during a period in which transpiration exceeded absorption. Despite the absence of a shock phase it seems appropriate to include these substances in the same group as lycomarasmin.

Reduced water uptake and loss during the shock phase cannot be attributed to vascular occlusion since water uptake subsequently increases. For a

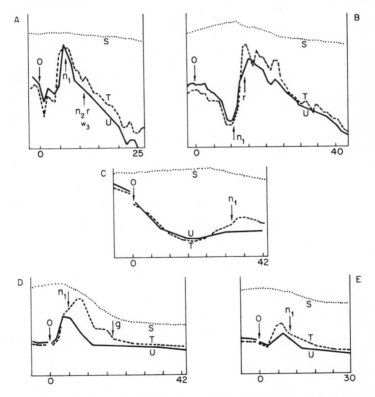

Fig. 13. Water relations in tomato shoots under treatment with A, lycomarasmin-iron complex, 10^{-3} M; B, alternaric acid, 10^{-6} M; C, lycomarasmic acid, 10^{-3} M; D, culmomarasmin; E, lycomarasmic acid-iron complex, 10^{-3} M. Adapted from Gäumann *et al.* (1952a, 1959). Key as for Fig. 12. Numerical values for ordinates [water uptake and loss (g/h) and shoot fresh weight (g)] have not been inserted as only the *relative* values are being considered.

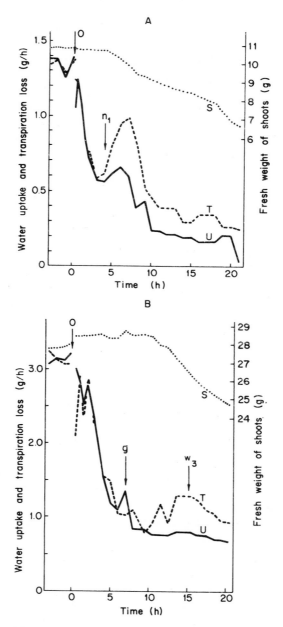

Fig. 14. Water relations of tomato shoots treated with fusaric acid solution, 10^{-2} M; A, at pH 7.5; B, at pH 4.3. Adapted from Gäumann *et al.* (1958). Key as for Fig. 12.

similar reason it cannot be attributed to permanent damage to leaf tissues resulting in loss of the tension attributable to their osmotic properties. In the absence of information on stomatal movement in these studies it may be surmised that the shock phase results from stomatal closure following an initial effect of translocated toxin on the mesophyll or epidermal cells, or both. La Rue (1930), quoted by Williams (1950), found that leaf epidermal cells are supplied with water directly from the tissues adjacent to the veins, and Heath (1938) demonstrated that stomatal movement is largely determined by the difference in turgor between guard cells and adjacent epidermal cells. Possibly, an initial effect of the toxin on the epidermal cells leads to stomatal closure.

Support for the view that the shock phase is a manifestation of rapid stomatal closure is provided by the fact that *in darkness*, when the stomata would already be closed, this phase in lycomarasmin toxicity is much reduced or virtually eliminated (Fig. 12). Such evidence is not conclusive, because the reduced uptake in darkness would also decrease the rate at which the toxin arrived at its site of action.

Eventually, however, it appears that the effects of toxins override the mechanisms maintaining stomatal closure both in light and, at the highest concentration, in darkness. The high rates of transpiration that develop after the shock phase presumably indicate that the stomata are wide open.

Gäumann and Jaag (1947) considered that lycomarasmin at concentrations of 10^{-3} M or higher completely destroyed the semipermeability of plasma membranes, allowing solutes to escape and thus eliminating osmotic effects and causing a loss of turgor and of capacity for water retention. The transient rise in transpiration was attributed to the release of water into the intercellular spaces. Gäumann and Jaag (1947) postulated that lycomarasmin at low concentrations (10^{-4}–10^{-5} M) increased permeability of cell membranes to water but not to solutes.

In the systems with which Gäumann and co-workers were working, the development of water deficits associated with very high transpiration rates appeared to be the *result* of cell damage rather than the *cause*. Excessive transpiration *accompanied* or *followed* cell damage; it did not precede it.

Nevertheless, it must be borne in mind that these systems, like most of those used to study toxic action, have two characteristics that distinguish them from diseased plants subject to toxic action of an invading pathogen:

1. The toxin is supplied uniformly to the vascular system of the whole shoot at a constant concentration that would generally exceed that found in the plant.
2. The supply of water (plus toxin) is unrestricted; the shoot under test is under negligible water stress, other than that induced by the treatment.

Hence, it is not to be expected that such systems will exactly reproduce the effects of *in vivo* toxin production (see Section IV,B,4,c).

The second pattern of disturbance in water relations induced by fungal and bacterial toxins, such as those investigated by Gäumann and co-workers, resembles the first in that application of the metabolite causes a reduction in water uptake and loss. The reduction sometimes occurs very rapidly, but not always. The pattern differs from the first in that any subsequent increase in the transpiration rate is small, and even if it occurs the rate of water uptake continues to decline and shows no significant increase at any time. It is exemplified (Fig. 15) by the actions of skyrin and baccatin, of certain bacterial "slime-substances," and of the "model" low molecular weight polysaccharide, inulin. Flaccidity of leaves or of stems and petioles, and sometimes longitudinal furrowing of stems, are characteristic symptoms. Leaf necrosis *may* develop later but is not a consistent feature.

This toxigenic pattern is suggestive primarily of vascular dysfunction, with leaf necrosis of secondary importance. Reference has already been made to vascular occlusion caused by fungal and bacterial metabolites and by products of host tissue degeneration. The bacterial slimes studied by Kunz (1953) doubtless reduce water movement by occluding vessels or by increasing viscosity of the vascular fluids. However, there is a further mechanism that may contribute to vascular dysfunction, particularly in young shoots, viz, collapse of tissues surrounding the vessels as a result of the action of toxins. Such a mechanism is suggested by the stem furrowing induced by skyrin and bacterial slimes. The phytotoxic antibiotic, patulin, similarly exerts its toxic action in stems and petioles, causing their collapse before leaf symptoms develop.

Stem furrowing is also a characteristic effect of fusaric acid introduced into the plant at an acid pH. The pattern of disturbance of water relations by fusaric acid at pH 4.3 thus resembles that of the "vascular" toxins of the second group, although some leaf toxicity appears to be superimposed on this pattern. This is indicated by a small and transient increase in water uptake and, at a later stage, a marked increase in transpiration (Fig. 14).

Gäumann's (1957) report that, with increasing pH, leaf injury by fusaric acid increased while stem injury declined was supported by Kern *et al.* (1957), who showed that ^{14}C-labeled fusaric acid accumulated predominantly in the stem at pH 4.3 and in the leaves at pH 7.0. These distributions were attributed to the fact that penetration of fusaric acid into cells is effected more rapidly by the undissociated form, so that under acid conditions the toxin would pass into the tissues surrounding the vascular system and be removed from the transpiration stream before reaching the leaves. With increasing pH, progressively larger amounts of toxin would reach the leaf tissues. Labeled toxin entering the leaves tended to be concentrated in the interveinal areas,

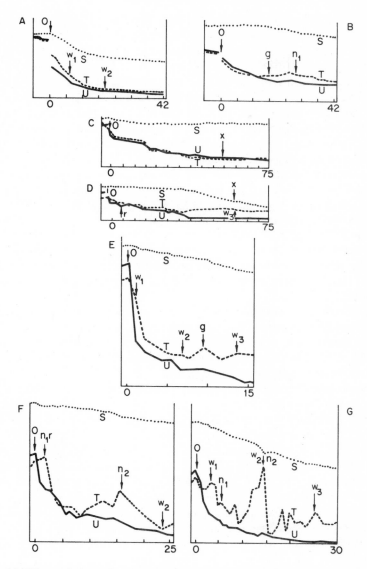

Fig. 15. Water relations of tomato shoots under treatment with A, skyrin; B, baccatin; C, inulin at 10^{-4} M; D, inulin at 10^{-4} M and lycomarasmin at 10^{-3} M; E, "slime-substance" from *Bacterium solanacearum*; F, "Fraction 4" toxin from *B. solanacearum*; G, a viable suspension of *B. solanacearum* in water. Adapted from Gäumann *et al.* (1959), Gäumann and Jaag (1950), and Kunz (1953). x = time at which treatments in C and D were discontinued, the test solution being replaced with water. Key to other symbols as for Fig. 12.

where necrosis was subsequently initiated. Studies by Sivadjian and Kern (1958) indicated that these interveinal zones of toxin accumulation developed a higher transpiration rate than the other parts of the leaf.

b. *Toxins and Cell Permeability.* The toxic mechanisms of fusaric acid have been investigated in considerable detail. Gäumann (1958) pointed out that, unlike the highly specific activity of the toxin of *Pseudomonas tabaci* in inhibiting methionine metabolism in the host, fusaric acid evidently can interfere in host cell physiology in a number of different ways, for example, by impairment of oxidative phosphorylation (by the pyridine ring) and of cytochrome oxidase activity (through the effects of the aliphatic side chain in the β position) and by interfering in other enzymic processes through its capacity to chelate manganese, copper, iron, calcium, and magnesium. The chelating properties of fusaric acid have been considered particularly by Lakshminarayanan (1955) and Kalyanasundaram and Subba-Rao (1957). Whatever the toxic mechanisms of fusaric acid may be, the central issue is whether its action on water relations of shoots can be interpreted in terms of the permeability changes that have been inferred from the macroscopic features of the induced syndrome.

Gäumann *et al.* (1952b) showed that a number of wilt toxins and antibiotics exerted a damaging effect on osmotic properties of plant cells. Some, such as fusaric acid and patulin, were active at concentrations as low as 10^{-7}–10^{-5} M. Others, active between 10^{-4} and 10^{-3} M, included alternaric acid and the lycomarasmin-iron complex.

The permeability effects of fusaric acid were studied by Bachmann (1956), working with *Rhoeo discolor* and *Spirogyra nitida*, and by Gäumann and Loeffler (1957), working with tomato stem pith cells. Effects on permeability to water of tomato pith cells became evident between 5×10^{-8} and 10^{-7} M. With increasing concentrations up to 10^{-6} M, permeability was progressively increased to a maximum of about 1.5 times normal. Further *increases* in concentrations to 10^{-5} M gave *reduced* increases in permeability. Beyond 10^{-5} M permeability was reduced, to about 60% of normal at 10^{-3} M. Epidermal cells of *Rhoeo discolor* gave a similar pattern of response, although the tissue was more sensitive than tomato stem pith cells (Fig. 16). Reduced permeability (described by Gäumann as "waterproofing") probably is attributable to precipitation of the plasma membranes.

Gäumann and Loeffler (1957) noted that increasing water permeability "leads to excessive transpiration of diseased plants as noted by Scheffer and Walker (1953) and to damage of the semi-permeability of the plasma membrane, rendering the protoplasts accessible to toxins, enzymes and other products of the pathogen." Evidence of changes in semipermeability had been provided by Linskens (1955), who showed that treatment with fusaric acid at

5×10^{-3} M and lycomarasmin at 2.5×10^{-3} M enabled substantially increased quantities of salts and amino acids to be washed from the surfaces of tomato leaves. Increased conductivity of leaf washings was detectable in 2.5–3 hours after application of the toxin, i.e., substantially earlier than visible symptoms would have occurred but probably coinciding with the end of the shock phase with lycomarasmin and with fusaric acid at pH 7.0.

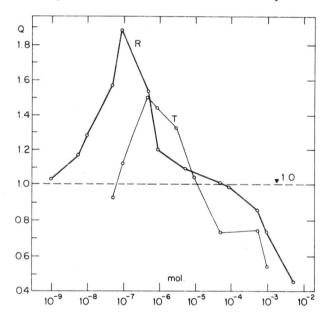

Fig. 16. Changes in water permeability in tomato stem pith cells (T) and *Rhoeo discolor* leaf epidermal cells (R) with various concentrations of fusaric acid. The permeability quotient (Q) represents the permeability relative to that of control tissues in water. From Gäumann (1958).

Permeability changes were modified by plant nutrition (Gäumann and Bachmann, 1958). Excessive nitrogen, phosphorus, or potassium tended to reduce sensitivity of *Rhoeo* cells to the effects of fusaric acid, though a high phosphorous level reversed the tendency to "waterproofing" at high concentrations of fusaric acid and led to renewed and drastic *increases* in permeability, a pattern of response that corresponded with that of *Spirogyra nitida* at normal levels of nutrition. Low levels of phosphorus and potassium had little effect on toxin sensitivity, but a low level of nitrogen nutrition completely eliminated the response to low concentrations of fusaric acid. This is of particular interest, because a common feature of wilt diseases is that host resistance is enhanced by low-nitrogen nutrition.

c. Two-Component Toxin Systems in Water Stress. There are reasonable indications that increased rates of transpiration in *Fusarium*-infected plants and shoots treated with *Fusarium* toxins may be attributable to increased permeability of mesophyll cells to water. However, this does not in itself lead to water deficit. As Ludwig (1960) has pointed out, the maximum increases in permeability induced by fusaric acid would increase water loss by a factor well within the range of day-to-day variation in the intact plant and would be compensated by increased uptake. Hence, under conditions of adequate water supply, there is no reason to suppose that toxin-induced increases in water permeability alone could lead to symptoms of water stress. It is clear that with an unrestricted and continuous supply of toxins, such as fusaric acid and lycomarasmin, the phase of increased transpiration with a balanced increase in uptake is of very limited duration and is rapidly succeeded by a phase in which solute losses from cells, accompanying disorganization of plasma membranes, lead to loss of osmotic tension in the system. As increasing numbers of cells enter this phase the capacity for water uptake shows a progressive decline. Initiation of this phase is marked by the divergence of the values for water uptake and loss. Rapid and simultaneous permeation of a shoot could be expected to result in its collapse within a relatively short time. However, this is not a realistic simulation of the conditions existing within the vascular tissues of a diseased plant, in which the following occurs:

1. The source of the toxin is initially small but increases progressively as the pathogen invades vascular tissues.
2. The distribution of the source may be limited to particular catenary sequences of vessels.
3. The toxin will be accompanied by other metabolites and possibly by host secretions or degradation products.
4. Some stress will already exist in the vascular system, to a degree depending on the availability of water in the soil and on factors affecting transpiration.

Reference was made earlier in Section IV,B,2 to the rather frequent reports of two-component toxin systems, commonly comprising a wilt-inducing polysaccharide and a necrosis-inducing toxin. These correspond essentially to the vessel-acting and leaf-acting categories of toxins noted above.

An example of a two-component system is provided by the work of Kunz (1953) on *Bacterium solanacearum*. Fractionation of a culture filtrate yielded a "slime substance," probably a polysaccharide, which induced flaccidity and stem furrowing (Fig. 15E), and a toxin ("fraction 4"), which caused leaf necrosis with increased transpiration (Fig. 15F). Some "overlapping" of the characteristic water-balance diagrams for the two types of activity possibly reflects incomplete separation. The action of the bacterium (Fig. 15G) evidently leads both to reduced uptake and excessive loss of water, although the *actual* water loss was less than in the healthy plant. A treatment of tomato shoots with a mixture of inulin and lycomarasmin (Gäumann and Jaag, 1950) resulted similarly in a continuous reduction in water uptake; water loss also declined at first, but later it increased without a corresponding rise in uptake. There was no obvious shock phase.

It is to be expected that a polysaccharide accompanying a toxin will (1) retard the rate of distribution of the toxin in the plant (2) increase the probability of water stress developing in the plant.

The argument advanced by Dimond and Edgington (1960) concerning the relative importance of vascular occlusion in the main stem and petioles is presumably equally applicable to retardation of flow by viscous polysaccharides. A much smaller proportion of the petiole vascular system needs to be rendered less efficient in order to induce stress than is required in the stem.

Hence, it may be postulated that the action of a two-component toxin system proceeds as follows: Fluid movement in a catena of infected vessels is retarded, partly by increase in drag caused by mycelium (this will be greatest in the lower parts of the system where "pioneer" hyphae have later become extensively branched, Fig. 17), partly by increased viscosity resulting from polysaccharide production by the pathogen, and possibly from production of gum by the host. Increased tension in the distal parts of the infected catena will be compensated by movement of water via lateral connections from non-infected sequences of vessels, so that the effects of the pathogen will be exerted for only a limited distance in advance. This distance will depend on the degree of retardation in infected vessels and on those structural characteristics of the xylem that determine the capacity for lateral movement.

Entry of the toxin complex into leaf-trace bundles of the transpiring leaf will add to the existing water stress (note that in studies of toxin effects on water relations of detached shoots there is no appreciable stress other than that induced by the treatment). A further increase of stress may arise as low concentrations of toxin entering the mesophyll and epidermal tissues cause increased permeability. However, as long as stomatal control is retained, the effect of excessive stress will be to induce stomatal closure. This, in turn, will reduce the movement of the toxin into the leaf. Recovery of full turgor will result in opening of the stomata, and the process will be repeated. However, Milthorpe and Spencer (1957) showed that reopening of the stomata at the end of a drought period was incomplete and that, in a series of cycles of alternating water deficiency and sufficiency, stomatal opening on recovery of turgor became progressively reduced. This is presumably a course of events associated with the phenomenon of "drought hardening." Presumably, also, such a mechanism enables leaves to show symptoms of desiccation while remaining turgid: Accumulation of toxic material in interveinal areas (see Section VI) of leaves drought hardened by pathogenic water stress leads to death and desiccation of the affected tissues, while tissues near the veins, receiving less toxin, remain turgid. Reports such as that of Blackhurst (1963) that verticillium wilt-resistant tomato "Loran Blood" was more drought resistant than the wilt-sensitive Ailsa Craig give further indications that water stress is a significant feature of vascular pathogenesis.

Clearly, the exact form of a syndrome induced by a two-component toxin system would be influenced to a considerable extent by the quantitative balance between the major toxic components. Predominance of the poly-saccharide component would lead essentially to a drought syndrome of greater or lesser intensity, with flaccidity in acute cases and drought hardening with growth reduction and the like in mild cases. Predominance of the necrotic factor could lead to sudden collapse and a water-soaked appearance of the leaves followed by desiccation, but in less acute cases various degrees of progressive chlorosis leading to necrosis are to be expected. Absence of poly-saccharide would result in much more rapid transport of the other toxic component.

Polysaccharides have been discussed as toxins, albeit acting via physical processes having the effect of reducing the rate of movement in infected vessels. It is at this point that "toxin" and "occlusion" concepts of wilt disease converge, because polysaccharides were discussed earlier in the context of vascular occlusion. A case in which tyloses apparently operate similarly in conjunction with necrosis-inducing toxins in the development of symptoms of a vascular disease is discussed in Section VI.

V. MECHANISMS COMPENSATING FOR VASCULAR DYSFUNCTION

In the context of water stress phenomena, vascular occlusions have been considered as factors deleterious to the host and contributing to pathogenesis. In some diseases, however, they may contribute to host resistance by limiting distribution of the pathogen and its metabolites in the plant (Talboys, 1964). Examples are provided by "gum barriers" limiting Stereum purpureum in certain plum varieties and tyloses restricting the distribution of Fusarium in sweet potato (McClure, 1950) and banana (Beckman, 1964). Treatment of elms with 2,3,6-trichlorophenylacetic acid induced tylosis and increased their resistance to Dutch elm disease (Smalley, 1962).

Hop verticillium wilt presents a paradoxical situation in that plants show-ing *acute* symptoms have virtually *no tyloses* though there is intense mycelial development in the vessels (Figs. 17 and 18), whereas plants showing *mild* symptoms often show *heavy tylosis* of both primary and secondary xylem in the roots (Fig. 10) and stem base (Fig. 11) (Talboys, 1958c). Intense tylosis is associated with very sparse proliferation of mycelium in the vascular system. Relatively minor damage is sustained by such plants because renewed or continued cambial growth results in formation of additional xylem in the stem (Fig. 11). If the rate of production of new xylem exceeds that at which existing xylem becomes occluded, the supply of water to the upper parts of the plant apparently is not seriously impaired. A consequence of this reaction is that the diameter of the lower parts of the stem may be more than twice that in a normal plant.

Because the vascular supply to a leaf in the primary axis arises from the primary xylem of the hop stem and forms no secondary xylem, it is clear that leaves arising from heavily tylosed regions of the stem will be subjected to water stress in addition to any direct toxic action of the limited amounts of the pathogen present in the vessels. Hence, symptom development and defoliation of the primary axis occur, to an extent determined by the vertical extent of tyloses in the stem. However, the habit of growth of the cultivated hop is such that a substantial part of its foliar surface is provided by the leaves on the flower-bearing lateral shoots arising in the axils of the primary leaves. Hence, a plant that has lost a considerable number of its primary leaves will not be adversely affected if its lateral shoots are adequately developed.

The vascular system of a lateral shoot undergoes secondary growth, and the outermost secondary wood is continuous with the outermost secondary vascular tissues in the stem. Thus, even though the vascular system supplying water to the primary leaves is blocked and defoliation occurs, the laterals continue to be supplied from the nonoccluded parts of the hyperplastic xylem tissues and remain free of disease symptoms (Fig. 19). In such circumstances the growth and cropping of the plant often are not seriously impaired. Occasionally, however, a period of adverse weather may result in a change of balance so that formation of tyloses " catches up " with xylem formation, and the plant collapses in a " quick-wilt " syndrome.

Tylosis followed by hyperplastic xylem formation appear to be characteristic wound reactions of the hop stem. An incision in the stem induces reactions that are histologically almost identical with those resulting from mild vascular infection. The evidence of Roberts and Fosket (1966) that wound-vessel formation can be stimulated by the combined action of IAA and gibberellic acid (GA) may provide a key to this response. The possible role of auxins in mechanisms of vascular occlusion has already been discussed.

Renewed xylem formation following pathogen-induced development of tyloses has been reported also by Schoeneweiss (1959) in white oak trees infected with the oak wilt pathogen. Here, also, the combination of tylosis and renewed xylem formation contributes to plant survival by limiting the spread of the pathogen and compensating for the resulting occlusion.

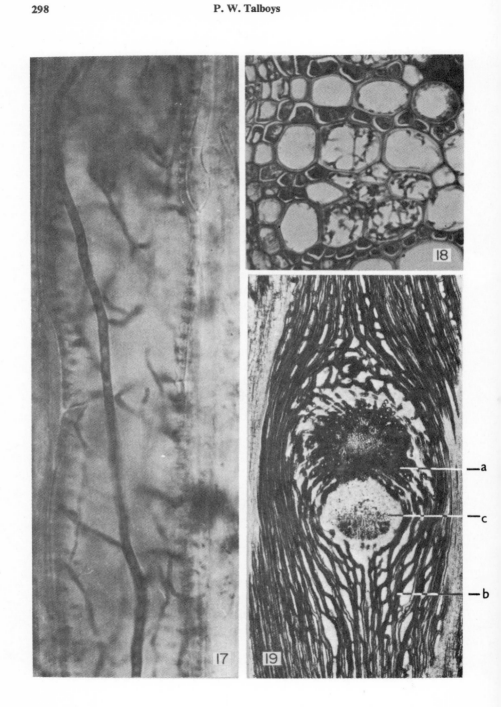

Reactions of this kind have not been reported in infected root systems. This probably reflects the fact that, whereas the stem vascular system provides the only pathway for water movement from roots to leaves, the root system provides numerous alternative pathways for the entry of water and its movement to the stem. Furthermore, root dysfunction frequently can be compensated by formation of new adventitious roots. This is a common feature of tomato verticillium wilt, and it has been suggested by Pegg and Selman (1959) that accumulation of IAA, produced either by the pathogen or as a host reaction to infection, induces adventitious root formation. Kalyanasundaram and Lakshminarayanan (1953) reported that culture filtrates of *Fusarium oxysporum* f. *vasinfectum* contained a factor that stimulated adventitious rooting in cotton shoots.

VI. ETIOLOGICAL DUALITY IN VASCULAR PATHOGENESIS

A given vascular disease can show substantial variation in its pattern of development. Differences in development may be related to the resistance of the host, to virulence and inoculum potential of the pathogen, and to environmental conditions including air temperature and humidity, soil temperature, and soil water and nutrient availability.

Thus, Bewley's (1922) description of tomato verticillium wilt (sleepy disease) with epinasty, yellowing, wilting, and necrosis differs from that of Selman and Pegg (1957), who noted acropetal yellowing and necrosis preceded by marked effects on growth, but no general wilting. Selman and Pegg suggested that the young plants used in their studies might be more toxin resistant than older plants.

Keyworth (1963) described two different syndromes in tomato fusarium wilt: A "normal" syndrome in the low-resistance "Bonny Best" variety on its own roots, showed a sequence of epinasty, leaflet flaccidity, and petiole flaccidity with progressive yellowing and necrosis of lamina tissues, whereas in scions of the more resistant "Pan America" on Bonny Best rootstocks,

---

Fig. 17. *V. albo-atrum* in vessel of young root, showing extensive branching from a "pioneer" hypha, and the complete absence of tyloses and gums, characteristic of acute disease development.

Fig. 18. *V. albo-atrum* in part of a single vascular bundle in the petiole of a hop leaf with typical interveinal necrosis, as in Fig. 2. Note the intense mycelial development and absence of tyloses and gums. Reproduced from Talboys (1958c).

Fig. 19. Tangential longitudinal section through the outer hyperplastic xylem of a hop stem with the mild verticillium wilt syndrome (as in Fig. 11). Continuity of the secondary xylem of the axillary shoot (a) with the hyperplastic xylem of the main axis (b) is evident; the leaf trace (c) has no connection with the hyperplastic xylem. Reproduced from Talboys (1958c).

there was chlorosis, initially marginal, with roughly circular necrotic patches and marginal necrosis developing later; the leaflets became desiccated, but the petioles remained rigid until the leaf died and abscised. The difference in disease expression in the two scion varieties was attributed to toxin hypersensitivity in the scion of the more resistant variety (Keyworth, 1964).

In banana fusarium wilt in Honduras a "yellowing" syndrome is characterized by strong yellowing and necrosis of *erect* leaves, which subsequently collapse. In a "nonyellowing" syndrome the petiole buckles and the green leaves "hang about the pseudostem as a green shroud for several days to a week before turning brown and dying" (Stover, 1959). The differences between the two banana wilt syndromes were attributed to variation in pathogenicity of isolates of *F. oxysporum* f. *cubense*. The chronic and acute forms of oil palm fusarium wilt that were distinguished by Prendergast (1957) appear to differ in a similar way, but in this case the syndrome differences have been attributed to environmental factors.

Variations in the syndrome pattern of a given vascular disease have generally been regarded, implicitly or explicitly, as *quantitative* variants of the same basic mechanism. Studies in hop verticillium wilt have shown, however, that in this disease there are qualitative differences between mild and acute syndromes.

Talboys (1958c) indicated that the acute form of the disease is characterized by intensive colonization of the vascular system by the pathogen, negligible tylosis, and the development of marginal and interveinal leaf necrosis, with persistent green coloring along and adjacent to the major veins. The mild syndrome is associated with extensive formation of tyloses and limited mycelial growth in vessels. Limited defoliation may follow leaf yellowing, and foliar necrosis, when it occurs, tends to be initiated in triangular sectors closely bounded by veins. Necrosis spreads from the veins rather than toward them as it does in the acute syndrome (Figs. 1 and 2).

The "tiger-stripe" foliar syndrome of acute disease is most easily interpreted in terms of a toxin-induced necrosis as postulated above (Section IV, B,c). The pattern is simulated by an uptake of an acid fuchsin solution. The dye accumulates initially in the interveinal areas, as indicated for fusaric acid by Kern *et al.* (1957) and Sivadjian and Kern (1958). Examination of leaves cleared after taking up a solution of acid fuchsin (Fig. 21) indicates that the dye moves out of the vascular system preferentially at the terminal tracheids and does not initially stain the mesophyll tissues adjacent to vessels. Similarly, the cells immediately in contact with the terminal tracheids are the first to show severe abnormalities during symptom development (Fig. 20). One factor contributing to preferential movement out of the terminal elements is probably their direct contact with the spongy mesophyll through which water exchange presumably is occurring at the highest rate.

Furthermore, the lower epidermis over the veins is without stomata. This could be expected to enhance the tendency for water and solutes to move more rapidly into the interveinal areas and for solutes to accumulate there. Toxin accumulation in a similar pattern will lead to necrosis as lethal concentrations are reached. Water stress would play only a subsidiary role in this syndrome although reduced water uptake appears to be a feature of the action of a toxin, probably a polysaccharide, produced *in vitro* by *V. albo-atrum* (Talboys, 1957).

The formation of tyloses appears to be a dominant factor in the mild syndrome of hop verticillium wilt, and there is no reason to doubt that yellowing of leaves and defoliation of the lower parts of the primary axis result from water stress induced by vascular dysfunction. Sectorial yellowing and necrosis are more complex. They are certainly a consequence of the presence of tyloses and mycelium in a few of the vessels supplying the leaf. The symptoms illustrated in Fig. 1 were associated with the presence of numerous tyloses in most of the vessels in one petiole bundle and a few vessels in an adjacent bundle. Small amounts of mycelium were also detectable, but movement of toxin was presumably limited by the tyloses. The network of vascular tissues in the leaf makes it difficult to induce water stress in a limited area of the lamina, but obstruction of the vessels on both sides of an interveinal area apparently can give rise to a stress in that area. The development of necrosis is associated with the death of cells adjacent to vessels (Fig. 22) rather than to those adjoining the terminal tracheids as in the "tiger-stripe" syndrome. This suggests that as a result of vascular occlusion the fungal metabolites are exerting a toxic action close to the site of production rather than at a distance.

The acute syndrome normally leads to death of at least the above-ground parts of the hops. Exceptionally, death can also result from development of the mild syndrome, in circumstances in which tylosis becomes very extensive and the compensating mechanism discussed in Section V becomes ineffective. There are, therefore, two entirely different sequences of events that can terminate in the death of the aerial parts of hops infected with *V. albo-atrum*.

Duality in hop wilt symptom patterns thus seems to be associated with a duality of mechanism: vascular occlusion predominating in one form of the disease and toxaemia in the other.

The question arises whether other diseases that show marked duality in symptom expression, such as the ones noted at the beginning of this section, also show qualitative differences in host-parasite interactions comparable with those occurring in hop verticillium wilt. A joint role of occlusion and toxic action has been proposed by Page (1959a) for banana fusarium wilt, but their relative importance in "yellowing" and "nonyellowing" syndromes was not elucidated.

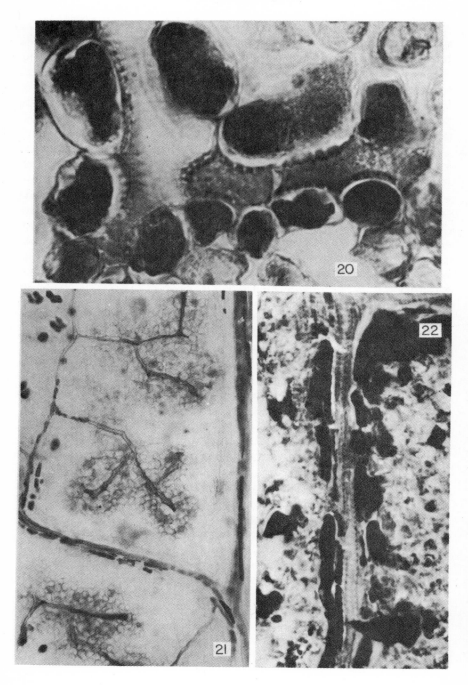

The conflicting evidence for vascular occlusion (Scheffer and Walker, 1953; Dimond and Waggoner, 1953c) and for toxin-induced wilting (Davis, 1954; Keyworth, 1964) in tomato fusarium wilt suggests that, in this disease also, different mechanisms may predominate in different circumstances.

VII. CONCLUSIONS

The tacit assumption that vascular diseases form a homogenous group because their symptoms all arise as a result of water stress is clearly not the whole truth. Many diseases are characterized by typical symptoms of water stress, with flaccidity (often initially reversible), yellowing, and defoliation or collapse occurring under acute stress or with dwarfing and drought hardening occurring under chronic stress. These tend to be associated with vascular occlusion by tyloses, gums, pathogen-produced polysaccharides, or the debris of enzymic degradation of host tissues.

However, numerous other syndromes are characterized by sharply defined patterns of foliar chlorosis and necrosis, initiated sometimes along the veins (vein clearing), sometimes in the interveinal and marginal area of the laminae, and by progressive desiccation without flaccidity. Such patterns are more readily interpreted in terms of the accumulation of pathogen produced (or pathogen induced) toxins, the action of which results in death of tissue *followed* by water loss, although even here some degrees of water stress induced by polysaccharides, tyloses, or gums, probably contributes to the ultimate symptom pattern.

Investigations on processes of symptom induction in vascular diseases have generally been made by plant pathologists. It is perhaps appropriate, therefore, that a plant pathologist should point out that such investigations have generally failed to take account of the extreme complexity of the systems under examination. Few investigators, e.g., have taken adequate account of stomatal movements during the course of their experiments. Studies of toxic effects on detached shoots have ignored the water stresses inherent in an intact system, and insufficient attention has been paid to the possibility of

---

Fig. 20. Tangential section through hop leaf with interveinal necrosis (as in Fig. 2) showing terminal tracheids with development of dark granular deposits in the adjacent mesophyll cells. One tracheid contains granular material, probably a gum.

Fig. 21. Cleared whole mount of hop leaf after short period of uptake of acid fuchsin. Stain is concentrated in the zone around the terminal tracheids but is not evident in the cells adjoining the vessels.

Fig. 22. Tangential section through necrotic area of a hop leaf with sectorial necrosis (as in Fig. 1), showing damage to cells adjacent to vessels; the effect shown in Fig. 20 is not evident.

multi-component systems operating in processes of induction of disease symptoms.

The work of Beckman *et al.* (1962) provides a lead to plant pathologists in the application of advanced techniques to this problem. It is to be hoped that plant physiologists also may find studies of water relations of diseased plants a challenging field for investigation.

REFERENCES

Anonymous (1963). Plant pathology: *Verticillium* in quince rootstocks. *Ann. Rept. East Malling Res. Sta. Kent* **1962**, 26.

Bachmann, E. (1956). Der Einfluss von Fusarinsäure auf die Wasserpermeabilität von pflanzlichen Protoplasten. *Phytopathol. Z.* **27**, 255.

Banfield, W. M. (1941). Distribution by the sap stream of spores of three fungi that induce vascular wilt diseases of elm. *J. Agr. Res.* **62**, 637.

Bartholomew, E. T. (1928). Internal decline (Endoxerosis) of lemons. VI. Gum formation in the lemon fruit and its twig. *Am. J. Botany* **15**, 548.

Bateman, D. F. (1966). Hydrolytic and trans-eliminative degradation of pectic substances by extracellular enzymes of *Fusarium solani* f. *phaseoli*. *Phytopathology* **56**, 238.

Beckman, C. H. (1956). Production of pectinase, cellulases, and growth-promoting substance by *Ceratostomella ulmi*. *Phytopathology* **46**, 605.

Beckman, C. H. (1964). Host responses to vascular infection. *Ann. Rev. Phytopathol* **2**, 231.

Beckman, C. H. (1966). Cell irritability and localization of vascular infections in plants. *Phytopathology* **56**, 821.

Beckman, C. H., and Halmos, S. (1962). Relation of vascular occluding reactions in banana roots to pathogenicity of root-invading fungi. *Phytopathology* **52**, 893.

Beckman, C. H., Kuntz, J. E., Riker, A. J., and Berbee, J. G. (1953). Host responses associated with the development of oak wilt. *Phytopathology* **43**, 448.

Beckman, C. H., Mace, M. E., Halmos, S., and McGahan, M. W. (1961). Physical barriers associated with resistance in *Fusarium* wilt of bananas. *Phytopathology* **51**, 507.

Beckman, C. H., Brun, W. A., and Buddenhagen, I. W. (1962). Water relations in banana plants infected with *Pseudomonas solanacearum*. *Phytopathology* **52**, 1144.

Bewley, W. F. (1922). Sleepy disease of tomato. *Ann. Appl. Biol.* **9**, 116.

Blackhurst, F. M. (1963). Induction of *Verticillium* wilt disease symptoms in detached shoots of resistant and of susceptible tomato plants. *Ann. Appl. Biol.* **52**, 79.

Blackhurst, F. M., and Wood, R. K. S. (1963). Resistance of tomato plants to *Verticillium albo-atrum*. *Trans. Brit. Mycol. Soc.* **46**, 385.

Blodgett, E. C. (1965). *Verticillium* wilt of stone fruit in Washington. *Washington State Univ. Agr. Expt. Sta., Sta. Circ.* **425**, 9 pp.

Brandes, L. W. (1919). Banana wilt. *Phytopathology* **9**, 339.

Brian, P. W., Elson, G. W., Hemming, H. G., and Wright, J. M. (1952). The phytotoxic properties of alternaric acid in relation to the etiology of plant diseases caused by *Alternaria solani* (Ell. & Mart.) Jones & Grout. *Ann. Appl. Biol.* **39**, 308.

Brooks, F. T., and Brenchley, G. H. (1931). Further injection experiments in relation to *Stereum purpureum*. *New Phytologist* **30**, 128.

Brooks, F. T., and Moore, W. C. (1926). Silver-leaf disease. *J. Pomol. Hort. Sci.* **5**, 61.

Brooks, F. T., and Storey, H. H. (1923). Silver-leaf disease. IV. *J. Pomol. Hort. Sci.* **3**, 117.

Brown, F., Hirst, E. L., and Jones, J. K. N. (1948). The structure of almond-tree gum. I. The constitution of the aldobionic acid derived from the gum. *J. Chem. Soc.* p. 1677.

Butler, E. J., and Jones, S. G. (1949). "Plant Pathology." Macmillan, New York.

Caroselli, N. E. (1954). *Verticillium* wilt of maple. *Dissertation Abstr.* **14**, 2186.

Chattaway, M. M. (1949). The development of tyloses and secretion of gum in heartwood formation. *Australian J. Sci. Research Ser. B* **2**, 227.

Clauson-Kaas, N., Plattner, P. A., and Gäumann, E. (1944). Über ein welkeerzeugendes Stoffwechselprodukt. *Ber. Schweiz. Botan. Ges.* **54**, 523.

Crandall, B. S., and Baker, W. L. (1950). The wilt disease of American persimmon, caused by *Cephalosporium diospyri*. *Phytopathology* **40**, 307.

Davis, D. (1954). The use of intergeneric grafts to demonstrate toxins in the *Fusarium* wilt disease of tomato. *Am. J. Botany* **41**, 395.

Deese, D. C., and Stahmann, M. A. (1962). Formation of pectic enzymes by *Verticillium albo-atrum* on susceptible and resistant tomato stem tissues and on wheat bran. *Phytopathol. Z.* **46**, 53.

Dimond, A. E. (1947). Symptoms of Dutch elm disease produced by *Graphium ulmi* in culture. *Phytopathology* **37**, 7.

Dimond, A. E. (1966). Pressure and flow relations in vascular bundles of the tomato plant. *Plant Physiol.* **41**, 119.

Dimond, A. E., and Edgington, L. V. (1960). Mechanics of water transport in healthy and *Fusarium* wilted tomato plants. *Phytopathology* **50**, 634.

Dimond, A. E., and Waggoner, P. E. (1953a). The physiology of lycomarasmin production by *Fusarium oxysporum* f. *lycopersici*. *Phytopathology* **43**, 195.

Dimond, A. E., and Waggoner, P. E. (1953b). On the nature and role of vivotoxins in plant disease. *Phytopathology* **43**, 229.

Dimond, A. E., and Waggoner, P. E. (1953c). The water economy of *Fusarium* wilted tomato plants. *Phytopathology* **43**, 619.

Dimond, A. E., and Waggoner, P. E. (1953d). The cause of epinastic symptoms in *Fusarium* wilt of tomatoes. *Phytopathology* **43**, 663.

Dimond, A. E., Waggoner, P. E., and Davis, D. (1954). Origin of symptoms in wilt diseases of plants. *Science* **120**, 777.

Dowson, W. J. (1922). On the symptoms of wilting of Michaelmas daisies produced by a toxin secreted by a Cephalosporium. *Trans. Brit. Mycol. Soc.* **7**, 283.

Esau, K. (1948). Anatomic effects of the viruses of Pierce's disease and phony peach. *Hilgardia* **18**, 423.

Fahmy, T. (1923). The production by *Fusarium solani* of a toxic excretory substance capable of causing wilting in plants. *Phytopathology* **13**, 543.

Feder, W. A., and Ark, P. A. (1951). Wilt-inducing polysaccharides derived from crown-gall, bean-blight and soft rot bacteria. *Phytopathology* **41**, 804.

Fergus, C. L., and Wharton, D. C. (1957). Production of pectinase and growth-promoting substance by *Ceratocystis fagacearum*. *Phytopathology* **47**, 635.

Foster, R. E. (1946). The first symptoms of tomato wilt: clearing of the ultimate veinlets in the leaf. *Phytopathology* **36**, 691.

Gäumann, E. (1957). Fusaric acid as a wilt toxin. *Phytopathology* **47**, 342.

Gäumann, E. (1958). The mechanisms of fusaric acid injury. *Phytopathology* **48**, 670.

Gäumann, E., and Bachmann, E. (1958). Über den Einfluss der Ernährung auf die Schadigung der Wasserpermeabilität der Protoplasten durch Fusarinsäure. *Phytopathol. Z.* **31**, 1.

Gäumann, E., and Jaag, O. (1946). Über das Problem der Welkekrankheiten bei Pflanzen. *Experientia* **2**, 215.

Gäumann, E., and Jaag, O. (1947). Die physiologischen Grundlagen des parasitogenen Welkens. I. *Ber. Schweiz. Botan. Ges.* **57**, 3.

Gäumann, E., and Jaag. O. (1950). Über das toxigene und das physikalisch induzierte Welken. *Phytopathol. Z.* **16**, 226.

Gäumann, E., and Loeffler, W. (1957). Über die Wirkung von Fusarinsäure auf die Wasserpermeabilität der Markzellen von Tomatenpflanzen. *Phytopathol. Z.* **28**, 319.

Gäumann, E., Kern, H., and Sauthoff, W. (1952a). Untersuchungen uber zwei Welketoxine. *Phytopathol. Z.* **18**, 405.

Gäumann, E., Naef-Roth, S., Reusser, P., and Ammann, A. (1952b). Über den Einfluss einiger Welketoxine und Antibiotica auf die osmotischen Eigenschaften pflanzlicher Zeller. *Phytopathol. Z.* **19**, 160.

Gäumann, E., Kern, H., Schüepp, H., and Obrist, W. (1958). Der Einfluss der Fusarinsäure auf den Wasserhaushalt abgeschnittener Tomatensprosse. *Phytopathol. Z.* **32**, 225.

Gäumann, E., Kern, H., and Obrist, W. (1959). Der Einfluss einiger Welketoxine auf den Wasserhaushalt abgeschnittener Tomatensprosse. *Phytopathol. Z.* **36**, 111.

Gilman, J. C. (1916). Cabbage yellows and the relation of temperature to its occurrence. *Ann. Missouri Botan. Garden* **3**, 25.

Goss, R. W. (1924). Potato wilt and stem-end rot caused by *Fusarium eumartii. Nebraska Agr. Expt. Sta. Res. Bull.* **27**.

Goss, R. W., and Frink, P. R. (1934). Cephalosporium wilt and die-back of the white elm. *Nebraska Univ. Agr. Expt. Sta. Res. Bull.* **70**.

Gothoskar, S. S., Scheffer, R. P., Walker, J. C., and Stahmann, M. A. (1953). The role of pectic enzymes in *Fusarium* wilt of tomato. *Phytopathology* **43**, 535.

Gottlieb, D. (1943). The presence of a toxin in tomato wilt. *Phytopathology* **33**, 126.

Gottlieb, D. (1944). The mechanism of wilting caused by *Fusarium bulbigenum* var. *lycopersici. Phytopathology* **34**, 41.

Green, R. J. (1954). A preliminary investigation of toxins produced *in vitro* by *Verticillium albo-atrum. Phytopathology* **44**, 433.

Griffiths, D. A., and Isaac, I. (1963). Wilt of lupin and sunflower caused by species of *Verticillium. Hort. Res.* **2**(2), 104.

Gruen, H. E. (1959). Auxins and fungi. *Ann. Rev. Plant Physiol.* **10**, 405.

Hall, W. C., and Morgan, P. W. (1964). Auxin-ethylene interrelationships. *Colloq. Intern. Centre Natl. Rech. Sci. (Paris)* **123**, 727.

Harris, H. A. (1940). Comparative wilt induction by *Erwinia tracheiphila* and *Phytomonas stewarti. Phytopathology* **30**, 625.

Harris, R. V. (1925). The blue stripe wilt of the raspberry. *J. Pomol. Hort. Sci.* **4**, 221.

Haskell, R. J. (1919). *Fusarium* wilt of potato in Hudson River valley, New York. *Phytopathology* **9**, 223.

Heath, O. V. S. (1938). An experimental investigation of the mechanism of stomatal movement, with some preliminary observations upon the response of ᵗhe guard cells to "shock." *New Phytologist* **37**, 385.

Heitefuss, R., Stahmann, M. A., and Walker, J. C. (1960). Production of pectolytic enzymes and fusaric acid by *Fusarium oxysporum* f. *conglutinans* in relation to cabbage yellows. *Phytopathology* **50**, 367.

Henry, B. W., Moses, C. S., Richards, C. A., and Riker, A. J. (1944). Oak wilt: its significance, symptoms and cause. *Phytopathology* **34**, 636.

Hepting, G. H. (1939). A vascular wilt of the mimosa tree (*Albizzia julibrissin*). *U.S. Dept. Agr. Circ.* **535**, 10 pp.

Hepting, G. H. (1944). Sapstreak, a new killing disease of sugar maple. *Phytopathology* **34**, 1069.

Hirst, E. L. (1949). The occurrence and significance of the pentose sugars in nature, and their relationship to the hexoses. *J. Chem. Soc.* p. 522.

Hirst, E. L., and Jones, J. K. N. (1938). The constitution of damson gum. Part I. Composition of damson gum and structure of an aldobionic acid (glycuronosido-2-mannose) derived from it. *J. Chem. Soc.* p. 1174.

Hodgson, R., Peterson, W. H., and Riker, A. J. (1949). The toxicity of polysaccharide and other large molecules to tomato cuttings. *Phytopathology* 39, 47.

Hough, L., and Pridham, J. B. (1959). The composition of plum gums. *Biochem. J.* 73, 550.

Husain, A., and Dimond, A. E. (1958). The function of extracellular enzymes of Dutch Elm disease pathogen. *Proc. Natl. Acad. Sci. U.S.* 44, 594.

Husain, A., and Kelman, A. (1957). Presence of pectic and cellulolytic enzymes in tomato plants infected with *Pseudomonas solanacearum*. *Phytopathology* 47, 111.

Husain, A., and Kelman, A. (1958a). Relation of slime production to mechanism of wilting and pathogenicity of *Pseudomonas solanacearum*. *Phytopathology* 48, 155.

Husain, A., and Kelman, A. (1958b). The role of pectic and cellulolytic enzymes in pathogenesis by *Pseudomonas solanacearum*. *Phytopathology* 48, 377.

Hutchinson, C. M. (1913). Rangpur tobacco wilt. *Mem. Dept. Agr. India Bacteriol. Ser.* 1, 67.

Isaac, I. (1949). A comparative study of pathogenic isolates of *Verticillium*. *Trans. Brit. Mycol. Soc.* 32, 137.

Jenkins, W. A. (1948). Root rot disease-complexes of tobacco in Virginia. I. Brown root rot. *Phytopathology* 38, 528.

Johnson, J. (1921). *Fusarium* wilt of tobacco. *J. Agr. Res.* 20, 515.

Jones, J. K. N. (1939). The constitution of cherry gum. I. Composition. *J. Chem. Soc.* p. 558.

Jones, J. K. N. (1950). The structure of peach gum. I. The sugars produced on hydrolysis of the gum. *J. Chem. Soc.* p. 534.

Kalyanasundaram, R., and Lakshminarayanan, K. (1953). Rooting of cut shoots of cotton in culture filtrate of *Fusarium vasinfectum*. *Nature* 171, 1120.

Kalyanasundaram, R., and Subba-Rao, N. S. (1957). Temperature effect on the *in vivo* production of fusaric acid. *Current Sci. (India)* 26, 56.

Kalyanasundaram, R., and Venkata Ram, C. S. (1956). Production and systemic translocation of fusaric acid in *Fusarium* infected cotton plants. *J. Indian Botan. Soc.* 35, 7.

Kamal, M., and Wood, R. K. S. (1955). Role of pectic enzymes in the *Verticillium* wilt of cotton. *Nature* 175, 264.

Kamal, M., and Wood, R. K. S. (1956). Pectic enzymes secreted by *Verticillium dahliae* and their role in the development of the wilt disease of cotton. *Ann. Appl. Biol.* 44, 322.

Kern, H., and Kluepfel, D. (1956). Die Bildung von Fusarinsäure durch *Fusarium lycopersici in vivo*. *Experientia* 12, 181.

Kern, H., Sanwal, B. D., Flück, V., and Kluepfel, D. (1957). Die Verteilung der radioaktiven Fusarinsäure in Tomatensprossen. *Phytopathol. Z.* 30, 31.

Keyworth, W. G. (1942). *Verticillium* wilt of the hop (*Humulus lupulus*). *Ann. Appl. Biol.* 29, 346.

Keyworth, W. G. (1963). The reaction of monogenic resistant and susceptible varieties of tomato to inoculation with *Fusarium oxysporum* f. *lycopersici* into stems or through Bonny Best rootstocks. *Ann. Appl. Biol.* 52, 257.

Keyworth, W. G. (1964). Hypersensitivity of monogenic resistant tomato scions to toxins produced in Bonny Best rootstocks invaded by *Fusarium oxysporum* f. *lycopersici*. *Ann. Appl. Biol.* 54, 99.

Kiessig, R., and Haller-Kiessig, R. (1957). Beitrag zur Kenntnis einer infectiosen Welkekrankheit der Luzerne (*V. albo-atrum* R & B). *Phytopathol. Z.* 31, 185.

Kovachich, W. G. (1948). A preliminary anatomical note on vascular wilt disease of the oil palm (*Elaeis guineensis*). *Ann. Botany* [*N.S.*] **12**, 327.

Kozlowski, T. T. (1964). "Water Metabolism in Plants." Harper, New York.

Kozlowski, T. T., Kuntz, J. E., and Winget, C. H. (1962). Effect of oak wilt on cambial activity. *J. Forestry* **60**, 558.

Kuntz, J. E., and Riker, A. J. (1956). Oak wilt. *Wisconsin Univ. Agr. Expt. Sta. Bull.* **519**, 12 pp. illus.

Kunz, R. (1953). Die Wirkungsweise von *Bacterium solanacearum* E.F.S., dem Erreger der tropischen Schleimkrankheit des Tabaks, auf *Solanum lycopersicum* L. *Phytopathol. Z.* **20**, 89.

La Rue, C. D. (1930). The water supply of the epidermis of leaves. *Papers Mich. Acad. Sci.* **13**, 131.

Lakshminarayanan, K. (1955). Role of cystine chelation in the mechanism of *Fusarium* wilt of cotton. *Experientia* **11**, 388.

Lakshminarayanan, K., and Subramanian, D. (1955). Is fusaric acid a vivotoxin? *Nature* **176**, 697.

Leach, J. G., Lilly, V. G., Wilson, H. A., and Purvis, M. R. (1957). Bacterial polysaccharides: the nature and function of the exudate produced by *Xanthomonas phaseoli*. *Phytopathology* **47**, 113.

Leal, J. A., and Villanueva, J. R. (1962). Lack of pectic enzyme production by nonpathogenic species of *Verticillium*. *Nature* **195**, 1328.

Le Tourneau, D. (1957). The production of oligosaccharides by *Verticillium albo-atrum*. *Phytopathology* **47**, 527.

Linford, M. B. (1931a). Transpirational history as a key to the nature of wilting in the *Fusarium* wilt of peas. *Phytopathology* **21**, 791.

Linford, M. B. (1931b). Studies of pathogenesis and resistance in pea wilt caused by *Fusarium orthoceras* var. *pisi*. *Phytopathology* **27**, 797.

Linskens, H. F. (1955). Der Einfluss der toxigenen Welke auf die Blattausscheidungen der Tomatenpflanze. *Phytopathol. Z.* **23**, 89.

Ludwig, R. A. (1952). Studies on the physiology of hadromycotic wilting in the tomato plant. *Macdonald Coll. Tech. Bull.* **20**.

Ludwig, R. A. (1960). Toxins. *In* "Plant Pathology" (J. G. Horsfall and A. E. Dimond, eds.), Vol. **2**, pp. 315–351. Academic Press, New York.

McClure, T. T. (1950). Anatomical aspects of the *Fusarium* wilt of sweet potatoes. *Phytopathology* **40**, 769.

McDonnell, K. (1958). Absence of pectolytic enzymes in a pathogenic strain of *Fusarium oxysporum* f. *lycopersici*. *Nature* **182**, 1025.

Mace, M. E., and Solit, E. (1966). Interaction of 3-indoleacetic acid and 3-hydroxytyramine in *Fusarium* wilt of banana. *Phytopathology* **56**, 245.

McIntyre, G. A. (1964). Production of pectic enzymes by *Verticillium albo-atrum*. *Dissertation Abstr.* **25**, 779.

Matta, A. (1963). A comparison of pectolytic and cellulolytic activity and the pathogenicity on eggplant of some *Verticillium* isolates. *Phytopathol. Med.* **2**, 55.

Melhus, I. E., Muncie, J. H., and Ho, W. T. H. (1924). Measuring water flow interference in certain gall and vascular diseases. *Phytopathology* **14**, 580.

Milthorpe, F. L., and Spencer, E. J. (1957). Experimental studies of the factors controlling transpiration. III. The interrelations between transpiration rate, stomatal movement, and leaf-water content. *J. Expl. Botany* **8**, 413.

Naef-Roth, S., Kern, H., and Toth, A. (1963). Zur Pathogenese des Parasitogenen und Physiologischen Silberglanzes am Steinobst. *Phytopathol. Z.* **48**, 232.

Nelson, R. (1950). *Verticillium* wilt of peppermint. *Mich. State Univ. Agr. Expt. Sta. Tech. Bull.* **221**, 259 pp.

Nishimura, S. (1957). Pathochemical studies on water-melon wilt. V. On the metabolic products of *Fusarium oxysporum* f. *niveum. Nippon Shokubutsu Byori Gakkaiho* (*Ann. Phytopathol. Soc. Japan*) **22**, 215.

Norman, A. G. (1937). "The Biochemistry of Cellulose, the Polyuronides, Lignin, etc." Clarendon Press, Oxford.

Noviello, C., and Snyder, W. C. (1962). *Fusarium* wilt of hemp. *Phytopathology* **52**, 1315.

Orton, W. A. (1902). The wilt disease of cowpea and its control. *U.S. Dept. Agr. Bur. Plant Ind. Bull.* **17**, 9.

Page, O. T. (1959a). Observations on the water economy of *Fusarium*-infected banana plants. *Phytopathology* **49**, 61.

Page, O. T. (1959b). Fusaric acid in banana plants infected with *Fusarium oxysporum* f. *cubense. Phytopathology* **49**, 230.

Parker, K. G. (1959). *Verticillium* hadromycosis of deciduous tree fruits. *Plant Disease Reptr.* **Suppl. 255**.

Peace, T. R. (1962). "Pathology of Trees and Shrubs." Oxford Univ. Press (Clarendon), London and New York.

Pegg, G. F., and Selman, I. W. (1959). An analysis of the growth response of young tomato plants to infection by *Verticillium albo-atrum*. II. The production of growth substances. *Ann. Appl. Biol.* **47**, 222.

Pierson, C. F., Gothoskar, S. S., Walker, J. C., and Stahmann, M. A. (1955). Histological studies on the role of pectic enzymes in the development of *Fusarium* wilt symptoms in tomato. *Phytopathology* **45**, 524.

Plattner, P. A., and Clauson-Kaas, N. (1945). Über Lycomarasmin, den Welkstoff aus *Fusarium lycopersici* Sacc. *Experientia* **1**, 195.

Porter, C. L., and Green, R. J. (1952). Production of exotoxin in the genus *Verticillium. Phytopathology* **42**, 472.

Prendergast, A. G. (1957). Observations on the epidemiology of vascular wilt disease of the oil palm. *J. West Africa Inst. Oil Palm Res.* **2**, 148.

Prendergast, A. G. (1963). A method of testing oil palm progenies at the nursery stage for resistance to vascular wilt disease caused by *Fusarium oxysporum*, Schl. *J. West Africa Inst. Oil Palm Res.* **4**, 156.

Presley, J. T. (1950). *Verticillium* wilt of cotton, with particular emphasis on variation of the causal organism. *Phytopathology* **40**, 497.

Roberts, L. W., and Fosket, D. E. (1966). Interaction of gibberellic acid and indole-acetic acid in the differentiation of wound vessel members. *New Phytologist* **65**, 5.

Robinson, D. B., Larson, R. H., and Walker, J. C. (1957). *Verticillium* wilt of potato in relation to symptoms, epidemiology, and variability of the pathogen. *Wisconsin Univ. Agr. Expt. Sta. Res. Bull.* **202**.

Rosen, H. R. (1926). Efforts to determine the means by which the cotton-wilt fungus, *Fusarium vasinfectum*, induces wilting. *J. Agr. Res.* **33**, 1143.

Sadasivan, T. S., and Saraswathi-Devi, L. (1957). Vivotoxins and uptake of ions by plants. *Current Sci.* (*India*) **26**, 74.

Scheffer, R. P., and Walker, J. C. (1953). The physiology of *Fusarium* wilt of tomato. *Phytopathology* **43**, 116.

Schoeneweiss, D. F. (1959). Xylem formation as a factor in oak wilt resistance. *Phytopathology* **49**, 335.

Selman, I. W., and Pegg, G. F. (1957). An analysis of the growth response of young tomato plants to infection by *Verticillium albo-atrum*. *Ann. Appl. Biol.* **45**, 674.

Sequeira, L. (1963). Growth regulators in plant disease. *Ann. Rev. Phytopathol.* **1**, 5.

Sewell, G. W. F., and Wilson, J. F. (1964). Occurrence and dispersal of *Verticillium* conidia in xylem sap of the hop (*Humulus lupulus* L.). *Nature* **204**, 901.

Sivadjian, J., and Kern, H. (1958). Über den Einfluss von Welketoxinen auf die Transpiration von Tomatenblättern. *Phytopathol. Z.* **33**, 241.

Smalley, E. B. (1962). Prevention of Dutch elm disease by treatments with 2, 3, 6-trichlorophenyl acetic acid. *Phytopathology* **52**, 1090.

Smith, Erwin F. (1899). Wilt disease of cotton, watermelon and cowpea (*Neocosmospora* nov. gen.). *U.S. Dept. Agr. Bull.* **17**.

Spalding, D. H. (1960). Mechanisms of the stripe pathogen, *Cephalosporium gramineum*, in killing cultivated wheat. *Dissertation Abstr.* **21**, 421.

Spalding, D. H., Bruehl, G. W., and Foster, R. J. (1961). Possible role of pectolytic enzymes and polysaccharide in pathogenesis by *Cephalosporium gramineum* in wheat. *Phytopathology* **51**, 227.

Stewart, F. C. (1897). A bacterial disease of sweet corn. *N.Y. State Agr. Expt. Sta. (Geneva, N.Y.) Bull.* **130**.

Stover, R. H. (1959). Studies on *Fusarium* wilt of bananas. IV. Clonal differentiation among wild type isolates of *F. oxysporum* f. *cubense*. *Can. J. Botany* **37**, 245.

Struckmeyer, B. E., Beckman, C. H., Kuntz, J. E., and Riker, A. J. (1954). Plugging of vessels by tyloses and gums in wilting oaks. *Phytopathology* **44**, 148.

Subba-Rao, N. S. (1954). Fluorescence phenomenon in fusariose wilt of cotton. *J. Indian Botan. Soc.* **33**, 443.

Talboys, P. W. (1957). The possible significance of toxic metabolites of *Verticillium albo-atrum* in the development of hop wilt symptoms. *Trans Brit. Mycol. Soc.* **40**, 415.

Talboys, P. W. (1958a). Some mechanisms contributing to *Verticillium*-resistance in the hop root. *Trans. Brit. Mycol. Soc.* **41**, 227.

Talboys, P. W. (1958b). Degradation of cellulose by *Verticillium albo-atrum*. *Trans. Brit. Mycol. Soc.* **41**, 242.

Talboys, P. W. (1958c). Association of tylosis and hyperplasia of the xylem with vascular invasion of the hop by *Verticillium albo-atrum*. *Trans. Brit. Mycol. Soc.* **41**, 249.

Talboys, P. W. (1964). A concept of the host-parasite relationship in *Verticillium* wilt diseases. *Nature* **202**, 361.

Talboys, P. W., Bennett, M., and Wilson, J. F. (1961). Tolerance to *Verticillium* wilt diseases in the strawberry. *Ann. Rept. East Malling Res. Sta. Kent* **1960**, 94.

Taylor, J. B. (1963). The inactivation of *Verticillium albo-atrum* in apricot trees. *Phytopathology* **53**, 1143.

Thatcher, F. S. (1942). Further studies of osmotic and permeability relations in parasitism. *Can. J. Res.* **20**, 283.

Thomas, C. A. (1949). A wilt-inducing polysaccharide from *Fusarium solani* f. *eumartii*. *Phytopathology* **39**, 572.

Thomas, C. A. (1966). A study of the wilt disease of *Phaseolus* spp. caused by *Fusarium oxysporum*. Ph.D. Thesis, Univ. London, London, England.

Threlfall, R. J. (1959). Physiological studies on the *Verticillium* wilt disease of tomato. *Ann. Appl. Biol.* **47**, 57.

Trione, E. J. (1960). Extracellular enzyme and toxin production by *Fusarium oxysporum* f. *lini*. *Phytopathology* **50**, 480.

True, R. P., and Tryon, E. H. (1956). Oak stem cankers initiated in the drought year 1953. *Phytopathology* **46**, 617.

Van der Meer, J. H. H. (1925). *Verticillium*-wilt of herbaceous and woody plants. *Mededeel. Landbouwhoogeschool Wageningen* **28**(2), 1.

Van der Meer, J. H. H. (1926). *Verticillium* wilt of maple and elm seedlings in Holland. *Phytopathology* **16**, 611.

Waggoner, P. E., and Dimond, A. E. (1954). Reduction in water flow by mycelium in vessels. *Am. J. Botany* **41**, 637.

Waggoner, P. E., and Dimond, A. E. (1955). Production and role of extracellular pectic enzymes of *Fusarium oxysporum* f. *lycopersici*. *Phytopathology* **45**, 79.

Walker, J. C. (1957). "Plant Pathology," 707 pp. McGraw-Hill, New York.

Walter, J. M., May, C., and Collins, C. W. (1943). Dutch elm disease and its control. *U.S. Dept. Agr. Circ.* **677**, 12 pp.

Wardlaw, C. W. (1961). "Banana Diseases." Longmans, London.

Waterman, A. M. (1956). *Verticillium* wilt of yellow poplar (*Liriodendron tulipifera*). *Plant Disease Reptr.* **40**, 349.

Wheeler, H., and Luke, H. H. (1963). Microbial toxins in plant disease. *Ann. Rev. Microbiol.* **17**, 223.

White, J. G. (1954). Physiology of *Endoconidiophora fagacearum*, the fungus causing oak wilt, with special reference to growth and toxin production in synthetic media. *Dissertation Abstr.* **14**, 450.

White, J. G. (1955). Toxin production by the oak wilt fungus *Endoconidiophora fagacearum*. *Am. J. Botany* **42**, 759.

White, R. P. (1927). Studies on tomato wilt caused by *Fusarium lycopersici* Sacc. *J. Agr. Res.* **34**, 197.

Wilhelm, S., and Taylor, J. B. (1965). Control of *Verticillium* wilt of olive through natural recovery and resistance. *Phytopathology* **55**, 310.

Williams, W. T. (1950). Studies in stomatal behaviour. IV. The water relations of the epidermis. *J. Exptl. Botany* **1**, 114.

Wilson, C. L. (1961). Study of the growth of *Ceratocystis fagacearum* in wood with the use of autoradiograms. *Phytopathology* **51**, 210.

Wilson, C. L. (1963). Wilting of persimmon caused by *Cephalosporium diospyri*. *Phytopathology* **53**, 1402.

Wilson, C. L. (1965). *Ceratocystis ulmi* in elm wood. *Phytopathology* **55**, 477.

Winstead, N. N., and Walker, J. C. (1954). Production of vascular browning by metabolites from several pathogens. *Phytopathology* **44**, 153.

Wolf, F. T., and Wolf, F. A. (1948). A toxic metabolic product of *Fusarium oxysporum* var. *nicotianae* in relation to a wilting of tobacco plants. *Phytopathology* **38**, 292.

Wood, R. K. S. (1960). Pectic and cellulolytic enzymes in plant disease. *Ann. Rev. Plant Physiol.* **11**, 299.

Woolley, D. W. (1948). Studies on the structure of lycomarasmin. *J. Biol. Chem.* **176**, 1291.

Woolliams, G. E. (1966). Host range and symptomatology of *Verticillium dahliae* in economic, weed, and native plants in interior British Columbia. *Can. J. Plant Sci.* **46**, 661.

Yabuta, T. Kambe, K., and Hayashi, T. (1934). Biochemistry of the "bakanae"-fungus. I. Fusaric acid, a new product of the "bakanae" fungus. *Nippon Nogei-kagaku Kaishi* (*J. Agr. Chem. Soc. Japan*) **10**, 1059.

Young, H. C., and Bennett, C. W. (1921). Studies in parasitism. I. Toxic substances produced by fungi. *Mich. Acad. Sci. Ann. Rept.* **22**, 205.

Zentmyer, G. A. (1942). Toxin formation by *Ceratostomella ulmi*. *Science* **95**, 512.

Zentmyer, G. A. (1949). Verticillium wilt of avocado. *Phytopathology* **39**, 677.

AUTHOR INDEX

Numbers in italics refer to the pages on which the complete references are listed.

SUBJECT INDEX

A

Abscission of leaves, *see also* Defoliation, 48, 257, 260, 262, 300
Absorption, *see also* Water consumption
  minerals, 88, 89, 96, 114, 165–171, 263
  water, 24, 101, 208, 237, 263
Absorption lag, 206
Adenine, 117, 122
Adenosine monophosphate, 119
Adiabatic lapse rate, 60
ADP, 118
Advective energy, 4, 33, 74, 148
Aerodynamic equation, 5
Aerodynamic roughness, 15, 46, 67
Air embolism, 90
Air resistance, 60, 64, 67, 68, 71
Alanine, 120
Albedo, 14, 60
Alternaric acid, 285, 286, 292
Alternate bearing, *see* Biennial bearing
Amination, 117
Amino acids, 117, 119, 138, 139
Amino nitrogen, 138, 139
Amylase, 138
Anemometers, 64
Anhydrobiosis, 151
Annual ring, 205–233
Anthesis, 156
Antitranspirants, 101, 102, 108, 109
Apical meristems, 194, 195
Apparent evaporation, 31
Apoplast, 90, 95
Apricot verticillium wilt, *see* Verticillium wilt
Ascorbic acid, 109, 121
Asparagine, 117, 120
ATP, 116, 119, 121
Atrazine, 101
Autofluorescense, 122
Auxesis, 151
Auxin, 212, 213, 222, 223, 282

Available range, 101, 146, 147, 198, 203
Avocado verticillium wilt, *see* Verticillium wilt

B

Baccatin, 285, 290, 291
Bacterial wilt, 267, 279
Banana fusarium wilt, *see* Fusarium wilt
Basal area, 225, 226, 229, 230
Basiphil gum, 272
Benzoic acid, 94
Biennial bearing, 241
Boot stage, 19
Boundary layer, 102
Bound water, 121, 156
Bowen's ratio, 10, 13
Breeding, 99
Bud scales, 110
Bulk density, 237
5-butylpicolinic acid, 283

C

^{14}C, 92, 116, 117, 290
Caffein, 118
Callose, 122
Cambial growth, *see also* Growth, 114, 205–233, 297
Cambial initials, 220
Capillary fringe, 200
Canopy resistance, *see also* Resistance, 16
Carbowax, 110, 282
Catalase, 109
Catenary process, 95
CDP, 118
Cell enlargement, 208, 210, 212, 213, 234, 236
Cell thickening, 213, 214
Cellulase, 269
Cellulose, 116, 269
Cell wall plasticity, 116

325